'*Transitions in Energy Efficiency and Demand* provides an important contribution to the energy transition literature, correcting the usual bias towards energy supply. Drawing on case studies of innovation, it highlights demand-side innovations in system change; and by using sociotechnical approaches, the case studies avoid the trap of thinking of innovation simply in terms of technical fixes.'

Nick Eyre, *Professor, University of Oxford, UK*

'*Transitions in Energy Efficiency and Demand* is at the forefront of research on energy innovation and energy demand, providing new and in-depth insights into both technological and social change across a range of domains. Essential reading for scholars, policy makers, business leaders, students, and anyone else interested in a low-carbon, energy-efficient and low-demand energy transition.'

Marilyn Brown, *Professor, Georgia Tech, USA*

Transitions in Energy Efficiency and Demand

Meeting the goals enshrined in the Paris Agreement and limiting global temperature increases to less than 2°C above pre-industrial levels demands rapid reductions in global carbon dioxide emissions. Reducing energy demand has a central role in achieving this goal, but existing policy initiatives have been largely incremental in terms of the technological and behavioural changes they encourage. Against this background, this book develops a sociotechnical approach to the challenge of reducing energy demand and illustrates this with a number of empirical case studies from the United Kingdom. In doing so, it explores the emergence, diffusion and impact of low-energy innovations, including electric vehicles and smart meters. The book has the dual aim of improving the academic understanding of sociotechnical transitions and energy demand and providing practical recommendations for public policy.

Combining an impressive range of contributions from key thinkers in the field, this book will be of great interest to energy students, scholars and decision-makers.

Kirsten E.H. Jenkins is a Lecturer in Human Geography and Sustainable Development within the School of Environment and Technology, University of Brighton, UK.

Debbie Hopkins is jointly appointed by the Transport Studies Unit and the School of Geography and the Environment at the University of Oxford (UK) as a Departmental Research Lecturer.

Routledge Studies in Energy Transitions

Considerable interest exists today in energy transitions. Whether one looks at diverse efforts to decarbonize, or strategies to improve the access levels, security and innovation in energy systems, one finds that change in energy systems is a prime priority.

Routledge Studies in Energy Transitions aims to advance the thinking which underlies these efforts. The series connects distinct lines of inquiry from planning and policy, engineering and the natural sciences, history of technology, STS, and management. In doing so, it provides primary references that function like a set of international, technical meetings. Single and co-authored monographs are welcome, as well as edited volumes relating to themes, like resilience and system risk.

Series Editor: Dr. Kathleen Araújo, Boise State University and Energy Policy Institute, Center for Advanced Energy Studies (US)

Series Advisory Board
Morgan Bazilian, Colorado School of Mines (US)
Thomas Birkland, North Carolina State University (US)
Aleh Cherp, Central European University (CEU, Budapest) and Lund University (Sweden)
Mohamed El-Ashry, UN Foundation
Jose Goldemberg, Universidade de Sao Paolo (Brasil) and UN Development Program, World Energy Assessment
Michael Howlett, Simon Fraser University (Canada)
Jon Ingimarsson, Landsvirkjun, National Power Company (Iceland)
Michael Jefferson, ESCP Europe Business School
Jessica Jewell, IIASA (Austria)
Florian Kern, Institut für Ökologische Wirtschaftsforschung (Germany)
Derk Loorbach, DRIFT (Netherlands)
Jochen Markard, ETH (Switzerland)
Nabojsa Nakicenovic, IIASA (Austria)
Martin Pasqualetti, Arizona State University, School of Geographical Sciences and Urban Planning (US)

Mark Radka, UN Environment Programme, Energy, Climate, and Technology
Rob Raven, Utrecht University (Netherlands)
Roberto Schaeffer, Universidade Federal do Rio de Janeiro, Energy Planning Program, COPPE (Brasil)
Miranda Schreurs, Technische Universität München, Bavarian School of Public Policy (Germany)
Vaclav Smil, University of Manitoba and Royal Society of Canada (Canada)
Benjamin Sovacool, Science Policy Research Unit (SPRU), University of Sussex (UK)

How Power Shapes Energy Transitions in Southeast Asia
A Complex Governance Challenge
Jens Marquardt

Energy and Economic Growth
Why We Need a New Pathway to Prosperity
Timothy J Foxon

Accelerating Sustainable Energy Transition(s) in Developing Countries
The Challenges of Climate Change and Sustainable Development
Laurence L. Delina

Petroleum Industry Transformations
Lessons from Norway and Beyond
Edited by Taran Thune, Ole Andreas Engen and Olav Wicken

Sustainable Energy Transformations, Power and Politics
Morocco and the Mediterranean
Sharlissa Moore

Energy as a Sociotechnical Problem
An Interdisciplinary Perspective on Control, Change, and Action in Energy Transitions
Edited by Christian Büscher, Jens Schippl and Patrick Sumpf

Transitions in Energy Efficiency and Demand
The Emergence, Diffusion and Impact of Low-Carbon Innovation
Edited by Kirsten E.H. Jenkins and Debbie Hopkins

For more information about this series, please visit: www.routledge.com/Routledge-Studies-in-Energy-Transitions/book-series/RSENT

Transitions in Energy Efficiency and Demand

The Emergence, Diffusion and Impact of Low-Carbon Innovation

Edited by
Kirsten E.H. Jenkins
and Debbie Hopkins

First published 2019
by Routledge
2 Park Square, Milton Park, Abingdon, Oxon OX14 4RN

and by Routledge
52 Vanderbilt Avenue, New York, NY 10017

Routledge is an imprint of the Taylor & Francis Group, an informa business

© 2019 selection and editorial matter, Kirsten E.H. Jenkins and Debbie Hopkins; individual chapters, the contributors

The right of Kirsten E.H. Jenkins and Debbie Hopkins to be identified as the authors of the editorial matter, and of the authors for their individual chapters, has been asserted in accordance with sections 77 and 78 of the Copyright, Designs and Patents Act 1988.

The Open Access version of this book, available at www.taylorfrancis.com, has been made available under a Creative Commons Attribution-Non Commercial-No Derivatives 4.0 license.

Trademark notice: Product or corporate names may be trademarks or registered trademarks, and are used only for identification and explanation without intent to infringe.

British Library Cataloguing-in-Publication Data
A catalogue record for this book is available from the British Library

Library of Congress Cataloging-in-Publication Data
A catalog record has been requested for this book

ISBN: 978-0-8153-5678-3 (hbk)
ISBN: 978-1-351-12726-4 (ebk)

Typeset in Goudy
by Wearset Ltd, Boldon, Tyne and Wear

Contents

List of figures xii
List of tables xiii
Notes on contributors xiv
Preface xix

1 Introduction: new directions in energy demand research 1
KIRSTEN E.H. JENKINS, STEVE SORRELL,
DEBBIE HOPKINS AND CAMERON ROBERTS

PART I
Analytical perspectives 13

2 Of emergence, diffusion and impact: a sociotechnical
perspective on researching energy demand 15
FRANK W. GEELS, BENJAMIN K. SOVACOOL
AND STEVE SORRELL

3 A normative approach to transitions in energy demand: an
energy justice and fuel poverty case study 34
KIRSTEN E.H. JENKINS AND MARI MARTISKAINEN

PART II
The emergence and diffusion of innovations 51

4 Electric vehicles and the future of personal mobility in the
United Kingdom 53
NOAM BERGMAN

5 Experimentation with vehicle automation 72
DEBBIE HOPKINS AND TIM SCHWANEN

6 The United Kingdom smart meter rollout through an energy
  justice lens   94
  KIRSTEN E.H. JENKINS, BENJAMIN K. SOVACOOL
  AND SABINE HIELSCHER

7 Overcoming the systemic challenges of retrofitting residential
  buildings in the United Kingdom: a Herculean task?   110
  DONAL BROWN, PAULA KIVIMAA, JAN ROSENOW
  AND MARI MARTISKAINEN

PART III
Societal impacts and co-benefits   131

8 Exergy economics: new insights into energy consumption
  and economic growth   133
  PAUL BROCKWAY, STEVE SORRELL, TIM FOXON
  AND JACK MILLER

9 Energy-saving innovations and economy-wide rebound
  effects   156
  GIOELE FIGUS, KAREN TURNER AND
  ANTONIOS KATRIS

PART IV
Policy mixes and implications   175

10 Political acceleration of sociotechnical transitions: lessons
   from four historical case studies   177
   CAMERON ROBERTS AND FRANK W. GEELS

11 The challenge of effective energy efficiency policy in the
   United Kingdom   195
   JANETTE WEBB

12 Policy mixes for sustainable energy transitions: the case of
   energy efficiency   215
   FLORIAN KERN, PAULA KIVIMAA, KAROLINE ROGGE
   AND JAN ROSENOW

13 Managing energy and climate transitions in theory and practice:
   a critical systematic review of Strategic Niche Management   235
   KIRSTEN E.H. JENKINS AND BENJAMIN K. SOVACOOL

PART V
Conclusion 259

14 Conclusion: towards systematic reductions in
   energy demand 261
   KIRSTEN E.H. JENKINS, DEBBIE HOPKINS AND
   CAMERON ROBERTS

   Index 274

Figures

2.1	Multi-level perspective on transitions – illustrating the emergence, diffusion and impact of radical innovations as three consecutive circles	19
4.1	Timeline of the central vision	61
5.1	Local projects and emerging technical trajectories	75
6.1	Key components of the smart meter communication service and service providers	95
7.1	Key systemic challenges for driving retrofit uptake	115
7.2	The incumbent 'atomised market model' for residential retrofit	116
7.3	The Energiesprong Managed Energy Services Agreement (MESA)	117
8.1	Normalised time series of inputs and outputs to production in Portugal	140
8.2	Trends in primary to useful exergy efficiency in the UK, US and China (1971–2010)	143
8.3	Normalised trends in primary exergy, useful exergy and primary to useful exergy efficiency in China between 1971 and 2010	144
8.4	Primary exergy (top) and useful exergy (bottom) intensity of Portugal, UK, US, Austria and Japan, 1850–2010	145
8.5	Final exergy (top) and useful exergy (bottom) intensities in the EU-15, 1960 to 2010	146
8.6	Counterfactual simulations of UK economic output 1971–2013 using the MARCO-UK model	150
10.1	The positioning of the four case studies	178
13.1	Strategic Niche Management author demographics (n = 100)	243
13.2	Strategic Niche Management article methods (n = 45)	244
13.3	Distribution of Strategic Niche Management cases by technology (n = 24)	247

Tables

2.1	Key research debates with regard to sociotechnical transitions and low-energy innovation	29
4.1	Final sample of documents with visions of future transport in the UK including EVs	59
5.1	Examples of connected and autonomous vehicle (CAV) activities in Oxford and Greenwich	84
6.1	Sixty-seven anticipated short- and long-term benefits to smart meters in the UK	97
6.2	Estimated benefits to the smart meter implementation programme in the UK	99
7.1	Policy mix for achieving widespread comprehensive residential retrofit in the UK	123
8.1	Breakdown of end-uses, by useful exergy category and energy carrier group	141
8.2	Decomposing the drivers of useful exergy consumption in the UK, US and China over the period 1971–2010	143
9.1	Percentage change in key macroeconomic variables, relative to the baseline scenario, following a costless 10 per cent increase in household residential energy efficiency	160
9.2	Percentage change in key macroeconomic variables following a 5 per cent increase in Scottish household energy efficiency under alternative fiscal regimes	163
9.3	Changes in CO_2 emissions associated with a decreased spending in UK households use of UK EGWS outputs following a 10 per cent energy efficiency improvement	167
12.1	Relationship between policy development processes and the expected coherence and consistency of a policy mix	219
12.2	Overview of policy process theories and their application in transition studies	226
13.1	Strategic Niche Management guidelines and potential dilemmas	237
13.2	Content analysis coding framework	239
13.3	Indicative policy recommendations by analytical category	251
14.1	Six sociotechnical research debates and areas for future study	271

Contributors

Editors

Debbie Hopkins is Departmental Research Lecturer at the Transport Studies Unit and the School of Geography and the Environment, University of Oxford. She also holds a Junior Research Fellowship at Mansfield College, Oxford, and is a research affiliate of the Centre for Sustainability, University of Otago, New Zealand. In 2016, she co-edited *Low Carbon Mobility Transitions*, which built upon her research interests across climate change, low-carbon futures and sociotechnical transitions. At Oxford, she leads research into expectations of automation in freight, everyday experiences of UK truckers, and novel methodologies for researching mobile work.

Kirsten E.H. Jenkins is an early career Lecturer in Human Geography and Sustainable Development within the School of Environment and Technology (SET) at the University of Brighton. Prior to this, she was a Research Fellow in Energy Justice and Transitions within the Centre on Innovation and Energy Demand (CIED). Her background is as a sustainable development and human geography scholar with research interests that centre on energy justice, energy policy, and sustainable energy provision and use. She has published widely, serves as Managing Editor of *Energy Research & Social Science* and Associate Fellow of the Durham Energy Institute, and has worked on projects funded by the RCUK Energy Programme and ESRC.

Contributors

Noam Bergman is Researcher at the Science Policy Research Unit (SPRU), University of Sussex, UK. He has researched transitions to sustainability from a variety of perspectives, including low-carbon technologies such as microgeneration and electric vehicles, reorienting finance towards low-carbon development, and social innovations for sustainability, ranging from local networks to environmental activism. Prior to CIED his work includes involvement in the EU Project MATISSE on integrated sustainability assessment and a SuperGen consortium project on microgeneration, as well as a solo EDF-funded project on Sustainable Behaviour and Technology.

Paul Brockway is Senior Research Fellow at the School or Earth & Environment, University of Leeds, UK. His research addresses on an urgent global question: can we decouple energy use from economic growth to meet both climate and economic goals? He focuses on studying the interaction between energy use and society at the useful stage of the energy provision chain using exergy-based analysis, where exergy is 'available energy'. This emerging field is producing new insights into the significant role of energy in economic growth, and in turn the decoupling problem. He co-leads the international Exergy Economics research network (https://exergyeconomics.wordpress.com).

Donal Brown is Research Associate at CIED based at the SPRU at the University of Sussex and is also Sustainability Director at the Sustainable Design Collective. He holds a First-class BSc in Environmental Science and distinction in Climate Change and Policy MSc and is currently completing a PhD in domestic retrofit. He has worked for over ten years in all aspects of the construction industry. A sustainable energy and energy demand specialist in low-carbon housing, he researches and provides consultancy on energy efficiency, renewable energy and sustainable building solutions.

Gioele Figus is Research Associate in economics, working jointly between the Centre for Energy Policy and the Fraser of Allander Institute at the University of Strathclyde. He has expertise in modelling the impact of energy and environmental policy in the wide economy, through the development and use of large-scale economic models. He collaborates with the Scottish Government's centre of expertise on climate change where he is currently looking at integrating different large-scale modelling techniques for the analysis of energy efficiency and climate change actions. He has collaborated to different research projects funded by EPSRC, ESRC and UKERC.

Tim Foxon is Professor of Sustainability Transitions at SPRU, University of Sussex. His research explores the technological and social factors relating to the innovation of new energy technologies, the co-evolution of technologies and institutions for a transition to a sustainable low-carbon economy, and relations and interdependencies between energy use and economic growth. He is a member of the UK Energy Research Centre, the ESRC Centre for Climate Change Economics and Policy, CIED, and the new Centre for Research on Energy Demand Solutions. His has published 2 books, 60 academic articles, 15 book chapters and over 20 research reports. His book on the role of energy in past surges of economic development, and the implications for a low-carbon transition was published in October 2017.

Frank W. Geels is Professor of System Innovation and Sustainability at the University of Manchester. He is chairman of the international Sustainability Transitions Research Network (www.transitionsnetwork.org), and one of the world-leading scholars on sociotechnical transitions. Working in the field of innovation studies, he aims to understand the sociotechnical dynamics of sustainability transitions, analysing how firms, policymakers, consumers, public opinion and non-governmental organisations (NGOs) shape the emergence and diffusion of innovations. He has analysed sustainability transitions in electricity, transport and agro-food, and has published 5 books and 65 peer-reviewed articles. He was selected in the Thomson Reuters list of 'Highly Cited Researchers', identified as one of *The World's Most Influential Scientific Minds 2014*.

Sabine Hielscher is Research Fellow at the SPRU, University of Sussex and at Zentrum Technik und Gesellschaft (ZTG), Technische Universität Berlin. She is interested in the politics, processes and materialisations of civil society activities and the dynamics of everyday (sustainable) consumption patterns. Her recent work examines the expectations and visions behind large-scale 'smart' sociotechnical futures. Over the years, she has collaborated with a variety of civil society, public and business organisations, taking an interdisciplinary approach. She has worked on research projects funded by the AHRC, EPSRC, RCUK Energy Programme, European Commission and BMBF.

Antonios Katris is Research Associate at the Centre for Energy Policy, University of Strathclyde, UK. He was worked on the potential multiple benefits of energy efficiency at both the Scottish and UK level using economy-wide modelling. He also has some experience on the input–output analysis of international supply chains, primarily to identify points of significant carbon emissions and on topics of wider economic interest such as embodied labour cost. He has been involved as researchers on projects funded by EPSRC and ESRC.

Florian Kern is Head of the research field Ecological Economics and Environmental Policy at the Institute for Ecological Economy Research (IÖW) in Berlin, Germany, where he also leads on the cross-cutting themes 'Technology and Innovation' and 'Environmental Policy and Governance'. His research focuses on the governance of sociotechnical transitions towards sustainability. He is an associate editor of *Research Policy* and a member of the steering group of the Sustainability Transitions Research Network. Until May 2018 he was a Senior Lecturer at SPRU and Co-Director of the Sussex Energy Group and led a cross-cutting project on policy mixes as part of CIED.

Paula Kivimaa is Senior Research Fellow at the SPRU, University of Sussex and Senior Researcher in the Finnish Environment Institute (SYKE), Finland. She is an expert in sustainability transition and innovation studies, with a focus on the interface between policy and innovation and policy coordination. Her current research focuses on intermediaries, experiments and policy mixes in low-energy transitions.

Mari Martiskainen is Research Fellow at Sussex Energy Group, SPRU, University of Sussex. Her research centres around the transition to a fairer, cleaner and more sustainable energy world. She has worked with a range of conceptual approaches, including sustainability transitions, innovation intermediation, user innovation, power and politics. She has authored several articles in journals such as *Energy Research & Social Science, Environmental Innovation and Societal Transitions, Environment and Planning A* and *Research Policy*. She has written book chapters, conference proceedings and invited blog posts, and presents her research regularly to a range of audiences. In 2018, she was selected as a member of the Mayor of London Fuel Poverty Partnership.

Jack Miller was a postgraduate student with CIED between 2014 and 2017, conducting research into energy–economy interactions using the thermodynamic principles of exergy. He is now a researcher and adviser at the Parliamentary Office of Science & Technology, where he provides analysis of research evidence to UK parliamentarians. He has a background in energy policy and physics.

Cameron Roberts, University of Leeds, is Research Fellow studying the political economy of transport provision. His research background includes the history of science and technology, science and technology studies, and transitions theory. He has published on the role of politics and public story-lines in sociotechnical transitions, and on the deliberate acceleration of transitions to sustainability.

Karoline Rogge is Senior Lecturer in Sustainability Innovation and Policy at the SPRU and Co-Director of the Sussex Energy Group at the University of Sussex. In addition, she is Senior Researcher at the Fraunhofer Institute for Systems and Innovation Research (Fraunhofer ISI) in Karlsruhe, Germany. Her interdisciplinary research combines insights from innovation studies, environmental economics and political science to investigate the link between policy mixes and innovation, with a focus on low-carbon energy transitions. She received her PhD from ETH Zurich, Switzerland, with an empirical analysis of the innovation impact of the EU Emissions Trading System.

Jan Rosenow is Director of European Programmes at the Regulatory Assistance Project (RAP), a global team of highly-skilled energy experts. He is responsible for all aspects of leadership, management, and financial viability of RAP's work in Europe. He is also an Honorary Research Associate at Oxford University's Environmental Change Institute, an Associate Fellow at Sussex Energy Group of SPRU, University of Sussex and at the Free University of Berlin.

Tim Schwanen is Associate Professor of Transport Studies and Director of the Transport Studies Unit at the University of Oxford. His research interests are varied but many concentrate on radical and just transitions in the sociotechnical systems for the mobility of people, goods and information. He has published widely on these and other topics in journals in geography, urban studies, transport research and interdisciplinary science.

Steve Sorrell is Professor of Energy Policy in the SPRU, University of Sussex and Co-Director of CIED. He has undertaken a range of research on energy and climate policy, with particular focus on energy efficiency and resource depletion. This work is primarily informed by economics and has included case studies, econometric analysis and modelling. He has led several UK and international research projects and has acted as consultant to the European Commission, the UN, UK government departments, private sector organisations and NGOs.

Benjamin K. Sovacool, University of Sussex, is Professor of Energy Policy at the SPRU at the School of Business, Management, and Economics. There he serves as Director of the Sussex Energy Group and Director of the Center on Innovation and Energy Demand, which involves the University of Oxford and the University of Manchester. He works as a researcher and consultant on issues pertaining to energy policy, energy security, climate change mitigation and climate change adaptation. More specifically, his research focuses on renewable energy and energy efficiency, the politics of large-scale energy infrastructure, designing public policy to improve energy security and access to electricity, and building adaptive capacity to the consequences of climate change.

Karen Turner is Director of the Centre for Energy Policy (CEP) at the University of Strathclyde. Her main research interests lie in developing economy-wide modelling methods to explore the wider societal value delivered by, and whether 'the macroeconomic case' can be made for policy support of, a range of low-carbon energy policy solutions, with current focus on energy efficiency, hydrogen and Carbon Capture and Storage (CCS). She was Principle Investigator on the EPSRC Working with the End Use Energy Demand (EUED) Centres project titled 'Energy Saving Innovations and Economy-wide Rebound Effects' and has participated in and led a range of previous interdisciplinary research projects on energy policy relevant topics.

Janette Webb is Professor of Sociology of Organisations at the University of Edinburgh, UK. Her research is about energy and climate change. In collaboration with the EPSRC CIED, she is studying comparative European heat and energy efficiency policies and practices, with a particular focus on innovation in cities. Additional work is analysing local government energy developments in the UK and evaluating the Scottish Energy Efficiency Programme. She leads the Heat and the City research group and is a Co-Investigator in the UK Centre for Research on Energy Demand.

Preface

The research presented throughout this book represents the culmination of the work conducted in the first five years of the Centre on Innovation and Energy Demand (CIED), one of six End Use Energy Demand Centres funded by the Research Council United Kingdom (RCUK) Energy Programme. CIED sits at the forefront of research on the transition to a low-carbon economy, investigating new technologies and new ways of doing things that have the potential to transform the way energy is used and to achieve substantial reductions in energy demand. The research conducted by CIED is interdisciplinary, drawing on ideas from economics, history, innovation studies, sociology and geography. It is also multi-method, including qualitative and quantitative techniques ranging from historical and contemporary case studies, surveys, modelling to econometric analysis. Finally, it is practical, working with stakeholders to investigate the adoption of low-energy innovations relevant for a variety of sectors, including transport, industry, households and non-domestic buildings.

The authors featured in this book are either members of the core CIED groups at the Universities of Sussex, Manchester and Oxford, or members of the affiliate organisations, the Universities of Strathclyde and Edinburgh. All authors wish to give thanks to the RCUK for enabling the research presented within this book, as well as to the participants, collaborators, and Advisory Board members who have shaped this research. Both editors would also like to acknowledge and sincerely thank Cameron Roberts for his extensive commitment to the project, and Asa Morrison for his assistance with the formatting of the final volume.

1 Introduction

New directions in energy demand research

Kirsten E.H. Jenkins, Steve Sorrell,
Debbie Hopkins and Cameron Roberts

Introduction

Meeting the goal enshrined in the Paris Agreement of limiting global temperature increases to less than 2°C above pre-industrial levels demands rapid reductions in global carbon dioxide (CO_2) emissions. For example, the International Energy Agency (IEA) estimates that to provide a high likelihood (66 per cent probability) of meeting that target, cumulative global CO_2 emissions between 2015 and 2100 must be less than 880 Giga-tonnes (Gt) (IEA, 2017). For the energy sector alone, the IEA estimate a smaller 'carbon budget' of 790 Gt. To put this in perspective, global energy sector emissions stood at 32.5 Gt in 2017 – an increase of 1.4 per cent on the previous year and equivalent to ~4 per cent of the remaining budget (IEA, 2018). If emissions continue at this level, the budget will be exhausted in less than 25 years. Hence, to achieve the 2°C target, energy-related carbon emissions must fall very rapidly. The IEA estimate that emissions must fall by ~70 per cent by 2050[1] – implying a near complete decarbonisation of the electricity sector, retrofitting of the entire existing building stock, a major shift towards low-emission vehicles and an 80 per cent reduction in the carbon intensity of industrial sectors (IEA, 2017). By the end of the century, any residual anthropogenic CO_2 emissions would need to be balanced by CO_2 removals from the atmosphere.

There is no historical precedent for transforming energy systems at this scale and at this speed. Achieving this goal will require the rapid and extensive deployment of low-carbon technologies throughout all sectors of the global economy, with far-reaching implications for markets, infrastructures, institutions, social practices and cultural norms. What is more, emission reduction efforts will simultaneously have to address other concerns, including questions of social justice, energy access and energy security.

There is certainly some degree of political ambition to revolutionise the energy landscape. The 2016 Paris Agreement provides a strong basis for global mitigation efforts and these in turn have encouraged (and have been facilitated by) major improvements and cost reductions in renewable energy, electric vehicles, energy storage and other low-carbon technologies (UNFCC, 2015). Electricity from wind and solar is projected to be cheaper than fossil fuels by the mid-2020s, and global trends show a rapid uptake of these and other low-carbon technologies (Bloomberg New Energy Finance, 2018; IEA, 2017). Modern renewables now provide 10 per cent of global final energy demand and more than a quarter of global electricity generation, with a record 157 gigawatts (GW) being commissioned in 2017 (Frankfurt School–UNEP Centre/BNEF, 2018).

Yet despite these encouraging trends the rate of progress remains too slow, particularly in relation to improving energy efficiency and reducing energy demand. Global primary energy intensity (the ratio of primary energy consumption to GDP) fell by 1.2 per cent in 2017, but this is less than half the rate required to meet the 2°C target (IEA, 2017). While a business as usual scenario suggests a ~40 per cent increase in global primary energy demand by 2050, a 2°C scenario suggests practically no increase – unless negative emission technologies are deployed (IEA, 2017). There are a growing number of policy initiatives targeting energy demand, but many of these focus upon incremental technological improvements (e.g. insulation) and necessitate only modest changes in energy-related behaviour. But to meet the emission reduction targets, we must achieve *radical* changes in energy demand throughout all sectors of the global economy. Since only limited increases in global energy demand appear compatible with ambitious climate targets (Loftus *et al.*, 2015), developing countries must follow very different development paths than have been observed historically – leapfrogging to highly energy-efficient technologies and providing high levels of human welfare with much lower energy consumption that has been required in the past (Steckel *et al.*, 2013). And to allow space for increased energy demand in the developing world, there will need to be absolute reductions in energy demand in the developed world. Few countries have achieved this in the past, and it is likely to prove very challenging.

Reducing energy demand

The IEA estimates that improved energy efficiency and reduced energy demand could contribute up to half of the reductions in global carbon emissions over the next few decades (IEA, 2012a; IEA, 2015). In other words, changes in energy demand could contribute as much carbon abatement as all the low-carbon energy supply options combined. Similarly, the United Kingdom (UK) government has recognised that reducing energy demand can be a highly cost-effective approach to reaching climate targets, and positions both energy demand reduction and increased energy efficiency as core policy goals (DTI, 2003, 2007; DECC, 2011). But questions remain on how best to achieve these goals.

The demand for energy is driven by the demand for energy services, such as thermal comfort, illumination and mobility. Energy services form the last stage of an energy chain that begins with primary energy sources such as crude oil and nuclear power, continues through secondary energy carriers such as gasoline and electricity and then through end-use conversion devices such as boilers, furnaces, motors and lightbulbs. These conversion devices provide 'useful energy' such as low- and high-temperature heat, mechanical power and electromagnetic radiation, which in turn is preserved or trapped within 'passive systems' for a period of time to produce final energy services (Cullen and Allwood, 2010). So, for example, the heat delivered from a boiler (conversion device) is held within a building (passive system) for a period of time to provide thermal comfort (energy service).

It follows that there are three ways to reduce energy demand:

1. Improve conversion efficiencies and reduce transmission losses at all stages of the energy chain, including from primary to final energy (e.g. more efficient power stations) and from final to useful energy (e.g. more efficient boilers, engines and refrigerators).
2. Improve the ability of passive systems to trap energy for periods of time (e.g. more aerodynamic vehicles, better insulated buildings).
3. Reduce demand for energy services, such as heating, lighting and cooling, (e.g. lower internal temperatures, fewer overseas flights).

These changes can be achieved through a combination of retrofitting existing technologies (e.g. insulating a house), investing in new technologies (e.g. installing a condensing boiler) and changing energy-related behaviour (e.g. turning off lights when not in use). The latter in turn may involve either restraint (e.g. turning the thermostat down, giving up flying) or substitution by less energy-intensive services (e.g. shifting from cars to buses). Large improvements in energy efficiency are often associated with simultaneous shifts towards different energy carriers – such as replacing gas boilers with (more efficient) electric heat pumps or replacing gasoline cars with (more efficient) battery–electric vehicles. But much of the potential for reducing energy demand requires inter-linked changes in *all* of these areas. More fundamentally, radical reductions in energy demand are likely to require transitions to entirely *new systems* for providing energy services – such as intermodal transport, compact cities, and smart homes.

None of these options are straightforward and the complexity of the processes involved can easily be underestimated. Sorrell (2015) notes, for instance, that previous attempts to reduce energy demand have often proved unsuccessful; the assumptions on which policy interventions are based do not always reflect either the challenge involved or the factors shaping individual and organisational decision-making; and the complexity of economic systems can undermine the success of even well-designed interventions. There are numerous stumbling blocks on the road to energy demand reductions:

- **Reducing energy demand is complex:** Historically, economic growth has been closely linked to increased energy consumption, and few countries have achieved 'absolute decoupling' of primary energy consumption from gross domestic product (GDP) (see Chapter 8). The expectation that improved energy efficiency will lead to proportional reductions in energy demand can be misleading (Sorrell, 2009). The links between efficiency and demand are complex and rebound effects – in which consumers increase their consumption of energy services to take advantage of the fact that these services are now cheaper – can partly offset and sometimes completely eliminate the associated energy savings. In this regard, projections of the impact of policy instruments on energy demand often rely upon oversimplified assumptions (Wilhite et al., 2000; Sorrell, 2015).

- **Large-scale, rapid change is required:** Previous energy transitions (e.g. from localised wood use to centralised fossil fuels) have generally been long and arduous affairs (Smil, 2010). There may be some hope here, as past transitions have generally not been the result of deliberate government intervention (Geels et al., 2017). Yet the urgency of the climate change agenda means that we require larger, faster and more pervasive changes than have been achieved before, supported by policy efforts that have not existed in previous energy transitions (Sovacool, 2016). Such efforts will require substantial and sustained political commitment, combined with global cooperation in the face of powerful incentives to defect and free ride.

- **Energy demand is rising:** Even if the most optimistic forecasts for the upscaling of low-carbon energy supply are exceeded, increases in energy consumption will blunt their impact. Decarbonisation of energy supply must be combined with a break with the historically observed relationship between energy consumption and economic growth. If the rate of decarbonising energy supply is less than anticipated by the more optimistic scenarios, climate targets will only be achieved through greater efforts to reduce energy demand or the deployment of negative emission technologies. Given the uncertainties associated with the latter (Anderson and Peters, 2016), reducing energy demand must be a priority.

- **Societies are disinclined to change:** Energy demand is shaped by large-scale, capital-intensive and long-lived technologies and infrastructures (e.g. transport systems, buildings) that constrain the feasible rate of change. This inertia is reinforced by the entrenched habits and social practices that develop alongside these technologies and infrastructures, together with powerful political interests that resist change (Rosenbloom and Meadowcroft, 2014). For example, policies aimed at reducing automobile dependence face a backlash from motorists whose work and leisure patterns are built around the private car, and from motor and fossil fuel industries whose economic interests are threatened (Dudley and Chatterjee, 2011). For this reason, energy- and carbon-intensive forms of energy service provision continues to dominate and will be difficult to dislodge.

- **Carbon pricing is insufficient:** Carbon pricing can encourage reductions in energy demand and carbon emissions but is unlikely to be sufficient by itself. Carbon prices remain much lower than required to meet ambitious climate targets and attempts to raise them must overcome formidable political obstacles (Loftus et al., 2015). Carbon pricing can encourage organisations and individuals to pursue energy efficiency, but many are locked in to energy-inefficient systems and practices, with the costs of switching to more efficient systems frequently offsetting the financial benefits of lower energy consumption (see Gillingham et al., 2012). Moreover, the economic theories underpinning carbon pricing provide a poor guide to real-world individual and organisational behaviour (Brown, 2001; Wilson and Dowlatabadi, 2007).

- **Current policies neglect innovation:** Energy demand reduction requires rising energy/carbon prices alongside policies to reduce the economic barriers to improved energy efficiency (Sorrell et al., 2004). It requires interventions that encourage individuals and households to adopt *existing* energy-efficient technologies and practices, alongside support for *new* energy-efficient technologies throughout all stages of the innovation chain. But many policy measures are underrepresented in the current policy mix (e.g. innovation support) while others are confined to relatively incremental improvements (e.g. insulation). Thus, in the face of multiple barriers, current policy approaches appear insufficient.

Two things are clear from the preceding discussion. First, to reach our climate change targets, we must significantly reduce energy demand relative to business-as-usual scenarios, and possibly also in absolute terms. Second, the pathways to doing so defy simple or straightforward solutions. This brings us to the challenge of finding the most effective approach, and to the contribution of a 'sociotechnical' perspective on energy demand.

Perspectives on reducing energy demand: the sociotechnical approach

The challenge of reducing energy demand has been approached from many different theoretical perspectives including neoclassical economics (focusing on economic barriers to energy efficiency), social psychology (focusing on cognitive, emotional and affective influences on energy-related choices) and social practice theory (focusing on how habitual behaviour and social norms shape energy demand). Each approach offers valuable insights, but also has blind spots and weaknesses – particularly in relation to achieving more radical reductions in energy demand. This book therefore proposes a complementary *sociotechnical perspective* that can overcome some of these limitations. The sociotechnical approach is well established in the academic literature but has rarely been applied to energy demand.

A distinguishing feature of the sociotechnical approach is the expansion of the unit of analysis from individual technologies to the *sociotechnical systems* that provide energy services such as thermal comfort and mobility. Sociotechnical systems are understood as the interdependent mix of social and technical entities that function collectively to deliver specific energy services. They include physical artefacts (e.g. infrastructures, conversion technologies, passive systems), social arrangements (e.g. firms, supply chains, markets, regulations) and intangible elements such as skills, habits, routines, expectations and social norms (Geels, 2004). The sociotechnical system associated with electricity, for example, includes: physical artefacts such as power stations and transmission lines; social arrangements such as electricity markets, technical standards and industry associations; and intangible elements such as electrical engineering skills and the social practices associated with electricity provision and use (Hughes, 1983). Sociotechnical systems develop over many decades and the alignment and co-evolution of the different elements leads to mutual dependence and resistance to change (Geels, 2002). Since the configuration of sociotechnical systems shapes the level, nature and pattern of energy demand, significant reductions in energy demand requires not just changes in individual technologies, but far-reaching changes in the sociotechnical systems themselves. We term such changes *sociotechnical transitions*.

The sociotechnical perspective has its roots in the study of innovation but differs from more conventional approaches to innovation by: first, focusing on broader systems and processes of long-term change in those systems; and second, understanding innovation as both a technical and social process that necessitates complex relationships between a range of actors (including firms, researchers, policymakers and consumers). These actors develop strategies, make investments, learn, open up new markets and develop new routines. As an example, a sociotechnical account of transitions in the electricity system would include the changes within and interrelationships between: public policies and industry regulators; the strategies of generation, network and supply companies; the practices of electricity consumers; and the cognitive, normative and regulative rules that underpin different elements of the system (Geels, 2002; Hammond et al., 2013).

This book investigates how transitions in sociotechnical systems occur and their potential contribution to reducing energy demand. We assume that such transitions centre around particular *low-energy innovations* – defined as technologies or social practices that differ significantly from existing technologies and practices and have the potential to radically improve energy efficiency and/or reduce energy demand. An example would be the central role of heat pumps in a transition from gas to electric heating systems. We seek to make a distinctive contribution to the energy demand literature by developing a sociotechnical understanding of the emergence, diffusion and impact of such innovations. We aim to uncover the processes and mechanisms through which different types of low-energy innovation become (or fail to become) established, identify the role of different groups, explore the resulting impacts on energy demand and other social goals, and develop practical recommendations for both encouraging the diffusion of such innovations and maximising their long-term impact.

Our approach rests upon two assumptions. First, innovations must be situated and studied within broader sociotechnical systems, particularly when their diffusion is associated with fundamental changes in those systems (sociotechnical transitions). Second, to have a significant impact on energy demand, such innovations should be technologically radical, socially radical, or a combination of the two – what Dahlin and Behrens (2005) term 'systemically radical'. Radical innovations disrupt established sociotechnical systems – in this case the dominant energy- and carbon-intensive systems – and lead to far-reaching changes in the nature and functioning of those systems (see Chapter 2).

UK policy on energy demand

This book focuses primarily on the UK, one of a small number of countries that have made significant progress in reducing energy demand. Between 2001 and 2017, UK GDP grew by 31 per cent and population grew by 11.7 per cent, but primary energy demand fell by 19 per cent. These reductions have partly been achieved by the diffusion of low-energy innovations – such as energy-efficient lighting, appliances, boilers, electric motors and vehicles – and these in turn have been encouraged by policies such as building regulations, appliance standards and energy efficiency obligations. But demand reductions have also resulted from economic restructuring and the 'offshoring' of energy-intensive manufacturing to other countries. In this regard, reductions in UK energy use and emissions have been offset by increased energy use and emissions elsewhere. While such reductions may contribute to UK climate targets, they do little to address global climate change. Barrett et al. (2013) estimate, for instance, that while the UK's territorial greenhouse gas (GHG) emissions fell by 27 per cent between 1990 and 2008, it's 'consumption-based' emissions increased by ~20 per cent as a consequence of imported consumer goods displacing (more energy efficient) domestic production.

The UK has set long-term, legally binding targets for reducing GHG emissions and has established an independent Committee on Climate Change (CCC) to set intermediate targets and oversee progress. But the CCC (2018) warns that the UK is not on course to meet its 'carbon budgets' and that urgent action is required to both bring forward new policies and to reduce the risk of existing policies failing to deliver. Despite the UK's progress to date, new measures are urgently required to deliver deeper and faster improvements in energy efficiency, particularly in 'more difficult' sectors such as domestic heating (Shove, 2017; Staffell, 2017).

The UK has long history of energy efficiency policies and several of these have been very successful – including the series of obligations on energy suppliers to improve household energy efficiency (Mallaburn and Eyre, 2014) and the EU standards on the energy efficiency of domestic appliances. But there have also been notable failures, including the flagship Green Deal policy that was intended to deliver large-scale energy efficiency retrofits but was terminated only two years after its launch (Rosenow and Eyre, 2016). The IEA observes that UK energy efficiency policy has neglected security of supply and other concerns (IEA, 2006, 2012b; Kern et al., 2017), the CCC criticise the large-scale decline in investment after 2013 and the current dearth of policy initiatives, and Hardt et al. (2018) highlight the slowdown in the rate of efficiency improvement in industry and the limited scope for further energy savings through offshoring.

8 *Kirsten E.H. Jenkins et al.*

Overall then, the UK serves as both an exemplar of successful measures and a cautionary tale. While offshoring is clearly unsustainable in terms of reaching global emissions goals, the UK provides some good examples of what can be done, as well as what should be avoided. Several of these cases are covered in this book[2] and provide lessons that are relevant to a range of contexts.

About the book

This book is based upon research by the Centre on Innovation and Energy Demand (CIED), a five-year, social science research centre funded by the UK Research Councils. Focusing primarily on the UK, the book uses a sociotechnical approach to explore the challenge of reducing energy demand. The book includes theoretical discussions, literature reviews and a series of empirical case studies organised around the themes of *emergence*, *diffusion* and *impact* of low-energy innovations. The chosen cases include both new technologies (e.g. smart meters, vehicle automation and district heating) and new organisational arrangements (e.g. integrated policy mixes) that either have or could have significant impacts on energy demand. The book has the dual aim of improving the academic understanding of sociotechnical transitions and energy demand and providing practical recommendations for public policy.

Structure

This booked is structured around the themes of emergence, diffusion and impact – introduced in full in Chapter 2. We do not argue that all innovations follow a linear progression between these stages (which often overlap), but instead present them as a useful framework for conceptualising the innovation journey.

Emergence: The term *emergence* does not refer to the initial invention of new ideas (e.g. from scientific research), but the introduction of those ideas into society. Emerging technologies, behaviours, institutional arrangements and business models struggle to become established against more dominant systems and can easily fail. Before innovations can break through into broader markets, space needs to be created for learning and improvement, for the building of social networks and for stabilisation around a dominant configuration or design. The chapters on emergence examine these processes for specific low-energy innovations and uncover the conditions for their success.

Diffusion: Innovations spread when their performance improves and costs fall as a result of network, scale and learning economies; when public policies support their adoption; and when they become aligned with people's expectations and behaviours. Diffusion does not happen into an 'empty' world, but in the context of existing sociotechnical systems that provide barriers and active resistance. Many low-energy innovations are not intrinsically attractive to the majority of consumers since they are often (initially) more expensive and perform less well on key dimensions. The chapters on diffusion explore the mechanisms driving this process for selected low-energy innovations, and examine how infrastructures, business models, social norms, values and public policies need to change for such innovations to succeed.

Impact: The diffusion of low-energy innovations will only contribute to climate goals if they lead to significant reductions in economy-wide energy consumption. But research on innovation and sociotechnical transitions has paid relatively little attention to the ultimate impact of innovations on energy demand or other social goals. More generally, the links between economic growth, energy efficiency and energy consumption remain poorly understood. The chapters on impact therefore employ both orthodox and novel methods for estimating the historical impacts of low-energy innovations and for projecting their potential future impacts.

Chapters

Across each chapter, *Transitions in Energy Efficiency and Demand* moves from contextually-specific first principles through to empirical research in selected areas and, finally, to ideas for how these systems can be most effectively be changed. While each chapter is structured differently, they all include specific policy recommendations.

The first section of the book, 'Analytical perspectives' provides a conceptual and normative orientation to the problem of reducing energy demand. Chapter 2, provides a theoretical primer on the problems addressed by this book, and the potential contribution of the sociotechnical approach. This includes an overview of the 'multi-level perspective' on sociotechnical transitions and a survey of key debates relevant to emerge, diffusion and impact. Chapter 3 adds an ethical dimension to this discussion, considering the broader normative problems of energy provision through a case study of fuel poverty in the UK.

Chapter 4 begins the section on 'The emergence and diffusion of innovations' by considering visions of personal transport futures in the UK and the role of electric vehicles therein. It argues that policymakers would benefit from engaging with a variety of future visions, including the possibility of disruption and shocks and the failure to meet emission reduction targets. Chapter 5 then examines the experimentation with automated vehicles that is underway in several UK cities. It points to the highly managed processes of these experiments (e.g. where and when experiments occur, who is included/excluded, what counts as an experiment), which limit the opportunity for second-order learning and surprises. Chapter 6 explores the evolution of the UK smart meter rollout, including the obstacles faced and the potential implications for energy justice and consumer vulnerability. Lastly, Chapter 7 investigates the mammoth task of comprehensively upgrading UK residential buildings, highlighting the need for consistent and ambitious policy targets; the importance of new business models and finance mechanisms; and the role of intermediary actors in supporting policy implementation.

Moving on to the 'Societal impacts and co-benefits' section, Chapter 8 explores the importance of energy for economic growth and summarise a number of recent studies which suggest that efficiency improvements are key driver of growth and that the rebound effects from those improvements can be large. Chapter 9 is more forward-looking, using macroeconomic modelling to explore the economy-wide impacts of UK household energy efficiency improvements. They show how these can stimulate economic activity, leading to increased employment, investment and savings, and argue that a focus on rebound effects can obscure the wider economic and social benefits of improved energy efficiency.

The section on 'Policy mixes and implications' considers the policy frameworks for facilitating low-energy innovation. Chapter 10 uses a series of historical cases studies to investigate how policymakers can deliberately accelerate sociotechnical transitions – highlighting the importance of 'disarming' resistance from incumbent actors, popular support for the transition and the level of maturity of the core innovation. Chapter 11 goes on to discuss the challenge of delivering energy efficiency policy in the UK, arguing that political sensitivities about energy prices, neglect of the social benefits of energy efficiency and rigid adherence to neoclassical economic theory have hampered effective policy. This feeds directly into Chapter 12 on policy mixes for energy demand reduction. This chapter draws on the emerging policy mixes for energy transitions literature and highlights the comparative neglect of energy efficiency policy mixes. It goes on to summarise the empirical findings conducted as part of CIED with a view to both: (1) drawing out overall insights and avenues for future research and (2) establishing policy reflections on design principles for policy mixes in which energy efficiency plays a key role. Closing this section, Chapter 13 reviews the literature on Strategic Niche Management (SNM), identifying some lessons for both researchers and policymakers working towards low-energy transitions.

The conclusion (Chapter 14) summarises and elaborates the contributions of each chapter and develops a summative list of conceptual and policy principles for accelerating energy demand reduction. Taken together the chapters provide a comprehensive, sociotechnical account of the energy demand challenge and provide both new empirical results and practical suggestions for achieving meaningful change. We hope you enjoy!

Notes

1 This is a higher rate of reduction than assumed in many scenarios, since it excludes the possibility of temporarily overshooting the 2°C target and compensating subsequently through the use of negative emission technologies.
2 Although we also reference Denmark, Japan, Finland, New Zealand and the Netherlands, for example.

References

Anderson, K. and Peters, G. (2016) The trouble with negative emissions. *Science* 354(6390): 182–183.

Barrett, J., Peters, G., Wiedmann, T., Scott, K., Lenzen, M., Reolich, K. and Le Quuéré, C. (2013) Consumption-based GHG emissions accounting: A UK case study. *Climate Policy* 12(4): 451–470.

Bloomberg New Energy Finance (2018) New Energy Outlook 2018. Bloomberg, London, UK and New York, USA. Available at: https://about.bnef.com/new-energy-outlook/#toc-download.

Brown, M.A. (2001) Market failures and barriers as a basis for clean energy policies. *Energy Policy* 29(14): 1197–1207.

CCC (2018) Reduce UK Emissions: 2018 Progress Report to Parliament. Committee on Climate Change, London, UK. Available at: www.theccc.org.uk/publication/reducing-uk-emissions-2018-progress-report-to-parliament/.

Cullen, J.M. and Allwood, J.M. (2010) The efficient use of energy: Tracing the global flow of energy from fuel to service. *Energy Policy* 38(1): 75–81.

Dahlin, K.B. and Behrens, D.M. (2005) When is an invention really radical? Defining and measuring technological radicalness. *Research Policy* 34(5): 717–737.

DECC (2011) Planning Our Electric Future: A White Paper for Secure, Affordable and Low-carbon Electricity. Department of Energy and Climate Change, HM Government, London, UK.

DTI (2003) Energy White Paper: Our Energy Future – Creating a Low Carbon Economy. Department of Trade and Industry, London, UK.

DTI (2007) Meeting the Energy Challenge: A White Paper on Energy. HM Government, Department of Trade and Industry, London, UK.

Dudley, G. and Chatterjee, K. (2011) The Dynamics of Regime Strength and Instability: Policy Changes to the Dominance of the Private Car in the United Kingdom. In: Geels, F.W., Kemp, R., Dudley, G. and Lyons, G. (Eds) *Automobility in Transition? A Socio-Technical Analysis of Sustainable Transport*. Routledge, London, UK, 83–103.

Frankfurt School–Centre/BEFT (2018) Global Trends in Renewable Energy Investment. Frankfurt School of Finance & Management gGmbH, Frankfurt, Germany.

Geels, F.W. (2002) Technological transitions as evolutionary reconfiguration processes: A multi-level perspective and a case-study. *Research Policy* 31(8–9): 1257–1274.

Geels, F.W. (2004) From sectoral systems of innovation to socio-technical systems: Insights about dynamics and change from sociology and institutional theory. *Research Policy* 33(6–7): 897–920.

Geels, F., Sorrell, S., Schwanen, T., Jenkins, K. and Sovacool, B.K. (2018) Reducing demand through low-energy innovation: A sociotechnical review and critical research agenda. *Energy Research and Social Science* 40: 23–35.

Geels, F., Sovacool, BK., Schwanen, T. and Sorrell, S. (2017) The socio-technical dynamics of low-carbon transitions. *Joule* 1(3): 463–479.

Gillingham, K., Harding, M. and Rapson, D. (2012) Split incentives in residential energy consumption. *The Energy Journal* 33(2): 37–62.

Hammond, P.G., Howard, R.H. and Jones, I.J. (2013) The energy and environmental implications of UK more electric transition pathways: A whole systems perspective. *Energy Policy* 52: 103–116.

Hardt, L., Owen, A., Brockway, P., Heun, M.K., Barrett, J., Taylor, P.G. and Foxon, T.J. (2018) Untangling the drivers of energy reduction in the UK productive sectors: Efficiency or offshoring? *Applied Energy* 223(1): 124–133.

Hughes, T.P. (1983) *Networks of Power: Electrification in Western Society, 1880–1930*. John Hopkins University Press, Baltimore, Maryland, USA.

IEA (2006) Energy Policies of IEA Countries: The United Kingdom 2006 Review. International Energy Agency, Paris, France.

IEA (2012a) World Energy Outlook 2012. International Energy Agency, Paris, France.

IEA (2012b) Energy Policies of IEA Countries: The United Kingdom 2012 Review. International Energy Agency, Paris, France.
IEA (2015) World Energy Outlook Special Report 2015: Energy and Climate Change. International Energy Agency, Paris, France.
IEA (2017) Perspectives for the Energy Transition: Investment Needs for a Low-carbon Energy System. International Energy Agency, Paris, France.
IEA (2018) Global Energy & CO_2 Status Report 2017. International Energy Agency, Paris, France.
Kern, F., Kivimaa, P. and Martiskainen, M. (2017) Policy packaging or policy patching? The development of complex energy efficiency policy mixes. *Energy Research and Social Science* 23: 11–25.
Loftus, P.J., Cohen, A.M., Long, J.C. and Jenkins, J.D. (2015) A critical review of global decarbonization scenarios: what do they tell us about feasibility? *Wiley Interdisciplinary Reviews: Climate Change* 6: 93–112.
Mallaburn, P.S. and Eyre, N. (2014) Lessons from energy efficiency policy and programmes in the UK from 1973 to 2013. *Energy Efficiency* 7(1): 23–41.
Rosenbloom, D. and Meadowcroft, J. (2014) The journey towards decarbonization: Exploring socio-technical transitions in the electricity sector in the province of Ontario (1885–2013) and potential low-carbon pathways. *Energy Policy* 65: 670–679.
Rosenow, J. and Eyre, N. (2016) A post mortem of the Green Deal: Austerity, energy, efficiency, and failure in British energy policy. *Energy Research and Social Science* 21: 141–144.
Shove, E. (2017) What is wrong with energy efficiency? *Building Research and Information* 46(7): 1–11. DOI: 10.1080/09613218.2017.1361746.
Smil, V. (2010) *Energy Transitions: History, Requirements, Prospects*. Praeger/ABC-CLIO, Santa Barbara, USA.
Sorrell, S. (2009) Jevons' Paradox revisited: The evidence for backfire from improved energy efficiency. *Energy Policy* 37(4): 1456–1469.
Sorrell, S. (2015) Reducing energy demand: A review of issues, challenges and approaches. *Renewable and Sustainable Energy Reviews* 47: 74–82.
Sorrell, S., Schleich, J., O'Malley, E. and Scott, S. (2004) *The Economics of Energy Efficiency: Barriers to Cost-Effective Investment*. Edward Elgar, Cheltenham, UK.
Sovacool, B.K. (2016) How long will it take? Conceptualizing the temporal dynamics of energy transitions. *Energy Research and Social Science* 13: 202–215.
Staffell, I. (2017) Measuring the progress and impacts of decarbonising British electricity. *Energy Policy* 102: 463–475.
Steckel, J.C., Brecha, R.J., Jakob, M., Strefler, J. and Luderer, G. (2013) Development without energy? Assessing future scenarios of energy consumption in developing countries. *Ecological Economics* 90: 53–67.
UNFCCC (2015) *United Nations Framework Convention on Climate Change*. United Nations, New York, USA.
Wilhite, H., Shove, E., Lutzenhiser, L. and Kempton, W. (2000) The Legacy of Twenty Years of Energy Demand Management: We Know More about Individual Behaviour But Next to Nothing about Demand. In: Jochem, E., Sathaye, J., Bouille, D. (Eds) *Society, Behaviour, and Climate Change Mitigation. Advances in Global Change Research Vol. 8*. Springer, Dordrecht, Germany.
Wilson, C. and Dowlatabadi, H. (2007) Models of decision making and residential energy use. *Annual Review of Environment and Resources* 23: 169–203.

Part I
Analytical perspectives

2 Of emergence, diffusion and impact

A sociotechnical perspective on researching energy demand

Frank W. Geels, Benjamin K. Sovacool and Steve Sorrell

Introduction

Improvements in energy efficiency are widely expected to contribute more than half of the reductions in global carbon emissions over the next few decades (IEA, 2012) and are considered critically important to delivering the pledges made in the Paris Agreement (IEA, 2015). These improvements are expected to reduce energy demand below that projected in 'business as usual scenarios' and may also need to deliver absolute reductions in energy consumption. To provide a likely (66 per cent) chance of limiting global temperature increases to below 2°C, net global carbon emissions must peak by 2020 and fall to zero by approximately 2070 – an extraordinarily demanding target. In the near term (2040), this implies more than doubling the annual rate of energy efficiency improvement in appliances and the building stock (IEA, 2017). The rate and scale of change required is best described as revolutionary: there are few historical precedents for such accelerated efficiency improvements and existing policy initiatives have achieved only incremental progress towards that end (Geels et al., 2017).

To deliver such an ambitious target will require the rapid development and diffusion of multiple 'low-energy innovations' – innovations that differ significantly from existing technologies and practices and have the potential to improve energy efficiency and/or reduce energy demand. Many of these technologies, such as electric vehicles (EVs) or heat pumps, also involve a switch to low-carbon energy sources.

To date, policy efforts to improve energy efficiency and reduce energy demand have primarily been informed by neoclassical economics, behavioural economics and social psychology. These perspectives have numerous strengths, but also important limitations for understanding both the nature of the low-carbon challenge and the appropriate policy response to that challenge. In particular, they provide limited guidance on the emergence and diffusion of low-energy innovations and the associated processes of system transformation (Sorrell, 2015).

Neoclassical economics considers energy or carbon prices to be the critical variable in reducing energy demand, supported where appropriate by policies to reduce various economic barriers to energy efficiency, such as split incentives and asymmetric information (Brown, 2001; Sorrell *et al.*, 2004). However, for most consumers energy efficiency represents a secondary and largely invisible attribute of goods and services, thereby muting the response to price incentives. Factors such as comfort, practicality and convenience commonly play a much larger role in energy-related decisions, with energy consumption being dominated by habitual behaviour that is shaped by social norms (Shove, 2003). Moreover, carbon pricing is politically unpopular and energy efficiency remains a low political priority, resulting in a policy mix that is frequently ineffective (Kern *et al.*, 2017). Neoclassical economics also assumes rational decision-making by firms and individuals and thereby pays insufficient attention to the broader, non-economic determinants of decision-making (Stern *et al.*, 2016).

Insights from behavioural economics and social psychology can reveal the cognitive, emotional and affective influences on relevant choices and routines and suggest ways to 'nudge' people and organisations towards more energy-efficient choices and routines (Andrews and Johnson, 2016; Steg, 2016). But social–psychological research focuses overwhelmingly upon individual consumers and neglects the importance of interactions with other actors, organisational decision-making and economic and social contexts. More fundamentally, both economic and social psychology have an individualist orientation that underrates the significance of the collective and structural factors that shape behaviour, guide innovation and enable and constrain individual choice.

Thus, the dominant perspectives on reducing energy demand have a number of limitations and these limitations are reflected in the partial focus and frequent ineffectiveness of the current policy mix. Given this, we suggest a broader 'sociotechnical' perspective that more fully addresses the complexity of the challenges involved and which integrates insights from a number of social science disciplines, including innovation studies, science and technology studies, and history. We argue that reducing energy demand involves more than improving individual technologies or changing individual behaviours, but instead requires interlinked and potentially far-reaching changes in the broader 'sociotechnical systems' that deliver energy services, such as lighting, thermal comfort or mobility. We term these changes 'sociotechnical transitions'. These transitions are typically complex, protracted, multi-dimensional and path dependent, and the outcomes are difficult to predict. A sociotechnical transitions perspective acknowledges these characteristics and seeks to understand the transition process as a whole, rather than focusing upon individual technologies and behaviours.

Drawing from earlier work (Geels *et al.*, 2018), we have organised our discussion of sociotechnical systems and low-energy innovations under three research themes, namely: emergence, diffusion and impact. Although this is suggestive of a linear model of innovation, we think the distinction is useful since each theme encompasses very different analytical topics. *Emergence* and *diffusion* of radical low-energy innovations refer to different phases in decades-long transition processes (although the boundaries between them may be fuzzy). *Impact* refers to the ultimate effect of low-energy innovations on energy demand. Acknowledging complexities, we also identify crosscutting debates that span the three themes. The focus throughout is on theoretical and conceptual issues rather than specific empirical insights. Many of the debates are relevant to research on 'sociotechnical transitions' in general as well as to research on energy demand in particular.

The sociotechnical transitions approach

Numerous social scientific theories identify themselves as being 'sociotechnical', although they interpret that term in different ways. One recent review identified no less than 96 distinct theories that call themselves sociotechnical across more than a dozen disciplines (Sovacool and Hess, 2017). Nonetheless, there are some key distinctions that set our sociotechnical perspective apart from others, which we examine here.

Substantial reductions in energy demand will require *transitions* towards new or durably reconfigured sociotechnical systems for delivering heating, lighting, motive power, mobility and other energy services. For example, lower energy and lower carbon mobility may require: transforming the car fleet towards lightweight EVs; developing and diffusing associated technologies in materials, battery storage, controls and electric propulsion; establishing a national charging network; integrating this network with a smart transmission and distribution grid (including using EVs for electricity storage); developing new models for vehicle sharing and ownership; significantly expanding the share of public transport in total mobility; redesigning cities to encourage walking and cycling and so on.

Promising low-energy innovations provide the seeds for such transitions, but many of them initially have a very small market share and face uphill struggles against existing technologies and practices and the sociotechnical systems in which they are embedded. One implication is that current policy interventions (which revolve around more narrow dimensions such as cost structures, information provision and regulation) may be insufficient to bring about such non-marginal change. A second implication is that low-energy innovations should not be studied in isolation, but in the context of their compatibility with and struggles against existing sociotechnical systems. The specific framework we use to understand these issues is the Multi-Level Perspective (MLP), which we briefly summarise.

The MLP distinguishes three analytical levels (Geels, 2002; Geels and Schot, 2007; Rip and Kemp, 1998).

1 The incumbent *sociotechnical system* refers to the interdependent mix of technologies, industries, supply chains, consumption patterns, policies and infrastructures. These tangible system elements are reproduced by actors and social groups, whose perceptions and actions are shaped by formal rules (e.g. regulations, standards) and informal institutions (e.g. shared meanings, heuristics, rules of thumb, routines, social norms). The rules and institutions within a sociotechnical system are referred to as the sociotechnical *regime*. Owing to various 'lock-in' effects (Unruh, 2000), innovation in existing

systems is mostly incremental and path dependent, aimed at elaborating existing capabilities. Sources of lock-in include sunk investments (in skills, factories and infrastructures), economies of scale, increasing returns to adoption, and the momentum of established rules and institutions (Hughes, 1987). These reinforcing factors act to create stability in the incumbent system, and resistance to change.

2   *Niche innovations* refer to novelties that deviate on one or more dimensions from existing systems. The novelty may be a new practice (e.g. car-sharing), a new technology (e.g. battery–EVs), a new business model (e.g. energy service companies) or a combination of these. Many radical innovations initially have poor price/performance characteristics and are misaligned with – and obstructed by – the established sociotechnical system. Radical innovations therefore initially emerge in 'niches', which act as 'incubation rooms' that protect them against mainstream selection environments (Kemp et al., 1998). Examples are: particular application domains (e.g. the military), geographical areas, markets or subsidised programmes. Radical innovations are often developed by networks of 'fringe' actors, rather than by dominant firms (Van de Poel, 2000).

3   The *sociotechnical landscape* forms an exogenous environment beyond the direct influence of niche and regime actors but influencing them in various ways. This may be through gradual changes, such as shifts in cultural preferences, demographics and macro-political developments, or through short-term shocks such as macro-economic recessions and oil crises.

Niche actors are continually working on radical innovations (e.g. developing and improving technologies, opening up markets, finding customers, attracting investment, lobbying policymakers for support), but usually experience uphill struggles against existing systems, which are stabilised by multiple lock-in mechanisms. The MLP therefore suggests that transitions require the alignment of several processes within and between the three analytical levels:

> a) 'niche innovations gradually build up internal momentum (through learning processes, price/performance improvements and support from powerful groups), b) changes at the landscape level creates pressure on the regime, c) destabilisation of the regime creates windows of opportunity for niche innovations.
>
> (Geels and Schot, 2007, p. 400)

This combination of processes allows niche innovations to break through, and to trigger a series of broader changes in supply chains, infrastructures, policies, expectations and behaviours that ultimately transform the regime. The MLP has been illustrated and refined with historical case studies of transitions as well as contemporary applications.

Figure 2.1 schematically represents the MLP as a 'big-picture' understanding of transitions. The next three sections draw upon this framework to further

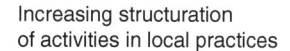

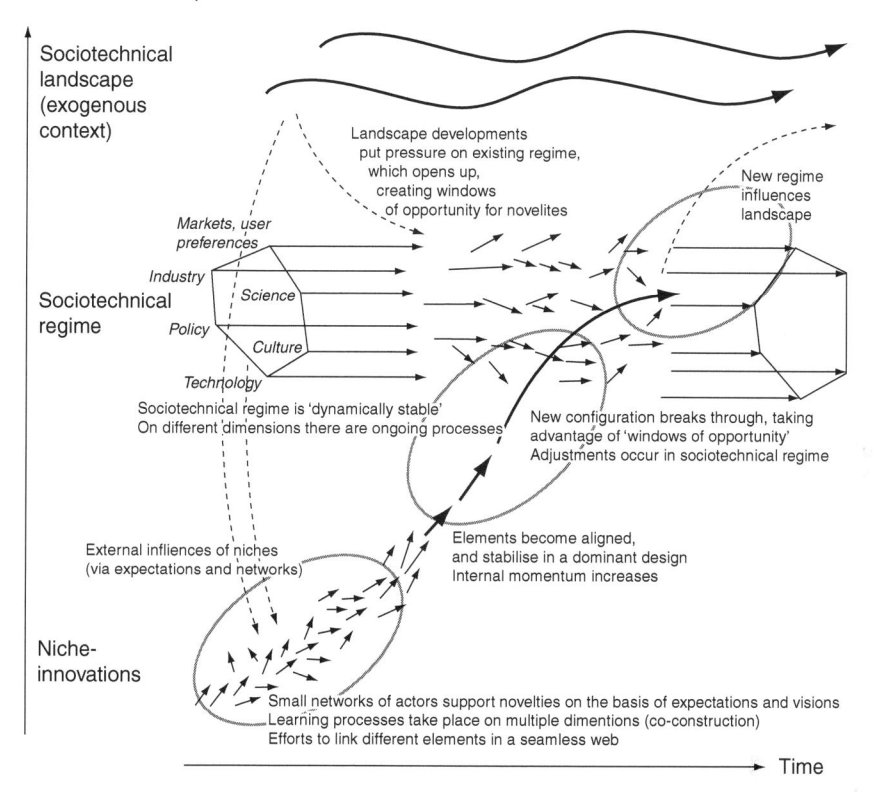

Figure 2.1 Multi-level perspective on transitions – illustrating the emergence, diffusion and impact of radical innovations as three consecutive circles.

Source: adapted from Geels (2002), with permission.

assess the processes through which low-energy innovations *emerge* and *diffuse*, together with their potential *impacts* on energy demand. In each section, we first provide a general *conceptualisation* of the relevant theme (emergence, diffusion, impact) and then highlight *research debates*. Although Figure 2.1 portrays these processes as three consecutive phases, real-world transitions deviate from the implied linearity.

Emergence of low-energy innovations

Sociotechnical research on *emergence* does not focus on the initial invention of new knowledge (e.g. from scientific research), but on the early introduction of innovations in real-world application domains, labelled 'niches' (Kemp *et al.*,

1998). The introduction of innovations tends to be difficult because the supportive sociotechnical contexts that allow innovations to thrive – e.g. networks of institutions, formalised and tacit knowledge, social norms and expectations, design standards, financial resources, supportive regulations and so forth – have yet to be established. A common manifestation of the absence of supportive contexts for innovations is the so-called 'valley of death' in innovation financing (Auerswald and Branscomb, 2003), where an emerging technology becomes too capital intensive for venture capital firms, while at the same time being too risky for project finance. Many novelties fail to cross this chasm or take a very long time to do so.

According to the Strategic Niche Management approach (Geels and Raven, 2006; Kemp et al., 1998; Schot and Geels, 2008; Smith and Raven, 2012), the creation of 'niches' or 'protective spaces' is a useful and important means of encouraging emerging innovations because they shield those innovations from the pressures imposed by the existing system and give them time to mature. Such protective spaces allow actors associated with innovations to address and reduce a wide range of uncertainties, including:

1 *Techno-economic uncertainties:* There may be competing technical configurations (EVs, for instance, may use lead acid, nickel metal hydride, lithium ion or zinc air batteries), each with different advantages and disadvantages.
2 *Finance and investment-related uncertainties:* Often it is difficult not only to obtain the funding that is necessary for technical development and practical experimentation, but also to evaluate the rationality of investments in innovations. To attract finance, product champions often make positive promises (Geels, 2002) and even expert analysts in technical areas often suffer from 'appraisal optimism' (Gilbert and Sovacool, 2016; Gross et al., 2013).
3 *Cognitive uncertainties:* Actors developing niche innovations often have different views and perceptions about technical specifications, consumer preferences, infrastructure requirements, future costs, and so forth (Sovacool et al., 2017). This 'interpretive flexibility' gives rise to debates, disagreements, discursive struggles and competing visions (Geels and Verhees, 2011; Goldthau and Sovacool, 2016).
4 *Social uncertainties:* The networks of actors developing niche innovations are often unstable and fluid. Actors may enter into partnerships for a few years, but then leave if difficulties arise or funding runs out (Olleros, 1986). Start-up or spin-off firms may be attracted by new opportunities, but then may also exit when economic ventures fail (as they often do in early phases).

To address these uncertainties, three core processes in the development of niche innovations have been identified in the literature (see Schot and Geels, 2008, for a summary):

- *Articulation of expectations and visions:* Expectations (defined as 'representations of future technological situations and capabilities' (Bakker et al., 2011)) are considered crucial for niche development because they provide

direction to learning processes, attract attention from policymakers, investors and other actors, and legitimate protection and nurturing (Borup et al., 2006; Melton et al. 2016; Van Lente, 2012).
- *Building of social networks*: This process is important to create a constituency behind an innovation, to facilitate interactions and knowledge exchange between relevant stakeholders, and to provide the necessary resources (e.g. venture capital, people, and expertise) for further development and subsequent diffusion (Kemp et al., 1998).
- *Learning processes* along multiple dimensions (Sengers et al., 2017), including: technical aspects and design specifications; market and user preferences; cultural and symbolic meanings; infrastructure and maintenance requirements; production processes; supply chains and distribution networks; regulatory standards; societal acceptability and environmental impacts.

Niches gain momentum if: first, visions and expectations become more precise and more broadly accepted; second, the alignment of various learning processes results in shared expectations and a 'dominant design'; and third, networks increase in size, including the participation of powerful actors that add legitimacy and expand resources (Schot and Geels, 2008). These processes of *stabilisation, acceptance and support* and *community building* tend to occur over sequences of concrete demonstration projects, experiences and trials (see Geels and Raven, 2006, for one conceptualisation of these processes).

Having summarised and characterised the niche innovation literature, we now identify two research debates that are relevant to the emergence of low-energy innovations.

The contribution of outsiders and incumbents to emergence

One debate relates to the role of new entrants relative to actors *within* incumbent regimes such as electric utilities and car manufacturers. The Strategic Niche Management (SNM) literature and the grassroots innovation approach (Seyfang and Haxeltine, 2012; Smith and Seyfang, 2013) often argue that start-ups, civil society organisations and 'grassroots' innovators tend to pioneer radical niche innovations because they are less 'locked in' and willing to think 'out of the box'. Incumbent actors, in contrast, focus on incremental innovations that fit easier with existing capabilities, capital investments and interests.

Recent work, however, has questioned this simple dichotomy, identifying many instances where incumbent actors develop radical niche innovations (Berggren et al., 2015; Geels et al., 2016). New entrants may also collaborate with incumbents in order to draw on their financial resources, technical capabilities and political connections. This may accelerate emergence but also entail some 'mainstreaming' and weakening of the more radical aspects of the innovation (Smith, 2007). The first research debate thus concerns the relative importance and roles of new entrants and incumbents in the emergence of low-energy niche innovations.

The role of visions and expectations in emergence

There are different views on sociotechnical visions and expectations, which underpins debates about their relative discursive (and material) strengths. Some (e.g. Loorbach, 2007) have an operational view and see visions as indicating long-term directions of transitions, which can then be explored with short-term projects that produce learning outcomes which can be used to adjust and fine-tune visions.

Sociologists of innovation, in contrast, have a more constructivist view that emphasises the 'performative' roles that visions and expectations play in early technological development (Bakker *et al.*, 2011; Borup *et al.*, 2006). Nightingale (1998) sees technological optimism and fantasy as an elemental part of the 'cognitive' dimension of innovation. Berkhout (2006) qualifies visions as strategic 'bids' for public support, which are an emergent property in all transitions. Van Lente (2012) further identifies three roles of expectations: they raise attention and legitimate the innovation as worthy of investment and support; they provide direction for search and learning processes; and they coordinate action in dispersed social networks.

There are also more critical views that emphasise 'hype-cycles' or cycles of 'promises-and-disappointment' and the strategic efforts by firms and engineers to push their innovations onto policy agendas by (over)promising their development speed and impact. Hype-cycle theory suggests that technologies move along a path from a trigger to a peak in expectations, then plummeting into a trough of disillusionment before eventually giving rise to a range of somewhat more modest expectations (Van Lente, 2012; Van Lente *et al.*, 2013). Hype-cycles are one mechanism behind the non-linearity of innovation journeys: climate change policies and low-carbon innovations can experience setbacks or accelerations, as well as repeated cycles of hype and disappointment, e.g. in low-carbon transport, where high hopes of battery–EVs (1990s) were followed by fuel cells and hydrogen (early 2000s), biofuels (early 200s), hybrid electric cars (mid-2000s) and battery–EVs again in the early 2010s (Melton *et al.*, 2016). This more critical view means that one should not necessarily assume that visions are a good basis for transition management. Visions can also be self-serving or keep more radical visions off the public agenda.

As a second research debate, researchers could fruitfully investigate how important these different roles of visions and expectations are in the emergence of low-energy innovations.

Diffusion of low-energy innovations

The widespread *diffusion* of low-energy innovations is necessary to achieve energy efficiency improvements on a substantial scale. However, large-scale diffusion in mass markets often means 'head-on' competition with incumbent sociotechnical systems, which are stabilised through the alignment of existing technologies with the business, policy, user and societal contexts. Therefore, the diffusion of low-energy innovations does not happen in an 'empty' world, but in the context of existing systems and incumbent actors who may pursue active resistance (Geels, 2014).

Another problem is that many low-energy innovations are not intrinsically attractive to the majority of consumers, since they are often (initially) more expensive and perform less well on key dimensions such as availability (early compact florescent lights and then LEDs) cost (early heat pumps or battery–EVs), or performance. Much of the recent policy interest in low-energy innovation is driven by public good concerns (e.g. sustainability, climate change) rather than by private interests (e.g. profit, utility), which implies that diffusion is unlikely to be driven solely by economic mechanisms, even with the help of carbon pricing. Policy support, cultural discourse and changing social practices are likely to be critically important factors as well, which means that a multi-dimensional approach is required.

The MLP conceptualises diffusion as entailing two interacting developments: (1) the creation of *endogenous momentum* of niche innovations; and (2) the *embedding* of niche innovations in wider contexts and environments.

Endogenous momentum arises gradually from the same processes that drive the emergence of innovations, namely: developing larger social networks with greater legitimacy and resources; aligning learning processes on multiple dimensions (technical, market, infrastructural, social political, cultural) often resulting in a 'dominant design'; and forming of clear and widely accepted visions of the future of the innovation. The gradual shift from the *emergence* phase to the *diffusion* phase is characterised by a *reversal*, with the innovation shifting from initial flexibility (when it is fluid and socially shaped) to 'dynamic rigidity' (Staudenmaier, 1989). Hughes (1987) describes the emerging *momentum* of new systems in terms of an increasing 'mass' of technical and organisational components, emerging directionality and system goals, and an increasing rate of perceptible growth. Thus, endogenous momentum is driven by multiple and reinforcing causal mechanisms including: expansion of social networks and bandwagon effects; positive discourses and visions; learning by doing; increasing returns to scale; network externalities; strategic games between firms (e.g. 'jockeying for position'); and increasing support from policymakers who see the innovation as a way of solving particular problems.

The diffusion of low-energy innovations also requires *embedding* within policy, social, business and user environments (Deuten *et al.*, 1997). This external fit may be difficult to foresee, as Rosenberg (1972, p. 14) noted more than 40 years ago: 'The prediction of how a given invention will fit into the social system, the uses to which it will be put, and the alterations it will generate, are all extraordinarily difficult intellectual exercises.' Achieving this fit may be especially difficult for more radical niche innovations that face a 'mismatch' with the existing sociotechnical system (Freeman and Perez, 1988). The process of societal embedding is conceptualised as a *co-construction* process that entails mutual adjustments between the innovation and wider contexts (Rip and Kemp, 1998). The degree of adjustment is a question for research: at one extreme end, the innovation is adjusted to fit in existing contexts, while, at the other end, the contexts are adjusted to accommodate the innovation (Smith and Raven, 2012).

The distinctive contribution of a sociotechnical approach to diffusion is to study the interaction between endogenous mechanisms and embedding in wider contexts. Although adoption decisions by individual consumers remain important, the sociotechnical perspective focuses upon the activities of a broader range of actors and the interrelationships between them. Within this literature, we highlight two debates that are relevant to the diffusion of low-energy innovations.

Political will and contextual pressures for deliberately accelerated diffusion

Ambitious climate targets will require much faster rates of technology diffusion than has been achieved in the past, potentially combined with the early retirement of existing technologies and infrastructure. Hence, a critically important debate relates to the best way to accelerate the diffusion of low-energy technologies. The German Advisory Council on Global Change (WBGU, 2011, p. 1) suggests that technical and policy instruments for low-carbon transitions are well developed, but that it is 'a political task to overcome the barriers of such a transformation, and to accelerate the change'. There are different views on the nature of this political task.

One view suggests that accelerated diffusion depends on 'leadership' or 'political will' or courage from policymakers (e.g. Figueres *et al.*, 2017; United Nations Climate Change Secretariat, 2016). But such a voluntarist orientation places very high hopes on the importance of politicians' own volition, and it reduces everything to politics. It also under-appreciates the fact that policymakers are locked-in by policy and wider sociotechnical regimes (Wilson, 2000) and tend towards risk avoidance (Howlett, 2014), especially for policies with diffuse and distant benefits.

A second view therefore suggests that it is more important to understand the *conditions* in which policymakers are more likely to introduce decisive policies (Meadowcroft, 2016). Conditions for accelerated diffusion may derive from external (landscape) shocks that change socio-political priorities and create a sense of urgency to accelerate deployment (Delina and Diesendorf, 2013; Sovacool, 2016). Pressure for stronger policies may also come from changes in public attention or discursive framings, social movement campaigns or from business interests that see commercial opportunities in low-carbon innovations (Raven *et al.*, 2016; Roberts and Geels, 2018).

Policy mixes for accelerated diffusion

Complementing the *political* focus above, there is ongoing discussion on *how* diffusion can be accelerated, which is especially relevant to low-carbon transitions and the time-sensitive problem of climate change (Grubler *et al.*, 2016; Kern and Rogge, 2016; Sovacool, 2016). The mainstream climate mitigation literature (IPCC, 2014) has identified a range of options where strengthened *policies* could help accelerate low-carbon transitions, such as R&D subsidies, feed-in tariffs, carbon pricing, performance standards and removing fossil fuel subsidies. But views differ on what combination of policy instruments is likely to be most effective and the manner in which they should be sequenced (Meckling *et al.*, 2017).

In line with the systemic approach to innovation taken throughout this chapter, we also take a systemic view of policy and policymaking. As Kern *et al.* (2017) identify, much of the policy advice literature still focuses on individual policy instruments, pairwise instrument interactions or intended policy mixes, neglecting the analysis of complex, real-world mixes, their development over time, and their consistency and coherency. We agree with Kivimaa and Kern (2016) about the need for a comprehensive *policy mix*, rather than individual, isolated instruments that tend to operate in a non-predictable and non-synergetic matter. As an example, Givoni *et al.*'s (2013) exploration of the transport sector illustrates that the deliberate and careful combination of mutually supportive policy packages may result in more effective and efficient outcomes through increasing public and political acceptability and the likelihood of implementation. It is important therefore to look at the whole system of policy instruments, to identify positive and negative interactions between policies, and overall characteristics of policy mixes (such as their coherence, consistency, comprehensiveness or credibility) and investigate how these hinder or stimulate the emergence, diffusion and impact of low-energy innovations.

Impact of low-energy innovations

Comprehending the *impacts* of low-energy innovations on improved efficiency or demand is central to public policy: energy efficiency improvements are considered to be the most promising, fastest, cheapest and safest means to mitigate climate change, as well as providing broader benefits, such as improved energy security, reduced fuel poverty, and increased economic productivity (Ryan and Campbell, 2014). However, compared to the large body of work on emergence and diffusion, the analysis of the impacts of low-energy innovations has received much less attention from sociotechnical researchers. Authors often emphasise the limitations of linear, deterministic approaches to projecting impacts; the frequency with which expectations of impacts are confounded by real-world experience (Gilbert and Sovacool, 2016); and the challenges associated with both anticipating impacts *ex ante* and measuring them ex post (McDowall and Geels, 2017).

Quantification of impacts is difficult within complex social systems but may nevertheless be feasible for more incremental kinds of innovations within restricted spatial and temporal boundaries, e.g. the adoption of condensing boilers and the retrofitting of loft and cavity wall insulation (Dowson *et al.*,

2012; Hamilton et al., 2013). In these examples, sufficient data exists for the historical impacts of these changes to be measured and the relevant systems are sufficiently stable for the future impacts to be modelled.

But establishing the historical or potential future impact of more radical innovations over longer periods of time presents much greater difficulties. For example, commonly used modelling tools may not capture all of the relevant mechanisms (McDowall and Geels, 2017), and certain types of outcomes may be difficult or impossible to anticipate (such as Brexit in the UK or the election of President Trump in the United States). The impacts of any change within a complex system are necessarily mediated through multiple interdependencies, time-delayed feedback loops, path dependencies and threshold effects. More fundamentally, the basic concept of 'impact' is problematic from a sociotechnical perspective, because of its connotations of technological determinism – with technology impacting on society in a linear and straightforward fashion (Rosenberg, 1995).

Hence, for radical and systemic innovations it is difficult to establish causality, assess historical impacts and project future 'impacts'. While historical analysis can provide rich descriptions of the co-evolutionary processes involved, the primary lesson is the contingent nature of impacts and our limited ability to anticipate them in advance. In this context, authors in the sociotechnical tradition have focused more upon transition processes than on the ultimate (environmental) impacts of those transitions.

Against this background, we identify two important research debates that are relevant to the impact of low-energy innovations.

Rebound effects of low-energy innovations

First, there is a critical debate on the rebound effects from low-energy innovations and the extent to which these may undermine their anticipated climate benefits (Sorrell and Dimitropoulos, 2008). Such effects result from mechanisms operating at different levels, across geographical scales and over different time periods, but only some of these are amenable to quantification. Moreover, attention to date has focused almost exclusively upon economic mechanisms to the neglect of other co-determinants.

As an illustration, consider the following example from transport systems: (a) fuel-efficient cars make travel cheaper, so people may choose to drive further and/or more often, thereby offsetting some of the energy savings; (b) joint decisions by consumers and producers may channel the benefits of improved technology into larger and more powerful cars, rather than more fuel-efficient cars; (c) drivers may use the savings on fuel bills to buy other goods and services which necessarily require energy to provide; (d) the energy embodied in new technologies (e.g. lightly materials) may offset some of the energy savings, especially when product lifetimes are short; (e) reductions in fuel demand translate into lower fuel prices which encourages increased fuel consumption, together with changes in incomes, prices, investments and industrial structures throughout the economy; and (f) more fuel-efficient vehicles deepen the lock-in to the sociotechnical system of car-based transportation, with associated and reinforcing changes in infrastructure, institutions, regulations, supply chains and social practices.

Rebound is therefore an emergent property of a complex system. A growing body of research is exploring mechanisms *a-b*, and to a lesser extent mechanisms *c-e* in transport and other areas (e.g. Chitnis and Sorrell, 2015; Stapleton *et al.*, 2016), but this research excludes non-economic mechanisms, tends to be confined to the short to medium term and stops short of assessing the impacts of broader changes in the relevant systems. Nevertheless, such studies indicate significant departures from anticipated impacts. There is a need to apply the relevant techniques to other innovations, contexts, datasets and time periods, and to extend the analysis to include broader psychological, social, institutional and other factors that either offset, reinforce or contribute additional rebounds – for example, the phenomena of 'moral licensing' (Harding and Rapson, 2013; Tiefenbeck *et al.*, 2013). More fundamentally, methods need to be found or refined to investigate the longer-term impacts of sociotechnical transitions, and to evaluate the sustainability claims of proponents of particular low-energy innovations. Without more careful investigation of their impacts, such claims may rest more on hope than on evidence.

Frameworks for evaluating broader impacts

Low-energy innovations may have wider impacts that go beyond energy demand reductions. Although this section is less of a debate, it highlights two research approaches (energy justice and exergy economics) that provide frameworks to assess these broader impacts.

The material and social transformations associated with low-energy innovations may involve contestations over what is *just, equitable* and *right*. Thus, there is a need for frameworks that explore questions of ethics and justice, including concern for where, how and with whom new technologies are socially embedded. Without a focus on justice, a low-energy revolution may fail to acknowledge the burden of not having enough energy, where some individuals lack access, are challenged by under-consumption and poverty, and face health burdens and shortened lives as a consequence of restricted energy choices (Jenkins *et al.*, 2016; Sovacool *et al.*, 2016). While some policies may have positive implications for equity and social justice (e.g. insulation retrofits for low-income households) others may have negative impacts – for example, subsidies for EVs, solar-photovoltaic (PV) and whole house retrofit often disproportionately benefit wealthier households.

Also important are broader debates about energy use, energy efficiency and economic growth. One emerging area of research that has the potential to throw new light on these issues is 'exergy economics' (Ayres and Warr, 2010). This approach hinges upon the thermodynamic concept of *exergy* and the use of second-law (systems move towards entropy and disorder) rather than first-law (energy can be neither created nor destroyed) measures of thermodynamic efficiency. The focus is on the *useful exergy inputs* into national economies – where useful exergy is the exergy outputs of end-use conversion devices, such as the mechanical drive from an engine or the high-temperature heat from a furnace. The core claim of researchers in this area, backed up by an increasing volume of empirical research, is that *useful exergy drives economic activity* (Ayres and Voudaris, 2014; Kümmel et al., 2015; Laitner, 2015). Increases in useful exergy, in turn, can be achieved by using more primary energy, shifting towards higher-quality energy carriers (e.g. from coal to oil) and improving energy efficiency. This claim directly contradicts orthodox economics that ignores the distinction between energy and exergy and ascribes little or no role to energy in explaining economic growth. A key implication of this emerging perspective is that energy efficiency improvements by producers may significantly boost economic growth, and that the economic importance of those improvements has been undervalued (Brockway et al., 2017). Clearly, a deeper understanding of this question is important to understand the broader economy-wide impacts of low-energy innovations.

Having discussed six research debates across emergence, diffusion and impact, we end with a brief discussion of research and policy implications.

Conclusion and policy implications

A sociotechnical approach focuses upon how radical innovations lead to new sociotechnical systems through the co-construction of multiple elements. Informed by detailed case studies, this perspective sheds new light on how sociotechnical systems evolve, stabilise and transform through the alignment of developments on multiple levels. The themes of emergence, diffusion and impacts are useful heuristic devices through which to understand the sociotechnical transitions required to accelerate improvements in energy efficiency and the means through which these can be encouraged. This introductory chapter has explored the sociotechnical conceptualisation of emergence, diffusion and impact and identified several research debates within each theme. These themes and debates are summarised in Table 2.1, together with the chapters in the book that relate to them.

In conclusion, the innovation, energy, transport and climate policy communities need to improve their analytical understanding of sustainable energy transitions and of the policy approaches to encouraging such transitions, especially in the area of energy efficiency and energy demand. There is a need to adopt a broader view of the process, which takes into account learning and experimentation, the multiple conditions necessary for systemic change and the coalitions of interests that can block or support emerging niche innovations. Analysts and policymakers should look beyond carbon pricing as a policy panacea and recognise that disagreement and contestation are normal dimensions of low-carbon transitions that need to be accommodated rather than ignored. The topic of sustainable energy transitions is too empirically rich and socially complex to be left to economists and social psychologists alone.

Table 2.1 Key research debates with regard to sociotechnical transitions and low-energy innovation

Theme	Debate	Relevant chapter(s)
Emergence	The role of visions and expectations in innovation emergence.	Electric vehicles and ride-sharing (Bergman, Ch. 4), automated mobility (Hopkins and Schwanen, Ch. 5).
	The relative role of outsiders and incumbents in emergence.	Smart meters (Jenkins *et al.*, Ch. 6), low-energy housing retrofits (Brown *et al.*, Ch. 7).
Diffusion	Political will or conditions for deliberately accelerated diffusion.	Managing transitions (Jenkins and Sovacool, Ch. 13), Historical evidence (Roberts and Geels, Ch. 10).
	Policy mixes for accelerated diffusion.	Policy challenges (Webb, Ch. 11) and policy mixes (Kern *et al.*, Ch. 12).
Impacts	The rebound effects from low-energy innovations.	Energy innovation and rebound effects (Figus *et al.*, Ch. 9).
	Frameworks for examining broader impacts.	Energy justice (Jenkins and Martiskainen, Ch. 3), exergy economics (Brockway *et al.*, Ch. 8).

Source: the authors.

Acknowledgements

This chapter borrows from, and extends the arguments initially presented in, 'Reducing energy demand through low carbon innovation: A sociotechnical transitions perspective and thirteen research debates', published in *Energy Research and Social Science* 40 (June 2018), pp. 23–35. The authors are appreciative to the Research Councils United Kingdom (RCUK) Energy Program Grant EP/K011790/1 'Centre on Innovation and Energy Demand', the Danish Council for Independent Research (DFF) Sapere Aude Grant 4182–00033B 'Societal Implications of a Vehicle-to-Grid Transition in Northern Europe', which have supported elements of the work reported here. Any opinions, findings, and conclusions or recommendations expressed in this material are those of the authors and do not necessarily reflect the views of RCUK Energy Program or the DFF.

References

Andrews, R.N.L and Johnson, E. (2016) Energy use, behavioural change, and business organizations: Reviewing recent findings and proposing a future research agenda. *Energy Research and Social Science* 11: 195–208.

Auerswald, P.E. and Branscomb, L.M. (2003) Valleys of death and Darwinian seas: Financing the invention to innovation transition in the United States. *The Journal of Technology Transfer* 28(3–4): 227–239.

Ayres, R.U. and Voudouris, V. (2014) The economic growth enigma: Capital, labour and useful energy? *Energy Policy* 64: 16–28.

Ayres, R.U. and Warr, B. (2010) *The Economic Growth Engine: How Energy and Work Drive Material Prosperity*. Edward Elgar Publishing, Cheltenham, UK.

Bakker, S., Van Lente, H. and Meeus, M. (2011) Arenas of expectations for hydrogen technologies. *Technological Forecasting & Social Change* 78(1): 152–162.

Berggren, C., Magnusson, T. and Sushandoyo, D. (2015) Transition pathways revisited: Established firms as multi-level actors in the heavy vehicle industry. *Research Policy* 44(5): 1017–1028.

Berkhout, F. (2006) Normative expectations in systems innovation. *Technology Analysis & Strategic Management* 18(3–4): 299–311.

Borup, M., Brown, N., Konrad, K. and Van Lente, H. (2006) The sociology of expectations in science and technology. *Technology Analysis & Strategic Management* 18(3–4): 285–298.

Brockway, P.E., Saunders, H., Heun, M.K., Foxon, T.J., Steinberger, J.K., Barrett, J.R. and Sorrell, S. (2017) Energy rebound as a potential threat to a low-carbon future: Findings from a new exergy-based national-level rebound approach. *Energies* 10(1): 51–75. DOI: 10.3390/en10010051.

Brown, M.A. (2001) Market failures and barriers as a basis for clean energy policies. *Energy Policy* 29(14): 1197–1207.

Chitnis, M. and Sorrell, S. (2015) Living up to expectations: Estimating direct and indirect rebound effects for UK households. *Energy Economics* 52(1): 100–116.

Delina, L. and Diesendorf, M. (2013) Is wartime mobilisation a suitable policy model for rapid national climate mitigation? *Energy Policy* 58: 371–380.

Deuten, J.J., Rip, A. and Jelsma, J. (1997) Societal embedding and product creation management. *Technology Analysis & Strategic Management* 9(2): 131–148.

Dowson, M., Poole, A., Harrison, D. and Susman, G. (2012) Domestic UK retrofit challenge: Barriers, incentives and current performance leading into the Green Deal. *Energy Policy* 50: 294–305.

Figueres, C., Schellnhuber, H.J., Whiteman, G., Rockström, J., Hobley, A. and Rahmstorf, S. (2017) Three years to safeguard our climate. *Nature* 546(7660): 593–595.

Freeman, C. and Perez, C. (1988) Structural Crisis of Adjustment, Business Cycles and Investment Behaviour. In: Dosi, G., Freeman, C., Nelson, R., Silverberg, G. and Soete, L. (Eds) *Technical Change and Economic Theory*. Pinter, London, UK, 38–66.

Geels, F.W. (2002) Technological transitions as evolutionary reconfiguration processes: A multi-level perspective and a case-study. *Research Policy* 31(8–9), 1257–1274.

Geels, F.W. (2014) Regime resistance against low-carbon transitions: Introducing politics and power into the multi-level perspective. *Theory, Culture and Society* 31(5): 21–40.

Geels, F.W. and Raven, R.P.J.M. (2006) Non-linearity and expectations in niche-development trajectories: Ups and downs in Dutch biogas development (1973–2003). *Technology Analysis & Strategic Management* 18(3/4): 375–392.

Geels, F.W. and Schot, J.W. (2007) Typology of sociotechnical transition pathways. *Research Policy* 36(3): 399–417.

Geels, F.W. and Verhees, B. (2011) Cultural legitimacy and framing struggles in innovation journeys: A cultural-performative perspective and a case study of Dutch nuclear energy (1945–1986). *Technological Forecasting & Social Change* 78(6): 910–930.

Geels, F.W., Berkhout, F. and Van Vuuren, D. (2016) Bridging analytical approaches for low-carbon transitions, *Nature Climate Change* 6(6): 576–583.

Geels, F.W., Schwanen, T., Sorrell, S., Jenkins, K. and Sovacool, B.K. (2018) Reducing energy demand through low carbon innovation: A sociotechnical transitions perspective and thirteen research debates. *Energy Research & Social Science* 40: 23–35.

Geels, F.W., Sovacool, B.K., Schwanen, T. and Sorrell, S. (2017) Sociotechnical transitions for deep decarbonization. *Science* 357(6357): 1242–1244.

Gilbert, A.Q. and Sovacool, B.K. (2016) Looking the wrong way: Bias, renewable electricity, and energy modeling in the United States. *Energy* 94: 533–541.

Givoni, M., Macmillen, J., Banister, D. and Feitelson, E. (2013) From policy measures to policy packages. *Transport Reviews* 33(1): 1–20.

Goldthau, A. and Sovacool, B.K. (2016) Energy technology, politics, and interpretative frames: Shale gas fracking in Eastern Europe. *Global Environmental Politics* 16(4): 50–69.

Gross, R., Heptonstall, P., Greenacre, P., Candelise, C., Jones, F. and Castillo, A.C. (2013) Presenting the Future: An Assessment of Future Costs Estimation Methodologies in the Electricity Generation Sector, UKERC Report. UK Energy Research Centre, London, UK.

Grubler, A., Wilson, C. and Nemet, G. (2016) Apples, oranges, and consistent comparisons of the temporal dynamics of energy transitions. *Energy Research and Social Science* 22: 18–25.

Hamilton, I.G., Steadman, P.J., Bruhns, H., Summerfield, A.J. and Lowe, R. (2013) Energy efficiency in the British housing stock: Energy demand and the Homes Energy Efficiency Database. *Energy Policy* 60: 462–480.

Harding, M. and Rapson, M. (2013) Do voluntary carbon offsets induce energy rebound? A conservationist's dilemma. University of California, Davis, USA.

Howlett, M. (2014) Why are policy innovations rare and so often negative? Blame avoidance and problem denial in climate change policy-making. *Global Environmental Change* 29: 395–403.

Hughes, T.P. (1987) The Evolution of Large Technological Systems. In: Bijker, W.E., Hughes, T.P. and Pinch, T. (Eds) *The Social Construction of Technological Systems: New Directions in the Sociology and History of Technology*. MIT Press, Cambridge, MA, USA, 51–82.

IEA (2012) World Energy Outlook 2012. International Energy Agency, Paris, France.

IEA (2015) World Energy Outlook Special Report 2015: Energy and Climate Change. International Energy Agency, Paris, France.

IEA (2017) Perspectives for the Energy Transition: Investment Needs for a Low-carbon Energy System. International Energy Agency, Paris, France.

IPCC (2014) Summary for Policymakers. In: Climate Change 2014: Mitigation of Climate Change. Contribution of Working Group III to the Fifth Assessment Report of the Intergovernmental Panel on Climate Change (Edenhofer, O., Pichs-Madruga, R., Sokona, Y., Farahani, E., Kadner, S., Seyboth, K., Adler, A., Baum, I., Brunner, S., Eickemeier, P., Kriemann, B., Savolainen, J., Schlömer, S., von Stechow, C., Zwickel, T. and Minx, J.C. (Eds)). Cambridge University Press, Cambridge, UK and New York, USA.

Jenkins, K., McCauley, D., Heffron, R., Stephan, H. and Rehner, R. (2016) Energy justice: A conceptual review. *Energy Research and Social Science* 11: 174–182.

Kemp, R., Schot, J. and Hoogma, R. (1998) Regime shifts to sustainability through processes of niche formation: The approach of strategic niche management. *Technology Analysis and Strategic Management* 10(2): 175–196.

Kern, F. and Rogge, K.S. (2016) The pace of governed energy transitions: Agency, international dynamics and the global Paris agreement accelerating decarbonisation processes? *Energy Research & Social Science* 22: 13–17.

Kern, F., Kivimaa, P. and Martiskainen, M., (2017) Policy packaging or policy patching? The development of complex energy efficiency policy mixes. *Energy Research and Social Science* 23: 11–25.

Kivimaa, P. and Kern, F. (2016) Creative destruction or mere niche support? Innovation policy mixes for sustainability transitions. *Research Policy* 45: 205–217.

Kümmel, R., Lindenberger, D. and Weiser, F. (2015) The economic power of energy and the need to integrate it with energy policy. *Energy Policy* 86L 833–843.

Laitner, J.A. (2015) Linking energy efficiency to economic productivity: Recommendations for improving the robustness of the U.S. economy. *WIREs Energy and Environment* 4(3): 235–252.

Loorbach, D. (2007) *Transition Management: New Mode of Governance for Sustainable Development.* International Books, Utrecht.

McDowall, W. and Geels, F.W. (2017) Ten challenges for computer models in transitions research: A commentary on Holtz et al. *Environmental Innovation and Societal Transitions* 22: 41–49.

Meadowcroft, J. (2016) Let's get this transition moving! *Canadian Public Policy* 42(1): S10–S17.

Meckling, J., Sterner, T. and Wagner, G. (2017) Policy sequencing toward decarbonization. *Nature Energy* 2(12): 918.

Melton, N., Axsen, J. and Sperling, D. (2016) Moving beyond alternative fuel hype to decarbonize transportation, *Nature Energy* 1(3).

Nightingale, P. (1998) A cognitive model of innovation. *Research Policy* 27(7): 689–709.

Olleros, F. (1986) Emerging industries and the burnout of pioneers. *Journal of Product Innovation Management* 1(1): 5–18.

Raven, R., Kern, F., Verhees, B. and Smith, A. (2016) Niche construction and empowerment through socio-political work. A meta-analysis of six low-carbon technology cases. *Environmental Innovation and Societal Transitions* 18: 164–180.

Rip, A. and Kemp, R. (1998) Technological Change. In: Rayner, S. and Malone, E.L. (Eds) *Human Choice and Climate Change, Volume 2.* Battelle Press, Columbus, OH, USA, 327–399.

Roberts, C. and Geels, F.W. (2018) Conditions and intervention strategies for the deliberate acceleration of socio-technical transitions: Lessons from a comparative analysis of two historical case studies in Dutch and Danish heating. *Technological Forecasting and Social Change*, under review.

Rosenberg, N. (1972) Factors affecting the diffusion of technology. *Explorations in Economic History* 10(1): 3–33.

Rosenberg, N. (1995) Innovation's uncertain terrain. *The McKinsey Quarterly* 3: 170–185.

Ryan, L. and Campbell, N. (2012) Spreading the Net: The Multiple Benefits of Energy Efficiency Improvements, IEA working paper. International Energy Agency.

Schot, J.W. and Geels, F.W. (2008) Strategic niche management and sustainable innovation journeys: Theory, findings, research agenda and policy. *Technology Analysis & Strategic Management* 20(5): 537–554.

Sengers, F., Wieczorek, A.J. and Raven, R. (2017) Experimenting for sustainability transitions: A systematic literature review, *Technological Forecasting and Social Change*, forthcoming. DOI: 10.1016/j.techfore.2016.08.031.

Seyfang, G. and Haxeltine, A. (2012) Growing grassroots innovations: Exploring the role of community-based initiatives in governing sustainable energy transitions. *Environment and Planning C: Government and Policy* 30(3): 381–400.

Shove, E. (2003) *Comfort, Cleanliness, and Convenience: The Social Organization of Normality.* Berg Publishers, Oxford, UK.

Smith, A. (2007) Translating sustainabilities between green niches and sociotechnical regimes. *Technology Analysis & Strategic Management* 19(4): 427–450.

Smith, A. and Raven, R.P.J.M. (2012) What is protective space? Reconsidering niches in transitions to sustainability. *Research Policy* 41(6): 1025–1036.

Smith, A. and Seyfang, G. (2013) Constructing grassroots innovations for sustainability. *Global Environmental Change* 23(5): 827–829.

Sorrell, S. (2015) Reducing energy demand: A review of issues, challenges and approaches. *Renewable and Sustainable Energy Reviews* 47: 74–82.

Sorrell, S. and Dimitropoulos, J. (2008) The rebound effect: Microeconomic definitions, limitations and extensions. *Ecological Economics* 65(3): 636–649.

Sorrell, S., O'Malley, E., Schleich, J. and Scott, S. (2004) *The Economics of Energy Efficiency: Barriers to Cost-Effective Investment*. Edward Elgar, Cheltenham, UK.

Sovacool, B.K. (2016) How long will it take? Conceptualizing the temporal dynamics of energy transitions. *Energy Research and Social Science* 13, 202–215.

Sovacool, B.K. and Hess, D.J. (2017) Ordering theories: Typologies and conceptual frameworks for sociotechnical change. *Social Studies of Science* 47(5): 703–750.

Sovacool, B.K., Heffron, R.J., McCauley, D. and Goldthau, A. (2016) Energy decisions reframed as justice and ethical concerns. *Nature Energy* 1: 1–6.

Sovacool, B.K., Kivimaa, P., Hielscher, S. and Jenkins, K. (2017) Vulnerability and resistance in the United Kingdom's smart meter transition, *Energy Policy* 109: 767–781.

Stapleton, L., Sorrell, S. and Schwanen, T. (2016) Estimating direct rebound effects for personal automotive travel in Great Britain. *Energy Economics* 54: 313–325.

Staudenmaier, J.M. (1989) The Politics of Successful Technologies. In: Cutliffe, S.H. and Post, R.C. (Eds) *In Context: History and the History of Technology: Essays in Honor of Melvin Kranzberg*. Lehigh University Press, Bethlehem, PA, USA, 150–171.

Steg, L. (2016) Values, norms, and intrinsic motivation to act proenvironmentally. *Annual Review of Environment and Resources* 41: 277–292.

Stern, P.C., Sovacool, B.K. and Dietz, T. (2016) Towards a science of climate and energy choices. *Nature Climate Change* 6: 547–555.

Tiefenbeck, V., Staake, T., Roth, K. and Sachs, O. (2013) For better or for worse? Empirical evidence of moral licensing in a behavioral energy conservation campaign. *Energy Policy* 57: 160–171.

United Nations Climate Change Secretariat (2016) Climate Action Now: Summary for Policymakers.

Unruh, G.C. (2000) Understanding carbon lock-in. *Energy Policy* 28(12): 817–830.

Van de Poel, I. (2000) On the role of outsiders in technical development. *Technology Analysis & Strategic Management* 12(3): 383–397.

Van Lente, H. (2012) Navigating foresight in a sea of expectations: Lessons from the sociology of expectations. *Technology Analysis and Strategic Management* 24(8): 769–782.

Van Lente, H., Peine, A. and Spitters, C. (2013) Comparing technological hype cycles: Towards a theory. *Technological Forecasting & Social Change* 80(8): 1615–1628.

WBGU (2011) *World in Transition: A Social Contract for Sustainability*. German Advisory Council on Global Change, Berlin, Germany.

Wilson, C.A. (2000) Policy regimes and policy change. *Journal of Public Policy* 20(3): 247–274.

3 A normative approach to transitions in energy demand
An energy justice and fuel poverty case study

Kirsten E.H. Jenkins and Mari Martiskainen

Introduction

Meeting the emissions targets enshrined in the Paris Agreement will necessitate low-carbon, sustainability-oriented transitions across multiple sociotechnical domains, including electricity and heat, industry and buildings, and transport, to name a few. As acknowledged by a sociotechnical approach, this requires not only the rapid transformation of our physical energy systems, but of the societies that create and use them. Indeed, in the case of transitions in energy demand, it will influence who uses which energy source, how and when. Electrification is projected to influence patterns of mobility, changing the ways in which we drive and fuel our vehicles (see Bergman, Chapter 4; Bergman *et al.*, 2017); the introduction of smart metering in the United Kingdom (UK) is enabling remote reading, with implications for user practice and social vulnerabilities, (see Jenkins *et al.*, Chapter 6; Sovacool *et al.*, 2017); and residential retrofit will alter the performance of our homes (see Brown *et al.*, Chapter 7), for instance.

While these energy transitions are promised with the best in mind – often manifesting as a push towards low-carbon energy production forms and energy efficiency measures – it is inevitable that there will be winners and losers. Pertinently, this includes those unable to access or afford them. In this vein, making sure that all voices are represented in transitions plans and their actualisation is undoubtedly a question of social justice, equity and fairness.

As outlined by Jenkins *et al.* (2018, p. 67), failure to adequately engage with questions of justice throughout the transition process is dangerous given that it 'may lead to aggravated poverty, entrenched gender bias and non-participation as outcomes or by-products of 'blinkered" decision-making'. Indeed, without a focus on justice, transitions may fail to acknowledge two sides of the debate surrounding energy demand: (1) the burdens of having too much energy or too many energy services, including subsequent waste, over-use and pollution, and (2) not having sufficient energy services, where some individuals lack access, are challenged by under-consumption and poverty, and may face health burdens and shortened lives as a consequence of restricted energy choices (Sovacool *et al.*, 2016). Thus, our starting assertion is that justice is, and must be, central to transitions debates and planning, even in the context of demand reduction. This is not always the case, however.

Scholars and analysts frequently envision the process by which sustainability transitions take place to be one of disruptive change through transformative innovation (Markard et al., 2012; Schot and Steinmuller, 2016). As a result, proponents of transformative change suggest that by involving stakeholders from the outset, it can present more all-encompassing, robust solutions to sustainability challenges. For instance, Linnenluecke et al. (2017) identify that plans for transformational change recognise that environmental challenges present opportunities to meet the (currently unmet) needs of those at the 'bottom of the pyramid' – including the poorest of the poor (see also Bezboruah and Pillai, 2013; McAlpine et al., 2015; Tebo, 2005). Within the transitions literature, then, there appears an emerging concern for particular individuals who are seen to deserve more socially just outcomes.

Yet despite ongoing debates about ethics and justice across many fields of academic literature, one social element missing from transitions frameworks is explicit, practice-oriented engagement with the energy justice concept and related approaches to justice concerns (Jenkins et al., 2018); an omission that, arguably, is mirrored in practice. Indeed, beyond the walls of academia, and despite the broadening utilisation of the transitions concept, it is increasingly acknowledged that the 'socio-' or social element is frequently missing in the transitions literature and transition plans, including failures to recognise their social justice and equity implications (see Goldthau and Sovacool, 2012; Jamieson, 2014; Markowitz and Shari, 2012; Newell and Mulvaney, 2013; Sovacool et al., 2016; Swilling and Annecke, 2012).

As an illustration, the 2009 UK government's Low Carbon Transition Plan (DECC, 2009) (the last document of its type) includes only passing reference to justice concepts as it mentions 'fairness', which is characterised as the fair distribution of costs only without attention to wider aspects of accessibility, fairness in the decision-making process, or the inclusion of all stakeholders and users.[1] For this reason, Eames and Hunt (2013, p. 58) note that even 'a 'low-carbon' transition has the potential to distribute its costs and benefits just as unequally [as historical fossil-based transitions] without governance mindful of distributional justice' or, as is argued throughout, issues of justice as recognition and procedural justice too.

This chapter serves a dual purpose. First, it reiterates and reaffirms the need for socially just transitions approaches in energy demand scholarship and explores the role of the three-tenet energy justice framework in this. In so doing, it introduces a potentially new audience to energy justice scholarship. Second, through a case study of fuel poverty, it begins to explore both the dangers of failing to acknowledge justice outcomes and, conversely, what doing so may practically look like. We close with policy-relevant recommendations towards the integration of energy justice thinking in demand reduction efforts.

Normative approaches to transitions in demand

Arguably, there are (at least) two weaknesses in our attempts to understanding the normativity of energy demand challenges to date. First, a large proportion of the research on normative approaches to energy demand – that relating to how things should or ought to be – is conducted from an individual's perspective, considering personal normativity and moral choice. As we aim to reduce energy demand, this is a clear advantage. Yet while they capture some concerns of normativity, they do not allow an understanding of how such normative approaches might be embedded in and emerge from transitions in energy demand. Further, individuals may find it difficult to change as they are locked into unsustainable infrastructure, including living in sub-standard housing stock or working on the minimum wage, thereby confining choice. Moreover, with these limitations in mind, a focus on individuals does not provide opportunities for the forms of systematic change we require.[2] For that, we need a broader outlook.

As a second weakness is that we, as scholars, have yet to agree on who we are concerned about. A large proportion of justice scholarship in the energy demand domain focuses on vulnerable consumers – those 'left behind' without regular or affordable access to energy, and this can vary considerably depending on context. Indeed, a range of work has emerged that considers particular vulnerable groups including the elderly or the unwell (Thomson et al., 2017a), those living in adverse housing conditions (Healy and Clinch, 2004), or individuals living with disabilities (Snell et al., 2015).

Yet work by Chatterton et al. (2016) shows that there are in fact only a few (typically well off) minority groups that place the greatest burden on energy networks due to their high energy consumption.[3] Therefore, they argue that we should target the less vulnerable in order to make meaningful progress with demand reduction. As an illustration, they state that

> in order to make these reductions, rather than assuming a need for an 80 reduction across all of society, it makes sense to at least examine the potential for reducing consumption in that sector of society that is consuming greater than 30% more energy per household than the average.
> (Chatterton et al., 2016, p. 87)

In this case then, the measures needed to reach each group would trade off. Indeed, these different approaches raise contrasting normative questions around who we should engage with, how and to what end.

While any one approach is inevitably limited in its ability to resolve a problem entirely – especially one as large and complex as injustice – the remainder of this chapter positions the energy justice framework as one way of further considering and potentially overcoming these challenges above challenges.

In line with the sociotechnical approach, the energy justice framework is one that forcefully reminds us that energy dilemmas are about more than merely hardware (Sovacool and Dworkin, 2015). The concept has emerged amid the realisation that our energy structures require widespread reform, and out of a growing interest in the justice implications of energy consumption and energy's societal impacts (Hall, 2013). Against the background of the environmental and climate justice literatures, and in light of this surrounding context, energy justice thus aims 'to provide all individuals, across all areas, with safe, affordable and sustainable energy' (McCauley et al., 2013, p. 1).

As part of its growing popularity, energy justice is increasingly characterised as an analytical tool – one that, for Heffron et al. (2015), can achieve a just balance in the energy trilemma, which they typify as economics (energy finance), the environment (climate change mitigation) and politics (energy security). As one example, Heffron et al. (2015, p. 172) develop an energy justice metric, which is designed to connect with economists through quantitative analysis of energy justice, allowing it to be evaluated in monetary terms. This approach produces three results: (1) an individual–country energy justice metric and (2) an energy justice metric for each type of energy generation source, e.g. nuclear power, both of which allow (3) the cost of energy justice to then be factored in to an economic model calculation in the form of a cost–benefit analysis. Sovacool and Dworkin (2015, p. 436) state in relation to such models, that energy justice thus 'presents a useful decision-making tool that can assist energy planners and consumers in making more informed energy choices' as well as serving as 'an important analytical tool for energy researchers striving to understand how values get built into energy systems or to resolve common energy problems'. In this regard, the energy justice concept shares some commonalities with the political reach of the transitions concept and sociotechnical approach more broadly.

The energy justice tenet framework

A range of tenet frameworks have emerged within the energy justice field. McCauley et al. (2013) use three – distributional justice, procedural justice and justice as recognition – whereas others dismiss the inclusion of recognition as a tenet, including Sidortsov and Sovacool (2015) who instead focus on distributional justice, procedural justice and cosmopolitanism as core concepts. In addition, Heffron and McCauley (2017) consider restorative justice and Sovacool et al. (2016) add the eight concerns of availability, affordability, due process, intragenerational equity, intergenerational equity, sustainability, transparency and accountability, and responsibility. In keeping with McCauley et al. (2013), however, this chapter utilises the framework of three core tenets, distributional justice, procedural justice and justice as recognition, including justice as recognition as the third tenet based on the works of Fraser (2014).

In a change from the norm, the order in which the tenets are typically used – distribution, procedure, recognition – is altered and, instead, justice as recognition is in second place. This leaves the structure of distributional justice, justice as recognition and procedural justice. This approach builds upon the work of Jenkins et al. (2016) who argue for a reordering of the tenets on the logic that if injustice is to be tackled, you must (a) identify the concern – distribution,

(b) identify who it affects – recognition, and only then (c) identify strategies for remediation – procedure. This is, in effect, a 'what, who and how' approach to tackling energy justice concerns, with the intention that energy justice can exist as a solution-based framework that not only characterises injustices but can also help tackle them. To this end, the following paragraphs not only introduce the conceptual background to these tenets, but also provide real-world case studies of their meaning in practice.

Distributional justice: fuel poverty in the UK and beyond

The first tenet of energy justice is distributional justice. Energy justice is an inherently spatial concept that includes both the physically unequal allocation of environmental benefits and ills and the uneven distribution of their associated responsibilities (Walker, 2009, p. 615): for example, exposure to risk. Thus, energy justice can appear as a situation where 'questions about the desirability of technologies in principle become entangled with issues that relate to specific localities' (Owens and Driffill, 2008, p. 4414), and represents a call for the distribution of benefits and ills on all members of society regardless of income, race etc. (Bullard, 2005; Heffron et al., 2015). To illustrate the application of distributional justice, we move here to the illustrative example of fuel poverty both in the UK and briefly, beyond.

In consumption terms, distributional justice is typically discussed in relation to the issues of affordability, availability and sustainability (Sovacool and Dworkin, 2015). This explains the ready application of energy justice literature to the issue of fuel poverty (see Fuller and McCauley, 2016; McCauley et al., 2013; Sovacool, 2015; Walker and Day, 2012), which was originally defined as having to spend more than 10 per cent of a households' income on energy bills (Boardman, 1991). Distributional justice in this early scenario reflected a concern for the unequal allocation of energy resources and the sometimes-prohibitive costs of them. Yet while the 10 per cent definition was useful (and indeed, is still a widely used indicator in many countries) it was deemed by many to be too rigid to enable for all the complexities linked to fuel poverty to be taken into account (Thomson et al., 2017b).

Throughout the history of the fuel poverty concept, then, the 10 per cent definition has been further divided into three main causes: (1) living in an energy-inefficient home, (2) having a low income and (3) facing high energy bills. Indeed, fuel poverty, or energy poverty,[4] has evolved to become a concern for the inability to 'attain a socially and materially necessitated level of domestic energy services' (Bouzarovski and Petrova, 2015, p. 31) – a wider definition than a simple proportion of income that considers the intersections between a home, its occupants and the energy system, all of which will vary depending on cultural and contextual factors. Nonetheless, despite this broadening context, what is being discussed here is what energy resources we do or do not have access to – a manifestation of energy outcomes that distributional justice can reveal and help us to understand.

To clarify, given that distributional justice is concerned with the distribution of goods and services across society, it follows that fuel poverty is a clear example; the majority of individuals enjoy better access to energy services than a minority few and the resultant injustices across social groups (e.g. according to their housing type or geography) leaves some groups more vulnerable than others. This is, first and foremost, a failure of not only energy policy, but also policies on housing, social issues and welfare standards. Second, it is something that we have to account for as we consider not only how to alleviate fuel poverty equitably, but how energy demand policy may further compound or enable this (Gillard et al., 2017).

The reported incidence rates of fuel poverty are high. In England, 11.1 per cent of the population were thought to be in fuel poverty in 2016, a figure that corresponds to approximately 2.55 million households (BEIS, 2018). This is despite positive progress towards the interim 2020 fuel poverty target, which presents a legal obligation for as many fuel poor homes as 'reasonably practicable' to be raised to Band E in the first instance, with Band C as the 2030 goal[5] (White et al., 2014). Rates in Scotland, Wales and Northern Ireland are even more concerning. In 2016, 26.5 per cent of the total Scottish population, or 649,000 homes, were deemed fuel pool and of those, 183,000 were extremely fuel poor (Scottish Government, 2018). The latest figures for Northern Ireland are 42 per cent and for Wales 23 per cent of households living in fuel poverty[6] (Department for Communities, 2018; Welsh Government, 2018). In this regard, distributional inequities are rife.

While numerous advocates have stated that improving energy efficiency and reducing demand can reduce fuel poverty (e.g. Sorrell, 2015), we must be careful of two caveats: first, that fuel poverty is not experienced the same way everywhere and second, that it can go beyond simple maldistribution.

The UK's framing of fuel poverty has been influential in other developed economies, including researchers in New Zealand (Day et al., 2016; Howden-Chapman et al., 2012; O'Sullivan et al., 2012; Viggers et al., 2013) and Europe, where concerns have appeared around a household's ability to heat their home, manifestations of damp and mould and energy bill debts (Day et al., 2016; Healy and Clinch, 2004; Thomson and Snell, 2013), and increasingly overheating. For example, a recent European-funded initiative – EU Energy Poverty Observatory (EPOV) – brings together researchers, academics, policymakers and practitioners across the EU, aiming to understand and address energy poverty in Europe in a more coherent way. EPOV seeks to illustrate, that (a) our assets are not distributed equally and (b) that through this maldistribution, issues of inequity emerge. It follows that when planning to overhaul energy infrastructure for low-carbon goals, we should also anticipate and mitigate the potential impacts of our choices with a view to reducing unequal burdens across a wide range of contexts, including attention to the potential externalisation of UK energy policy 'ills'.

Second, access cannot always be evenly distributed, at least not without major grid infrastructural changes. Access to ground source heat pumps or localised district heating systems that may lower localised energy costs is restricted, for instance, and although we use an old reference, some island communities in the UK still do not have mains gas supply for heating (Barbour and Twidell, 1981). Thus, Walker and Bulkeley (2006) and Eames and Hunt (2013) note that unequal distribution is not always unjust. Instead, it is often the 'fairness' of the processes surrounding infrastructural development that is important (Walker and Bulkeley, 2006, p. 4), and as such claims for distributional justice require that evidence of inequality are combined with an argument for fair treatment (Eames and Hunt, 2013). Throughout this chapter such arguments are taken to manifest as calls for justice as recognition and procedural justice too.

Justice as recognition

Justice as recognition is taken to be a means of engaging with the questions of 'who' is energy justice for, and, importantly, who is responsible for its provision. It appears as a concern for 'how people are involved in environmental decision-making, or "who (and what) is given respect"' (Eames, 2011). Drawing on Fraser (1999), Schlosberg (2007, p. 18) conceptualises the concerns around justice as recognition as three separate issues: (1) practices of cultural domination, (2) patterns of non-recognition (invisibility of individuals and their concerns), and (3) disrespect through stereotyping and disparaging language: misrecognition. Within this context, justice as recognition is more than tolerance, and requires that individuals must be fairly represented, that they must be free from physical threats, and that they must be offered complete and equal political rights (Schlosberg, 2003).

The process of cultural domination may include, as one of innumerate potential examples, 'the process of disrespect, insult and degradation that devalue some individuals and some places' identities in comparison to others' (Walker 2009, p. 615). In this context, justice as recognition calls for the respect of difference, and a move to prevent one group dominating others (Martin et al., 2013). Further, justice recognition also represents a call to acknowledge diversity within and between environmental justice movements (Hall et al., 2013). Thus, it includes calls to recognise the divergent perspectives of different ethnic, racial and gender differences (Fraser, 1999). Justice as recognition also appears as non-recognition, the invisibility of individuals and their concerns, as exemplified by the often-cited issue of fuel poverty. Finally, concerns may also arise not over a failure to recognise, but as misrecognising, a distortion of individual's views that may appear demeaning or contemptible (Schlosberg, 2003).

Fuel poverty has been shown to affect a household's health, wellbeing and quality of life. For example, living in a cold home has been linked to weight gain in babies and young children, while damp and mouldy homes can cause breathing problems and respiratory illnesses such as asthma (Guertler and Royston, 2013). The lack of heating and insulation can mean that those facing fuel poverty often live in very cold homes, being only able to heat certain parts, or room/s, which limits their use. It is also likely that with warming temperatures, overheating and being able to cool a home will become concerns too.

For those living in fuel poverty, their home becomes a place of discomfort, ill health and even death. In this respect, the health impacts of fuel poverty are very significant. It is estimated that each year, the English National Health Service (NHS) faces a bill of £1.36 billion for treating illnesses linked to fuel poverty (Public Health England, 2014). Mental health problems, respiratory problems like asthma and circulatory illnesses such as heart disease have been linked to cold and damp homes (Marmot Review Team, 2011). Indeed, as perplexing as it sounds, in a country of seemingly high development status, individuals die each year due to cold homes (Jolin, 2014). Fuel poverty can also have an impact on children's education. Living in a cold home can make completing homework difficult and illnesses due to cold and damp homes mean children have to take more time off school (Guertler and Royston, 2013). This comes as acknowledgement that different individuals can experience vulnerability to fuel poverty in different ways, depending on their personal circumstances (Middlemiss and Gillard, 2015).

In his exploration of affordable warmth and justice, Sovacool (2015) highlights concern for a particular group in society – those unable to access affordable heat, who often become visible through the effects listed above. Yet Walker and Day (2012, p. 71) state that fuel poverty 'can be read as a lack of recognition of the needs to certain groups, and, more fundamentally, as a lack of equal respect accorded to their wellbeing'. We seek to highlight here that fuel poverty is both visible and invisible and that one of the largest challenges is establishing who is being affected. Indeed, while identifying those who face fuel poverty is key towards recognising the problem and addressing it, it is not always easy to identify those who are suffering. Often, there is stigma attached to being fuel poor (Hards, 2013), so much so that households avoid seeking help for fear of being seen not being able to cope with certain aspects of their life. Furthermore, there are also those who may not realise that they are in fact living in fuel poverty, a factor that has recently been recognised by the 'Being Warm Being Happy' (2017) research project, which examines fuel poverty among adults with learning disabilities.

Beyond specific social groups that are statistically more likely to be affected, England's leading fuel poverty charity National Energy Action (NEA) identifies that fuel poverty can occur for anyone (NEA, 2017). Previous research has recognised that those considered vulnerable groups, i.e. children, the elderly and individuals with disabilities or long-term health conditions, can be especially susceptible to the impacts of fuel poverty (Public Health England, 2014). However, fuel poverty often becomes an issue at a time of crisis, when an individual's circumstances change dramatically and they become unable to 'function according to the dominant expectations of present-day energy markets' (Martiskainen et al., 2018, p. 29). In this regard, Sovacool (2015) outlines that the issue of fuel poverty intersects with procedural justice, as affected households have neither the time nor the means to participate in energy decision-making that may rectify injustices.

As we move towards policy for reducing fuel poverty and for reducing energy demand, it becomes paramount to both understand the links between the two and to gain a more nuanced recognition of energy needs and their link to vulnerability within particular groups (Gillard et al., 2017); a justice as recognition concern. This includes ongoing reflexivity around who is being negatively impacted by policy choices and who is responsible for those outcomes. In this context, Gillard et al. (2017, p. 55) identify that 'a recognition-based approach can help to identify the particulars of energy injustice for different groups and strengthen the political response'.

Procedural justice

The last tenet in the reordered tenet framework is procedural justice, or the 'how' of energy justice. Procedural justice concerns access to decision-making processes that govern the distributions outlined above, and manifests as a call for equitable procedures that engage all stakeholders in a non-discriminatory way (Bullard, 2005; Walker, 2009). It states that all groups should be able to participate in decision-making, and that their contributions should be taken seriously throughout. It also requires participation, impartiality and full information disclosure by government and industry (Davies, 2006), and the use of appropriate and sympathetic engagement mechanisms (Todd and Zografos, 2005). It is concerned, then, about the fairness of decision-making processes, or justice in 'doing', and emerges as a claim for representational space and free speech (Sayer, 2011; Sze and London, 2008). For Walker (2012) these requirements can be split in to four key rights:

1 access to information, what type of information and who it is provided by;
2 access to and meaningful participation in decision-making;
3 lack of bias on the part of decision-makers;
4 access to legal processes for achieving redress.

Procedural justice manifestations include, as an illustration, questions arising around how and for whom community renewables projects are developed (Walker and Devine-Wright, 2008), and the ethics of the emergent voluntarism debate, where communities volunteer to host facilities (Butler and Simmons, 2013).

Several programmes provided by the UK government, non-governmental organisations (NGOs) and charitable organisations have addressed fuel poverty over the years (see for example Rosenow et al., 2013; Sovacool, 2015). Addressing fuel poverty has been, for example, one of key drivers for building-related energy efficiency policies (Kern et al., 2017). However, despite all the efforts and government pledges to eradicate fuel poverty, it still exists widely in the UK and the number of fuel poor households has been rather stable in recent years.

The government's rhetoric outlined in 2001 was to end the problem of fuel poverty for vulnerable households by 2010. In 2002, this was clarified to reflect the goal of ending fuel poverty for vulnerable and non-vulnerable households living in social housing by 2010. Fuel poverty in other households was to be targeted after progress in these groups, with a target that by November 2016, English citizens should not be living in fuel poverty (DEFRA, 2003). This has changed to instead improving a reasonable proportion of fuel poor housing. In 2015, the UK government published a Fuel Poverty Strategy (DECC, 2015) for the first time in 14 years. This set out a new target for England: 'The fuel poverty target is to ensure that as many fuel poor homes as is reasonably practicable achieve a minimum energy efficiency rating of Band C, by 2030' (DECC, 2015, p. 12). This change in rhetoric was also coupled with the reduction in the number of households that had received energy efficiency measures under the various government schemes (for detailed analysis on UK policy mix changes, see Kern et al., 2017).

Fuel poverty is not an easy problem to solve. As fuel poverty has causes and implications that go beyond energy, addressing it will also require considering issues beyond energy policy (e.g. housing, social and health policy). Recently, scholars in the field of energy research have highlighted that fuel poverty needs to be rethought as a complex issue (e.g. Baker et al., 2018) However, there is a danger that those facing fuel poverty are still rather invisible in the quarters where energy, or social, policy decisions are made. Furthermore, as Sovacool (2015) has highlighted, households affected by fuel poverty may not have the time nor the means to participate in energy policy decision-making. As Gillard et al. (2017) identify, this can be either through a lack of capability, or a lack of trust. While there are several charitable organisations and community initiatives – such as Energy Cafés, which provide locally-targeted advice for those facing fuel poverty (see Martiskainen et al., 2018), these are often limited in scope, stop-start in nature and have to rely on external support and volunteer effort. The lack of sufficient government effort to get to the root of the problem indicates how those living in fuel poverty in the UK have become the invisible, yet accepted, losers in the UK's energy system.

Conclusions and policy recommendations: towards just energy transitions

As a result of our reflections, this section makes both conceptual and empirically founded conclusions and policy recommendations. First, as a larger conceptual claim and in line with previous work (Jenkins et al., 2018), we present the argument that it is within the overarching process of sociotechnical change that issues of energy justice emerge, where inattention to social justice issues can cause injustices, or via their inclusion can provide a means to solve them. Thus, we argue for greater engagement with the energy justice approach in both academia and in practice, where it can be used as a means to guide ethically sound decision-making. This comes partly as acknowledgement that social science perspectives on energy transitions are under-represented in academic scholarship (Guy and Shove, 2000; Sovacool, 2015; Wilhite et al., 2000) and, that where they do exist, the economic and geopolitical aspects of energy take precedence (Edberg and Tarasova, 2016).

As an illustration of why this is necessary in the context of energy demand, we have sought to highlight both the dangers of failing to appreciate justice outcomes and the factors that doing so may reveal, from patterns of poor housing infrastructure to vulnerable groups, inadequate processes, and even affluent parties who, through a process of redistribution may not only lower fuel poverty rates, but also decrease energy demand in the process. The target here is not only to make sure that demand processes are sustainable but, as a fundamental part of that, that they are fair.

We have highlighted that fuel poverty is an urgent issue that needs to be addressed as an energy, social and health issue, and at all levels of government, local authorities and health authorities. In this context, it is possible to give a series of case-specific, policy-relevant recommendations that illustrate the potential mobilisation of energy justice approaches. We base these both on our own analysis, and as a synthesis of recommendations based across relevant academic literature.

First, we identify that government and businesses must identify those who may be vulnerable and then both ascertain and make provision for vulnerable customers through targeted subsidies, exemptions and efficiency measures. This recommendation has bearing on energy efficiency policy in particular, where questions are raised about the equity and design of its implementation. In this context, justice principles can provide guidelines for policy interventions by ensuring that energy efficiency schemes reach households in a way that meets their specific needs (Gillard et al., 2017).

Second, as acknowledgement that fuel poverty goes beyond simple affordability to an outcome of living in an inefficient home, we require strong and consistent policies to upgrade the national housing stock (see also Chapter 7), combined with consumer engagement programmes to enable participatory justice throughout this process. At the same time, we must be careful not to reinforce structural or social inequalities e.g. through stereotyping recruitment practices (Gillard et al., 2017).

Third, in terms of achieving procedural justice, funded Energy Cafés can act as a triage service, bringing together local authorities, health workers, community organisations and individuals in a trusted setting, providing advice and ensuring that energy needs are met (Martiskainen et al., 2018). This becomes especially important when you consider that the more individuals that are successfully engaged and take on energy efficiency schemes, the greater the potential success of transition pathways. These and similar ventures require grant funding in order to provide a continued service, train fuel poverty advisors and transfer learning to others.

Fourth, we must acknowledge the impact of energy pricing and subsidies and their knock-on effects especially on low income and other potentially vulnerable or low-income consumers (see Chapter 6 on smart metering, as an example). This comes as acknowledgement that despite some positive policy measures (including the English Warm Front Scheme which was partially merged into the Energy Company Obligation), increasing prices for electricity and natural gas have rapidly increased compared to incomes, thereby mitigating some positive effects (Sovacool, 2015). In this regard, we need socially justice energy policy mixes (see Chapter 12).

Through these measures and others in the domain of energy demand, we have the potential to both acknowledge and embed normativity in energy demand reduction efforts.

Acknowledgements

This chapter borrows from, and significantly extends the arguments initially presented in, 'Humanizing sociotechnical transitions through energy justice: An ethical framework for global transformative change', published in *Energy Policy* 117 (June 2018), pp. 66–74.

Notes

1 Rawls (1985) describes that 'justice' can be divided into the principles of liberty and equality, with a further subdivision of the latter including the principle of 'fair equality of opportunity'. In this regard, they can be seen as fundamentally interlinked concepts.
2 Unless, of course, they are persuasive on a large scale, though there is limited evidence of this level of effectivity to date.
3 See Chatterton *et al.* (2016) for an exploration of who 'high consumers' may be considered as in the UK context, including the 'energy decadent', where individual circumstances allow high consumption through choice.
4 We acknowledge here that there is a muddied history around 'fuel' and 'energy' poverty distinctions. In line with Bouzarovski and Petrova (2015, p. 33), we present the case that both can be considered under the same umbrella – 'a set of domestic energy circumstances that do not allow for participating in the lifestyles, customs and activities that define membership of society'. This definition arises despite the fact that energy poverty arguably has a tradition in developing world electrification and fuel poverty in developed world originated as a concern for unaffordable warmth within the home.
5 An Energy Performance Certificate (or EPC) is required for properties when constructed, sold or let. The EPC provides details on the energy performance of the property and what you can do to improve it. It is banded between A (the highest) and G (the lowest).
6 Note that different indicators are used in England, and Scotland, Wales and Northern Ireland for measuring fuel poverty. England uses a 'low income high costs' indicator, which states that a household is in fuel poverty if their fuel bills are above national average and, were they to pay those costs, their residual income would be below the official poverty line. Scotland, Wales and Northern Ireland use the 10 per cent of indicator, i.e. a household is in fuel poverty if they have to spend more than 10 per cent of their income on fuel bills.

References

Baker, K., Mould, R. and Restrick, S. (2018) Rethink fuel poverty as a complex problem, *Nature Energy* 3: 610–612.

Barbour, D. and Twidell, J. (1981) Energy use on the island of North Ronaldsay, Orkney. *Energy for Rural and Island Communities*: 39–51.

Being Warm Being Happy (2017) Being Warm Being Happy Homepage. Available online at: https://beingwarmbeinghappy.org/.

BEIS (2018) Annual Fuel Poverty Statistics Report (2016 data). Department for Business, Energy and Industrial Strategy, London, UK.

Bergmans, N., Schwanen, T. and Sovacool, B.K. (2017) Imagined people, behaviour and future mobility: Insights from visions of electric vehicles and car clubs in the United Kingdom. *Transport Policy* 59: 165–173.

Bezboruah, K.C. and Pillai, V. (2013) Assessing the participation of women in microfinance institutions: evidence from a multinational study. *Journal of Social Service Research* 39(5): 616–628.

Boardman, B. (1991) *Fixing Fuel Poverty: Challenges and Solutions*. Belhaven Press, London, UK.

Bouzarovski, S. and Petrova, S. (2015) A global perspective on domestic energy deprivation: Overcoming the energy poverty-fuel poverty binary. *Energy Research and Social Science* 10: 31–40. DOI: 10.1016/j.erss.2015.06.007.

Bullard, R.D. (2005) Environmental Justice in the 21st Century. In: Dryzek, J. and Schlosberg, D. (Eds) *Debating the Earth*. Oxford University Press, Oxford, UK.

Butler, C. and Simmons, P. (2013) Framing Energy Justice in the UK: The Nuclear Case. In: Bickerstaff, K., Walker, G. and Bulkeley, H. (Eds) *Energy Justice in a Changing Climate: Social Equity and Low-carbon Energy*. Zed Books, London, UK.

Chatterton, T.J., Anable, J., Barnes, J. and Yeboah, G. (2016) Mapping household direct energy consumption in the United Kingdom to provide a new perspective on energy justice. *Energy Research and Social Science* 18: 71–87.

Davies, A. (2006) Environmental justice as subtext or omission: Examining discourses of anti- incineration campaigns in Ireland. *Geoforum* 37: 708–724.

Day, R., Walker, G. and Simcock, N. (2016) Conceptualising energy use and energy poverty using a capabilities framework. *Energy Policy* 93: 255–264.

DECC (2009) The UK Low Carbon Transition Plan: National Strategy for Climate and Energy. Department of Energy and Climate Change, London, UK.

DECC (2015) Cutting the Cost of Keeping Warm – A Fuel Poverty Strategy for England. March 2015. HM Government, London, UK.

DEFRA (Department of Environment, Food and Rural Affairs) (2003) The UK Fuel Poverty Strategy: 1st Annual Progress Report. HM Government, London, UK.

Department for Communications (2018) Fuel Poverty. Available at: www.communities-ni.gov.uk/topics/housing/fuel-poverty.

Eames, M. (2011) Energy, innovation, equity and justice. Energy justice in a changing climate: Defining an agenda. InCluESEV Conference, London

Eames, M. and Hunt, M. (2013) Energy Justice in Sustainability Transitions Research. In: Bickerstaff, K., Walker, G. and Bulkeley, H. (Eds) *Energy Justice in a Changing Climate: Social Equity and Low-Carbon Energy*. Zed Books, London, UK.

Edberg, K. and Tarasova, E. (2016) Phasing out of phasing in: Framing the role of nuclear power in the Swedish energy transition. *Energy Research and Social Science* 13: 170–179.

Fraser, N. (1999) Social Justice in the Age of Identity Politics. In: Henderson, G. and Waterstone, M. (Eds) *Geographical Thought: A Praxis Perspective*. Routledge, Oxford, UK.

Fraser, N. (2014) *Justice Interrupts*. Routledge, London, UK.

Fuller, S. and McCauley, D. (2016) Framing energy justice: Perspectives from activism and advocacy. *Energy Research and Social Science* 11: 1–8.

Gillard, R., Snell, C. and Bevan, M. (2017) Advancing an energy justice perspective on fuel poverty: Household vulnerability and domestic retrofit policy in the United Kingdom. *Energy Research and Social Science* 29: 53–61.

Goldthau, A. and Sovacool, K.B. (2012) The uniqueness of the energy security, justice, and governance problem. *Energy Policy* 41: 232–240.

Guertler, P. and Royston, S. (2013) Fact-file: Families and Fuel Poverty. Association for the Conservation of Energy, London, UK. Available at: www.ukace.org/wp-content/uploads/2013/02/ACE-and-EBR-fact-file-2012-02-Families-and-fuel-poverty.pdf.

Guy, S. and Shove, E. (2000) *A Sociology of Energy, Buildings and the Environment: Constructing Knowledge, Designing Practice*. Oxford, Routledge, UK.

Hall, S.M. (2013) Energy justice and ethical consumption: Comparison, synthesis and lesson drawing. *Local Environment* 18(4): 422–437.

Hall, M.S., Hards, S. and Bulkeley, H. (2013) New approaches to energy: Equity, justice and vulnerability: An introduction to the special issue. *Local Environment: The International Journal of Justice and Sustainability* 18(4): 413–421.

Hards, S.K. (2013) Status, stigma and energy practices in the home, *Local Environment* 18(4): 438–454.

Healy, J.D. and Clinch, J.P. (2004) Quantifying the severity of fuel poverty, its relationship with poor housing and reasons for non-investment in energy-saving measures in Ireland. *Energy Policy* 32(2): 207–220.

Heffron, R.J. and McCauley, D. (2017) The concept of energy justice across the disciplines. *Energy Policy* 105: 658–667.

Heffron, R.J., McCauley, D. and Sovacool, B.K. (2015) Resolving society's energy trilemma through the Energy Justice Metric. *Energy Policy* 87: 168–176.

Howden-Chapman, P., Viggers, H., Chapman, R., O'Sullivan, K., Barnard, L.T. and Lloyd, B. (2012) Tackling cold housing and fuel poverty in New Zealand: A review of policies, research, and health impacts. *Energy Policy* 49: 134–142.

Jamieson, D. (2014) *Reason in a Dark Time: Why the Struggle against Climate Change Failed – and What it Means for Our Future*. Oxford University Press, Oxford, UK.

Jenkins, K., McCauley, D., Heffron, R., Stephan, H. and Rehner, R. (2016) Energy justice: A conceptual review. *Energy Research and Social Science* 11: 174–182.

Jenkins, K., Sovacool, B.K. and McCauley, D. (2018) Humanizing sociotechnical transitions through energy justice: An ethical framework for global transformative change. *Energy Policy* 117: 66–74.

Jolin, L. (2014) The scandal of Britain's fuel poverty deaths. *Guardian*. 11 September 2014. Available at: www.theguardian.com/big-energy-debate/2014/sep/11/fuel-poverty-scandal-winter-deaths.

Kern, F., Kivimaa, P. and Martiskainen, M. (2017) Policy packaging or policy patching? The development of complex energy efficiency policy mixes. *Energy Research and Social Science* 23: 11–25. DOI: 10.1016/j.erss.2016.11.002.

Liddell, C. and Morris, C. (2010) Fuel poverty and human health: A review of recent evidence. *Energy Policy* 38: 2987–2997.

Linnenluecke, M.K., Verreynne, M., de Villiers Scheepers, M.J. and Venter, C. (2017) A review of collaborative planning approaches for transformative change towards a sustainable future. *Journal of Cleaner Production* 124(4): 3212–3224.

Markard, J., Raven, R. and Truffer, B. (2012) Sustainability transitions: An emerging field of research and its prospects. *Research Policy* 41(6): 955–967.

Markowitz, E.M. and Shari, A.F. (2012) Climate change and moral judgment. *Nature Climate Change* 2: 243–247.

Marmot Review Team (2011) The Health Impacts of Cold Homes and Fuel Poverty. Written by the Marmot Review Team for Friends of the Earth. University College London, London, UK. Available at: https://friendsoftheearth.uk/sites/default/files/downloads/cold_homes_health.pdf.

Martin, A., Gross-Camp, N., Kebede, B., McGuire, S. and Munyarukaza, J. (2013) Whose environmental justice? Exploring local and global perspectives in a payments for ecosystems services scheme in Rwanda. *Geoforum* 54: 167–177.

Martiskainen, M., Heiskanen, E. and Speciale, G. (2018) Community energy initiatives to alleviate fuel poverty: the material politics of Energy Cafés. *Local Environment* 23(1): 20–35.

McAlpine, C.A., Seabrook, L.M., Ryan, J.G., Feeney, B.J., Ripple, W.J., Ehrlich, A.H. and Ehrlich, P.R. (2015) Transformational change: Creating a safe operating space for humanity. *Ecology and Society* 20(1): 56–61.

McCauley, D., Heffron, R., Stephan, H. and Jenkins, K. (2013) Advancing energy justice: The triumvirate of tenets. *International Energy Law Review* 32(3): 107–110.

Middlemiss, L. and Gillard, R. (2015) Fuel poverty from the bottom-up: Characterising household energy vulnerability through the lived experience of the fuel poor. *Energy Research and Social Science* 6: 146–154.

NEA (2017) UK Fuel Poverty Monitor 2016–2017. A review of progress across the nations. National Energy Action (NEA) and Energy Action Scotland (EAS). Available at: www.nea.org.uk/wp-content/uploads/2016/05/FPM_2016_low_res.pdf.

Newell, P. and Mulvaney, D. (2013) The political economy of the 'just transition'. *The Geographical Journal* 179(2): 132–140.

Owens, S. and Driffill, L. (2008) How to change attitudes and behaviours in the context of energy. *Energy Policy* 36(12): 4412–4418.

O'Sullivan, K., Howden-Chapman, P. and Fougere, G. (2012) Death by disconnection: The missing public health voice in newspaper coverage of a fuel poverty-related death. *K tuitui: the New Zealand Journal of Social Sciences* 7: 51–60.

Pereira, R., Barbosa, S. and Carvalho, F.P. (2014) Uranium mining in Portugal: A review of the environmental legacies of the largest mines and environmental and human health impacts. *Environmental Geochemistry and Heath* 36(2): 285–301.

Public Health England (2014) Fuel Poverty and Cold Home-related Health Problems. Health equity briefing 7. September 2014. Public Health England and UCL Institute of Health Equity, London, UK.

Rawls, J. (1985) Justice as fairness: Political not metaphysical. *Philosophy and Public Affairs* 14(3): 223–251.

Rosenow, J., Platt, R. and Flanagan, B. (2013) Fuel poverty and energy efficiency obligations – A critical assessment of the supplier obligation in the UK. *Energy Policy* 62: 1194–1203.

Sayer, A. (2011) Habitus, work and contributive justice. *Sociology* 45(1): 7–21.

Schlosberg, D. (2003) The Justice of Environmental Justice: Reconciling Equity, Recognition, and Participation in a Political Movement. In: Light, A. and de-Shalit, A (Eds) *Moral and Political Reasoning in Environmental Practice*. MIT Press, Cambridge, USA.

Schlosberg, D. (2007) *Defining Environmental Justice: Theories, Movements and Nature.* Oxford University Press, Oxford, UK.

Schot, J. and Steinmuller, E.W. (2016) Framing Innovation Policy for Transformative Change: Innovation Policy 3.0. Science and Policy Research Unit, University of Sussex, Brighton, UK.

Scottish Government (2018) High Quality Sustainable Homes: Safe. Available at: www. gov.scot/About/Performance/scotPerforms/partnerstories/HARO/Indicators/High-quality-sustainable#A1.

Sidortsov, R. and Sovacool, B.K. (2015) Left out in the cold: Energy justice and Arctic energy research. *Journal of Environmental Studies and Sciences* 5: 302–307.

Snell, C.J., Bevan, M. and Thomson, H. (2015) Justice, fuel poverty and disabled people in England. *Energy Research and Social Science* 10: 123–132.

Sorrell, S. (2015) Reducing energy demand: A review of issues, challenges and approaches. *Renewable and Sustainable Energy Reviews* 47: 74–82.

Sovacool, B.K. (2015) Fuel poverty, affordability, and energy justice in England: Policy insights from the Warm Front Program. *Energy* 93(1): 361–371.

Sovacool, B.K. and Dworkin, M.H. (2015) Energy justice: Conceptual insights and practical applications. *Applied Energy* 142: 435–444.

Sovacool, B.K., Heffron, R.J., McCauley, D. and Goldthau, A. (2016) Energy decisions reframed as justice and ethical concerns. *Nature Energy* 1: 16–24.

Sovacool, B.K., Kivimaa, P., Hielscher, S. and Jenkins, K. (2017) Vulnerability and resistance in the United Kingdom's smart meter transition. *Energy Policy* 109: 767–781.

Swilling, M. and Annecke, E. (2012) *Just Transitions: Explorations of Sustainability in an Unfair World*. UCT Press, South Africa.

Sze, J. and London, J.K. (2008) Environmental justice at a crossroads. *Sociology Compass* 2(4): 1331–1354.

Tebo, P.V. (2005) Building business value through sustainable growth. *Research-Technology Management* 48(5): 28–32.

Thomson, H. and Snell, C. (2013) Quantifying the prevalence of fuel poverty across the European Union. *Energy Policy* 52: 563–572.

Thomson, H., Snell, C.J. and Bouzarovski, S. (2017a) Health, well-being and energy poverty in Europe: A comparative study of 32 European countries. *International Journal of Environmental Research and Public Health* 14(6): 1–17.

Thomson, H., Bouzarovski, S. and Snell, C. (2017b) Rethinking the measurement of energy poverty in Europe: A critical analysis of indicators and data. *Indoor and Build Environment* 26(7): 879–901.

Todd, H. and Zografos. C. (2005) Justice for the environment: Developing a set of indicators of environmental justice for Scotland. *Environmental Values* 14(4): 483–501.

Viggers, H., Howden-Chapman, P., Ingham, T., Chapman, R., Pene, G., Davies, C., Currie, A., Pierse, N., Wilson, H., Zhang, J., Baker, M. and Crane, J. (2013) Warm homes for older people: Aims and methods of a randomised community-based trial for people with COPD. *BMC Public Health* 13: 176. DOI: 10.1186/1471-2458-13-176.

Walker, G. (2009) Beyond distribution and proximity: Exploring the multiple spatialities of environmental justice. *Antipode* 41(4): 614–636.

Walker, G. (2012) *Environmental Justice: Concepts, Evidence and Politics*. Routledge, London, UK.

Walker, G. and Bulkeley, H. (2006) Geographies of environmental justice. *Geoforum* 37(5): 655–659.

Walker, G. and Day, R. (2012) Fuel poverty as injustice: Integrating distribution, recognition and procedure in the struggle for affordable warmth. *Energy Policy* 49: 69–75.

Walker, G. and Devine-Wright, P. (2008) Community renewable energy: what should it mean?. *Energy Policy* 36(2): 497–500.

Welsh Government (2018) Fuel Poverty. Available at: https://gov.wales/topics/environmentcountryside/energy/fuelpoverty/?lang=en.

White, V., Hinton, T., Bridgeman, T. and Preston, I. (2014) Meeting the Proposed Fuel Poverty Targets: Modelling the Implications of the Proposed Fuel Poverty Targets using the National Household Model. Centre for Sustainable Energy Report for the Committee on Climate Change, London, UK. Available at: www.theccc.org.uk/wp-content/uploads/2014/11/CCC_ModellingProposedFuelPovertyTargets_FinalReport_Nov2014.pdf.

Wilhite, H., Shove, E., Lutzenhiser, L. and Kempton, W. (2000) The Legacy of Twenty Years of Energy Demand Management: We Know More about Individual Behaviour but Next to Nothing about Demand. In: Jochem, E., Sathaye, J. and Bouille, D. (Eds) *Society, Behaviour and Climate Change Mitigation*. Kluwer Academic Publishers, Springer, USA.

Part II
The emergence and diffusion of innovations

4 Electric vehicles and the future of personal mobility in the United Kingdom

Noam Bergman

Introduction

The future of transport, and specifically, the question of how to achieve sustainable personal mobility, is a much-debated topic, considering environmental issues such as air pollution, inequality of access to mobility and urban traffic congestion. Further, growing concerns over climate change have put pressure on the transport sector to reduce its greenhouse gas emissions. Transport accounts for approximately 25 per cent of the UK's CO_2 emissions, nearly two-thirds of which comes from cars and vans (CCC, 2014). There are also major concerns over air pollution and road congestion.

However, changes to the road transport system are considered especially challenging. In the UK, like other countries in the developed world, road-based transport centred on the private car for personal mobility is dominant. Discourses among transport professionals and policymakers portray unrestricted mobility as a right (Doughty and Murray, 2016). A UK government-issued review on low-carbon cars opened with the statement 'Road transport underpins our way of life' (King, 2007, p. 3), and ties road transport with economic growth and technological progress, an approach still evident in the recent Industrial Strategy (HMG, 2017). The UK has a large automotive industry, with a turnover of £77.5 billion, employing over 800,000 people, including 169,000 in manufacturing (SMMT, 2017), making change politically and economically challenging. Overall, the UK is locked into *automobility*, where privately owned car use is reinforced by infrastructure, regulations, institutions, vested interests, norms, cultures and practices (Paterson, 2007; Schwanen, 2015; Urry, 2004). A significant shift towards sustainable personal mobility therefore requires systemic change – a *sociotechnical transition*.

Various technological and cultural innovations exist with the potential to make transport more sustainable, such as low-emission vehicles, integrated transport, and car-sharing clubs. A prominent example is electric vehicles (EVs) – an innovation with technical potential to reduce emissions from road transport, if powered by low-emission electricity. However, consumer uptake has been very slow, and EVs have met resistance from within the traditional automotive industry and ambivalence from policymakers. That might be changing with a variety of EVs available on the market in different countries. In the UK, sales have risen steeply in the past few years, with 37,000 new EVs[1] registered in 2016, and 47,000 in 2017 – nearly 1.9 per cent of total sales (SMMT, 2018). While EVs might leave many aspects of automobility intact, they have the potential to significantly change the transport system through diverse electrified transport not limited to cars, causing tension with existing industry, infrastructure and driving and refuelling norms.

This chapter considers future visions of personal transport, as imagined by influential transport policy and automotive industry actors. It builds on a recent study (Bergman, 2017; Bergman et al., 2017) that looked at visions of the future of UK road transport, by analysing documents including forecasts, pathways and scenarios by diverse transport sector actors in order to explore how current challenges are perceived, what strategies these incumbent (i.e. regime) actors follow, and what role EVs might have. This chapter continues that research with a transitions theory perspective, complemented by concepts from institutional dynamics.

The research found that regime actors' visions reflect regime actor strategies: in most visions there is a similar trajectory of continued automobility, which offers little discontinuity or disruption compared to the present and recent past, but rather favours gradual change with conventional vehicles prominent in the medium term, and radical innovations institutionalised to reduce their threat. This chapter argues that the creation of visions is an act of 'institutional work' by incumbents, aimed at shaping the future to minimise change that could undermine their power.

We next consider useful concepts from transition theory and institutional dynamics, and the role of visions of the future in these, before introducing the challenge of automobility and sustainable transport through these lenses.

Theoretical background

Transitions

This chapter considers two aspects of transition theory: First, regime actors and their strategies, and the dynamics of regime–niche interactions, for example, whether niche–regime relationships are *competitive* or *symbiotic*; the latter occurs when radical innovations are seen as complementary to the interests of regime actors and they become absorbed in the prevailing sociotechnical system (Geels and Schot, 2007). Bakker (2014) considers when incumbent actors might strategically support an innovation, considering that newly emerging systems might align with their short- or long-term interests, and that *individual or collective expectations* might guide them to engage with the innovation. Positive collective expectations of an innovation, which might serve an actor's long-term interests, could lead to supporting a transition (and therefore the innovation) through influencing the configuration of the emerging system 'just in case the transition does take place' (Bakker, 2014, p. 65).

Second, potential future trajectories of the transport system, including personal mobility, could be viewed using the typology of *transition pathways* (Geels and Schot, 2007) described below. These pathways are archetypal, and a transition could involve elements of different pathways, or begin following one pathway and continue in another.[2]

When there is ongoing moderate landscape pressure on the regime, but niches are not sufficiently developed to pose a threat, regime actors can respond by modifying trajectories and innovation activities, possibly adopting ideas from niche innovations, with cumulative readjustments. The regime changes trajectory, but its basic architecture remains largely unchanged in a *transformation* pathway. In some instances, the adopted innovations trigger further changes in the architecture of the regime, with ongoing adoption leading to changes to rules, technologies, policy and user practices, all accumulating to a major *reconfiguration*. Struggles between policy actors and industry are possible in both these pathways (Smink et al., 2015).

If a large, sudden landscape change occurs and incremental change is insufficient, the regime struggles in competition between incumbents and newcomers. If developed niche innovations exist, one innovation can ultimately break through to replace the existing regime, in a *(technological) substitution*. If, on the other hand, niches are insufficiently developed, regime actors might lose faith leading to collapse, or de-alignment, of the regime. There is a period of competition between co-existing niches, leading to multiple innovation trajectories – but also uncertainty. Eventually, one niche will win out and a new regime will be formed, completing the *de-alignment and re-alignment* pathway.

In the absence of strong landscape pressure, the regime exhibits dynamic stability in a *reproduction* pathway. Even in reproduction, regime actors need to act in order to maintain power.

Visions

It is well established that visions of the future, and the expectations they generate, are central to the process of technological innovation. Visions can motivate engineers and designers to initiate projects (van Lente, 1993), be used to attract financial support for research and innovation (Fujimura, 2003), and raise interest from a wide range of stakeholders, increasing an innovation's legitimacy and uptake (Geels and Verhees, 2011; Schot and Geels, 2008). Studies like Levidow and Papaioannou (2013) suggest the importance of visions for innovation processes in personal transport, considering different innovation pathways for different visions.

Uncertainty plays an important role in sustainability transitions, due to long time horizons and large investments, in which societal change and other (landscape) effects occur in addition to innovation and technological development. This suggests visions could be especially important in the transitions approach (Budde et al., 2012).

Visioning is a deeply political technique. Visions created by regime actors are part of the sociotechnical system's 'culture', and therefore might influence expectations and strategies of all actors. Regime actors can use visioning exercises to suit their ends: delay pressures, neutralise some risks and threats (e.g. perceived uncertainties over incumbent technologies ability to adapt) while sensitising audiences to others (e.g. end-users' unfamiliarity with yet-to-be-proven technologies), and present the future as a more or less linear extrapolation of the (recent) past. This is one reason why genuinely new or unexpected events or systemic changes are rarely considered in visioning exercises. Meanwhile, niche actors can deploy visions to enhance their legitimacy by building networks and gathering support and resources for their innovations (Schot and Geels, 2008).

Institutional dynamics

Recent thinking on institutional dynamics (Fuenfschilling and Truffer, 2014; Geels, 2014; Smink et al., 2015) suggests that regime actors must work to uphold incumbent institutions, as these 'do not automatically persist. Instead, they need constant maintenance' (Fuenfschilling and Truffer, 2016, p. 301). Rather than reacting to developments, incumbents seek to proactively shape the institutions through which they interact with other social groups. This *institutional work* includes lobbying policymakers by incumbent firms and, vice versa, policymakers reaching out to incumbent service or technology providers; communication with the general public via advertising, press conferences and releases, and information/education campaigns; commissioning research and technical reports; the formulation of technical standards; and the shaping of discourse (Geels, 2014; Smink et al., 2015). In Multi-Level Perspective (MLP) terms, the dynamic stability of the regime is maintained by regime (incumbent) actors performing institutional work to maintain power in the present and shape the future to reproduce present power structures.

Pulling these strands together, this chapter suggests that *visioning is a powerful and effective strategy of institutional work*. First, because the formation of visions draws on the imagined future to justify action in the present (Anderson, 2010; McCormack and Schwanen, 2011). Second, because visioning exercises allow actors to 'craft' development trajectories that suit their agendas. A vision of the future with clear expectations (Geels and Verhees, 2011), framed as technological progress with market potential (Ruef and Markard, 2010) can help an innovation secure legitimacy and get the public onside (Walker et al., 2010).

Automobility

A powerful regime

The UK transport system is underpinned by the regime of *automobility*. This can be described by rules that favour privately owned cars; see car mobility as both a right and a necessity; link (car) mobility to economic development and technological progress; and see cars as (capable of) becoming green and clean through technological change (e.g. Schwanen, 2015). There are widely shared expectations of (relative) continuity of the system, and so far, the automobility regime has remained dynamically stable.

Regime actors include the automotive industry and transport bodies, but also many planners, consultants and policymakers because their livelihoods and/or the wider social or political–economic orders of which they are part depend on automobility's endurance (Cohen, 2012; Paterson, 2007; Unruh, 2000). Path dependencies also derive from the lifestyles of numerous households, and the sunk investments in road infrastructure and car manufacturers' production processes (Driscoll, 2014; Penna and Geels, 2015). Increasing returns and adaptive preferences, as well as 'interlinked networks of dependency' to the sociotechnical systems of housing development and fossil fuel extraction further entrench the automobility regime (Driscoll, 2014; Penna and Geels, 2015; Urry, 2004; Wells and Nieuwenhuis, 2012).

Potential for change

While the automobility regime is powerful and enduring, there are a range of landscape dynamics putting pressure on the regime in recent years, which might have weakened it in the long run, thereby opening a window of opportunity (Budde *et al.*, 2012). Increased concerns over climate change and air pollution have led to tighter regulation and legislation. Private car use and ownership have stabilised and even declined across much of the Global North since around 1990, especially among younger people and in cities – described as *peak car* (Goodwin and Van Dender, 2013). Cultural change has led to the image of the car shifting from an 'icon of modernity' to a more utilitarian perspective (Cohen, 2012). Finally, there seems to be a reduced commitment of policy-makers to automobility (Geels, 2012), e.g. actors at the European Level (EU) level have increasingly challenged the automobility regime over the last 20 years through regulations on CO_2 emissions and proposed roadmaps towards sustainable mobility (Weyer *et al.*, 2015). In conjunction with these trends, there is a differentiation of strategy among car manufactures and other incumbents (Budde *et al.*, 2012): different firms might make different choices in terms of investment in EVs, other technologies, or improved internal combustion engine vehicles (ICEVs).

EVs may benefit from this differentiation as a long-standing innovation that has until now remained a niche. They offer a technological solution, while seemingly minimising behaviour and cultural change, although major infrastructural and industry changes are implied – including a reduction in oil consumption. Until recently, EVs had little involvement from regime actors (van Bree *et al.*, 2010), with sales and performance too low to be considered a serious threat. However, this has changed in recent years. All major car manufacturers now produce EVs, suggesting that the industry has 'picked its winning technology' (Bakker and Farla, 2015). Moreover, recent sales across the developed world, while still modest, suggest uptake is accelerating beyond the demonstration phase (Bakker and Farla, 2015; Nykvist and Nilsson, 2015). There is undoubtedly increased hype around EVs, with the Organization of the Petroleum Exporting Countires (OPEC) increasing their 2040 forecast for plug-in EVs by nearly 500 per cent from 2015 to 2016, and the International Energy Agency more than doubling its central forecast for 2030 EV fleet size (BNEF, 2017). However, taking the long-term view of a transition, it is too soon to say if this is another hype cycle that will be followed by disappointment, or the beginning of a transition to electric transport.

What would a transition to EVs look like?

Looking at potential transitions to EVs in Germany, Augenstein (2015) finds discrepancies between visions of the future based on sustainable electric mobility, and strategies rooted in the current regime; battery EVs as a techno-fix could jeopardise 'deeper' sustainability transitions to lower car dependency. Augenstein concludes that EVs cannot simply replace ICEVs, as they are a radical innovation that does not fit the current mobility regime. Success depends on the emergence of 'new functionalities' (Geels, 2005) that ICEVs can't offer, such as EV-based energy storage, or even EVs as mobile power supply and others we cannot predict. This could imply redefining the role of the car in society, matching the *technological substitution* pathway.

Van Bree et al. (2010) construct future transport visions using transition pathways. Some of their scenarios envision government measures to reduce emissions as the main driver for change, forcing manufacturers to scale up low-carbon vehicles experiments. EVs winning out requires entrance of new actors and significant change to rules and practices (e.g. reduced driving range and overnight recharging), suggesting a transition pathway of regime *de-alignment and re-alignment*. Other scenarios envision consumer preferences and high fuel prices forcing manufacturers to change. EVs win out if fast recharging infrastructure is rolled out, with ongoing systemic changes suggesting regime *reconfiguration*.

These different perspectives suggest a transition to EVs would cause significant disruption to the automobility regime. It follows that regime actors would attempt to either prevent a transition to electric mobility, or reduce the disruption by performing maintaining institutional work, which can alter technology's design, function, practices and image during its diffusion process (Fuenfschilling and Truffer, 2016). In transition terms, regime actors are expected to work towards a *transformation* pathway if *reproduction* is no longer tenable; however, the systemic changes EVs would cause to the entire transport system might make this more of a *reconfiguration*. Lending a historical perspective, Dijk et al. (2015) suggest that when EVs emerged in European markets in the mid-1990s, the transport regime protected itself by transforming and adapting the potentially disruptive innovations of EVs, by favouring hybrid cars (regime sustaining) over battery–EVs (disruptive).

Table 4.1 Final sample of documents with visions of future transport in the UK including EVs

Document	Focus
Scope for the Transport Sector to Switch to EVs and PHVs (BERR and DfT, 2008) Market Outlook to 2022 for BEVs and PHEVs (Hazeldine et al., 2009) How to Avoid an Electric Shock: Electric Cars: From Hype to Reality (Dings, 2009)	Electric vehicles
Market Delivery of Ultra-Low Carbon Vehicles in the UK (Lane, 2011) Leading the Charge: Can Britain Develop a Global Advantage in ULEVs? (Straw and Rowney, 2013)	Low-carbon vehicles
Pathways to Future Vehicles: A 2020 Strategy (EST, 2002) Passenger Car Market Transformation Model (EST, 2007) The King Review of Low-carbon Cars: Part I (King, 2007) The King Review of Low-carbon Cars: Part II (King, 2008) Powering Ahead: The Future of Low-carbon Cars and Fuels (Kay et al., 2013)	Road transport
Fourth Carbon Budget: Reducing Emissions through the 2020s (CCC, 2010) The Carbon Plan: Delivering our Low Carbon Future (DECC, 2011) Fourth Carbon Budget Review: Technical Report (CCC, 2013) Meeting Carbon Budgets – 2014 Progress Report to Parliament (CCC, 2014)	UK economy
Future Energy Scenarios: UK Gas and Electricity Transmission (National Grid, 2015)	Gas and electricity
Intelligent Infrastructure Futures: The Scenarios – Towards 2055 (Curry et al., 2006)	UK futures

Source: the author.

60 Noam Bergman

This chapter builds on previous work in considering the connection between visions of the future of personal mobility in the UK as they are constructed and used by incumbent (regime) actors and transition theory, specifically transition pathways and regime–niche interactions.

Methodology

The main research method was analysis of documents that discuss the UK's transport future. Documents were found through online searches and references in reports and academic papers. These documents were written by, and for, a range of stakeholders in the UK transport sector, including government, industry, consultancies and other bodies. Only documents that explicitly discuss EVs as part of the UK's transport future, and which contain projections about the mid-term future (2020s through 2050s) were selected. Over 30 documents published in 2002–2015 were identified and 16 were selected for in-depth textual analysis, still giving a wide range of perspectives. These are listed in Table 4.1.

The documents were coded through quantitative content analysis. A priori coding included searches for projections of future transport and factors separating different scenarios, such as drivers and barriers for innovation and uptake. Further coding looked for emerging themes and narratives. Some qualitative analysis was also used to understand the tone and context and to infer actors' agendas and strategies.[3]

The documents appear to be dominated by visions from incumbents (regime actors). This might be partly an artefact of the research approach. However, this might also be (partly) a consequence of the deliberate attempts of incumbents to engage in institutional work seeking to absorb EVs into the deep structure that has ensured the survival of automobility since at least the Second World War (cf. Geels, 2011). Moreover, incumbents are more likely to possess the resources and skills to produce the sort of system-level visions that are considered here.

Findings: visions and trajectories

The analysed documents were produced by a range of actors and are different in style and substance. Some use model forecasts, some review scenarios from other sources, and some use heuristic scenarios or storylines, demonstrating what 'could be' achieved. Nonetheless, there are some clear common themes emerging in the future trajectories, which one might call a 'central vision'. This section explores the central vision, the factors and assumptions underlying it, and highlight trajectories which notably diverge from it. The timeline of the central vision is briefly described in Figure 4.1.

The central vision

The most striking feature of this near-consensus future is that almost all of the visions analysed are dominated by an assumption of continued automobility.

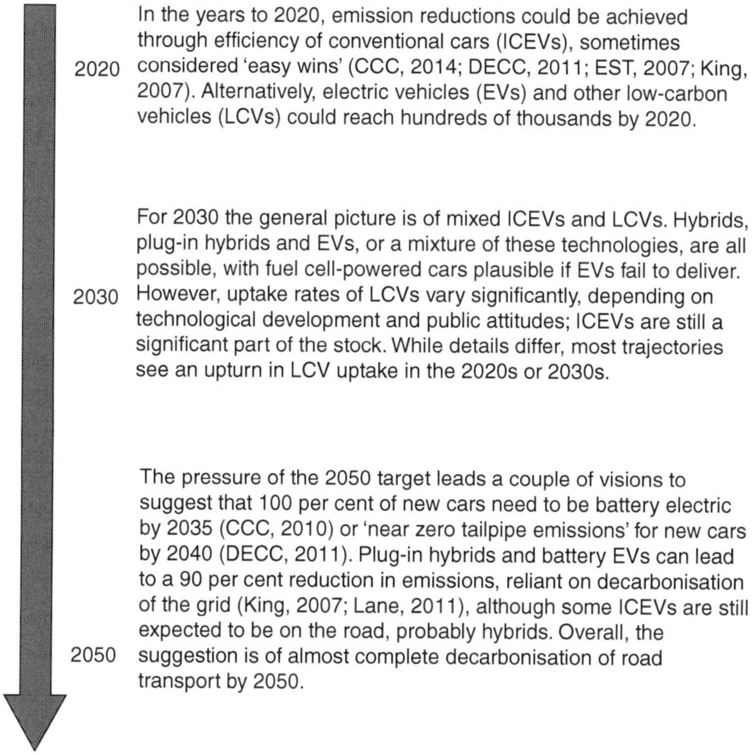

Electric vehicles and personal mobility 61

2020 — In the years to 2020, emission reductions could be achieved through efficiency of conventional cars (ICEVs), sometimes considered 'easy wins' (CCC, 2014; DECC, 2011; EST, 2007; King, 2007). Alternatively, electric vehicles (EVs) and other low-carbon vehicles (LCVs) could reach hundreds of thousands by 2020.

2030 — For 2030 the general picture is of mixed ICEVs and LCVs. Hybrids, plug-in hybrids and EVs, or a mixture of these technologies, are all possible, with fuel cell-powered cars plausible if EVs fail to deliver. However, uptake rates of LCVs vary significantly, depending on technological development and public attitudes; ICEVs are still a significant part of the stock. While details differ, most trajectories see an upturn in LCV uptake in the 2020s or 2030s.

2050 — The pressure of the 2050 target leads a couple of visions to suggest that 100 per cent of new cars need to be battery electric by 2035 (CCC, 2010) or 'near zero tailpipe emissions' for new cars by 2040 (DECC, 2011). Plug-in hybrids and battery EVs can lead to a 90 per cent reduction in emissions, reliant on decarbonisation of the grid (King, 2007; Lane, 2011), although some ICEVs are still expected to be on the road, probably hybrids. Overall, the suggestion is of almost complete decarbonisation of road transport by 2050.

Figure 4.1 Timeline of the central vision.
Source: the author.

Even the most recent documents do not take into account the 'peak car' phenomenon, but use long-standing, powerful discourse in which car-based road transport is central to mobility and tied to progress and economic growth. These visions assume that the UK's high travel demand will continue through 2020, 2030 or 2050, and that it will be met mostly through privately owned vehicles.

Many of the forecasts focus on what technology or fuel will power the cars of the future, as opposed to how or how much people will travel, or the broader question of the role the (private) car in the future. The challenge of reducing emissions is therefore seen primarily as a technical question of the distribution of different vehicle types. While the great reduction in emissions is seen as challenging, in almost all detailed scenarios the targets are met.

Another feature of the central vision is the link between automobility and a strong economy. Some documents explicitly link low-carbon vehicles (LCVs) to economic growth, suggesting the UK automotive industry can use its excellence in ICEV production to reach the forefront of LCV production. Further, it

62 *Noam Bergman*

is suggested that this engineering and manufacturing would deliver jobs and growth. Perhaps the most optimistic vision suggests the transition to LCVs has 'the potential to create jobs, rebalance the British economy towards manufacturing and exports, and promote sustainable economic growth in the UK' (Straw and Rowney, 2013, p. 8).

Factors affecting trajectories

There are many factors that are seen as affecting EV trajectories, including technology, public acceptance and uptake, policy, and economic factors.

Technological innovation is seen as a necessity for EVs to become mainstream in many visions, and most prominent is the battery – its performance, weight, price and reliability. Battery improvement is seen as essential for market penetration, as EVs have to compete both with ICEVs and with other low-carbon technologies. Battery cost is often portrayed as the most significant and least controllable factor affecting EV market penetration, but battery weight and lifetime, range and recharging times are also considered significant barriers.

While there is a common assumption that technological progress will act as a driver, *uncertainty* around *specific* technologies is commonly highlighted. This includes EVs, but also the rate of improvement in ICEVs and the rate of electricity decarbonisation; the collective effect is that the advantage of EVs is positioned as unknown (Dings, 2009; Lane, 2011). The earliest document (EST, 2002) is highly pessimistic about EVs, considering hydrogen more promising. There is more optimism later, suggesting EV rollout depends only on price coming down, although there is also acknowledgement of hype around EVs (e.g. Dings, 2009).

Technical issues and *public acceptance* issues overlap, for example, nearly every vision suggests that limited vehicle range and high upfront cost are crucial barriers to uptake. Lack of variety of car models and brands is seen as a barrier to uptake, suggesting EVs must mimic ICEVs in performance and choice, and meet expectations of comfort and speed, if they are to be widely purchased (CCC, 2014; EST, 2007; King, 2007). It appears that, with increasing confidence in EV technology over time, there has been a shift towards greater focus on public acceptance and awareness, with EVs' image seen as crucial to success. For example, Lane (2011) suggests manufacturers are rising to the challenge of developing low-carbon cars, leaving demand as the central obstacle.

Many of the documents portray people as fairly homogenous consumers having a *passive role* in any transition, limited almost exclusively to the choice of vehicle they purchase. Other behaviours, such as modal shift, trip reduction and eco-driving are not linked to EV trajectories; they appear to be seen as too marginal to significantly affect car sales. This is tied to the assumption that technological breakthroughs are needed if EVs are to succeed because of the presumed behavioural inertia and resistance from users, leaving *little room for adaptation among consumers*. The central vision seeks to replace ICEVs with LCVs with limited behaviour change or other disruption, indirectly delaying

Electric vehicles and personal mobility 63

LCVs until they can easily replace ICEVs. So, despite highlighting the need to create and sustain acceptance demand for EVs, consumers are not generally perceived as a driving force in the documents, in contrast with some scenarios of van Bree *et al.* (2010).

The role of government through *policy* and *legislation* has been described as 'the single biggest influence on the future of low-carbon cars and fuels' (Kay *et al.*, 2013, p. 123). Various market interventions to support EVs are discussed, including subsidies to reduce upfront price; other incentives for purchase, such as free parking or bus-lane use; investment in technologies through R&D funding; demonstration and commercialisation programmes, such as procurement of fleet vehicles; and supporting infrastructure.

Visions are also affected by *economic forecasts*. Most notably, the economic downturn in 2008 led to a drop in car manufacturing volumes, leading Hazeldine *et al.* (2009) to base their scenarios on severity of recession and speed of recovery, as these could affect development and deployment of EVs. The Committee on Climate Change pushed back its 2010 estimate (CCC, 2010) of EVs reaching cost-effectiveness in the mid-2020s to 2030 in a 2013 estimate (CCC, 2013).

These factors show that the central vision is not inevitable but relies on a variety of assumptions about the future. This makes it all the more important to contrast it with alternative visions, of which there are disappointingly few in these documents.

Specific trajectories of interest

A few visions have specific trajectories of interest, which challenge some of the assumptions above or stray significantly from the central vision. They both expose (often implicit) assumptions and agendas in the central vision and highlight different responses to the challenges the regime faces.

One vision (Kay *et al.*, 2013) contrasts an 'evolutionary' EV trajectory, which fits the central vision, with a 'revolutionary' trajectory. In the latter, new market entrants revolutionise vehicle design and manufacturing, introducing Information Computer Technology (ICT)-connected cars and compact, lighter (and therefore cheaper) designs suitable for shorter distances, leading to rapid uptake beginning in cities. The description suggests new functionalities, some still unknown, play a part. This is one of the only trajectories across all the analysed documents to suggest disruption, rather than incremental change, leading to greater institutional changes and potentially new norms around travel, a significant deviation from the central vision.

The Foresight work (Curry *et al.*, 2006), based on consultations with stakeholders from business, research and the public sector, is an outlier in its future visions. Its four scenarios are built around the two biggest uncertainties: whether technological progress will indeed deliver a low-carbon transport system, and whether people will accept intelligent infrastructure. Questioning the success of technology, independent from public acceptance, deviates from the consensus vision, and allows for futures that depart significantly from the present and recent past. Questioning public acceptance and according a significant, active role to the wider population also contrasts with the central vision, where user practices are principally seen as barriers to EV uptake.

Finally, the National Grid (2015) vision considers electricity use throughout the economy, including future projections of EVs. It differentiates trajectories by level of prosperity and level of 'green ambition'. This is the only document that considers the level of *social ambition* to decarbonise the economy, as opposed to actions by government, industry or consumers. Only one of their four scenarios, which has an environmentally engaged society and moderate economic growth, achieves the UK's renewable energy and emission reduction targets on time. This is the only document that explicitly constructs scenarios that fail to meet the targets, highlighting presumed success in many others.

Discussion and analysis

This section considers how the visions reflect actors' strategies, offering an analysis from a transition pathways perspective and an institutional work perspective, noting that some actors might consider LCVs inevitable and are acting to create a new consensus about the future.

Transition pathways in the visions

Considering the four archetypal transition pathways (Geels and Schot, 2007), the central vision conforms best to the *transformation* pathway. While change is seen as inevitable following landscape pressures in the form of policies and targets to reduce greenhouse gas emissions, it is envisioned as gradual, linear and hardly disruptive, with the main elements of automobility unchanged: transport is centred around privately owned cars; users are seen as consumers or drivers with high transport demand; large companies will shift over to making LCVs. While the arrival of new entrants is certainly possible, change is slow enough to allow many ICEV-producing incumbent firms to adapt. The transition dynamic is of regime actors adopting new technologies, gradually and almost seamlessly replacing ICEVs with LCVs (probably EVs). In other words, the regime reorients its trajectory. This is not surprising, considering the visions come from regime actors with strong vested interests.

Using the visions, regime actors craft possible futures in a way that makes the continuity of the sociotechnical regime more plausible. This is strengthened when visions by different actors converge towards a consensus or 'central vision'. Regime actors seek certainty over future policy, as it is questionable whether LCVs can reach a big market without government regulation and investment in infrastructure and R&D.

The transformation pathway sees incumbent manufacturers understanding the need for change, and over time 'gradually increasing use of electrification in the powertrain' (Kay *et al.*, 2013, p. 125) as they develop EVs from existing models, all without relinquishing power. This suggests extensive experimentation with alternative technologies and fuels largely carried out or controlled *by regime actors*, in contrast to outsiders; it could be seen as *niche absorption*, making the regime more fit for purpose in a changing landscape through acquiring new attributes and changing its trajectory (Haxeltine et al., 2008). A similar dynamic is described in van Bree et al. (2010, p. 537): 'Once carmakers become convinced that they can no longer address further tightening of regulations via adaptations of existing technology ... and that non-compliance will lead to substantial (financial) consequences, they scale up [fuel cell] and BEV experiments'.

In contrast, the 'revolutionary' trajectory (Kay et al., 2013) has elements of a *technological substitution*, where ICT-connected EVs taking off in cities leads to significant change in institutions and norms, disrupting the ICEV-based regime. The regime is found not fit for purpose when the landscape changes, and a more suitable niche innovation breaks through, changing not only the technology, but resulting in new travel norms and institutional changes.

In the Foresight visions (Curry et al., 2006), the 'Urban Colonies' scenario combines low-impact transport with strong public attitudes reducing acceptance of intelligent infrastructure, leading to a world where technological investment is focused on reducing environmental impact. Compact cities with a local focus reduce the need for travel, and transport is restricted to cleaner forms. This could be seen as a *reconfiguration* with extensive changes to infrastructure, institutions and travel norms, in contrast to the central vision.

The Foresight 'Tribal Trading' scenario (Curry et al., 2006) envisions a world after a severe energy shock, with the global economy damaged and infrastructure falling into disrepair. Transport has high environmental impact and is greatly reduced, with mobility no longer seen as a right. This trajectory has elements of *de-alignment and re-alignment*, where regimes collapse from strong pressure, and new ones emerge in a very different world. This contrasts sharply with the lack of disruption or discontinuity in the central vision.

In summary, disruptive change is unrealistically lacking in the central vision. The smooth transition does not take into account the radical nature of EVs which might rely on 'new functionalities' to succeed (Augenstein, 2015) – as opposed to a direct one-for-one replacement of ICEVs with EVs. It is important to appreciate that the regime of automobility is not entirely homogeneous or free from tensions; one disruption clearly lacking in the visions is change in fuel provision, as road transport is the biggest global source of oil demand. A shift to EVs would affect supply chain actors in the automobility system, requiring a significant reconfiguration of part of the regime. The lack of disruption is also evident in the paucity of scenarios in which emission targets are missed. Even economic disruption (recession) is seen as temporarily reducing car sales and delaying LCV development and deployment but does not challenge the regime. *This analysis suggests that the central vision is closest to the transformation pathway not because of inevitability or likelihood, but because it best suits incumbent actors' agendas.* This use of visions could be defined as institutional work, as described below.

66 *Noam Bergman*

Visions as institutional work

Creating visions can be seen as a strategy that regime actors use to maintain or restore institutional structures, by expressing expectations about the (long-term) future in order to influence the present (and near future). This section highlights some strategies deployed within these visions to neutralise the potentially disruptive nature of radical or niche innovations.

The visions show a varied, and sometimes contradictory, mix of imagined roles and responsibilities for *government and policy*. On the one hand, there are calls for non-meddling support for manufacturing, and on the other, recommendations that government work together with industry to develop the future of the automotive strategy in the UK. There are various calls for government to ensure funding for research and development (R&D) and not just basic research, and to guarantee long-term support for purchase in order to sustain demand, creating certainty for the industry. This suggests a recognition that creating a successful, sustained EV market is non-trivial and requires intervention and (financial) support. Uptake is seen as one of the biggest challenges and most documents identify ways to incentivise users through financial instruments, such as grants or subsidies for purchase, and some also suggest ways to raise awareness and improve EVs' image.

Both the call for government to tackle barriers to uptake and support the markets, and the focus on users and demand, arguably work to shift responsibilities for change away from manufacturers towards the state and civil society. These calls have a neoliberal character, minimising risk to the private sector and allowing industry actors their freedom without bearing full responsibility for emissions reduction targets, while potentially public and private sector grow closer. From an institutional perspective, this can be seen as *enabling work*, 'the creation of rules that facilitate, supplement and support institutions' (Lawrence and Suddaby, 2006, p. 230).

From a *regulation* perspective, documents from a range of actors call for gradually, but significantly, tightening emission targets. This would allow both improvement of ICEVs and increased uptake of LCVs, '[giving] the industry the required long-term security for investments in low-carbon car technology and infrastructure' (Dings, 2009, p. 7). There are calls for regulations to 'capture well-to-wheel (or even life cycle) emissions' (Kay *et al.*, 2013, p. xi). However, this could increase uncertainties around EVs since their well-to-wheel and life-cycle emissions depend on the electricity grid and the wider vehicle manufacture process, respectively. Institutionally, this could be seen as a *delay tactic*, favouring incumbents in the short term, while allowing them time to adapt.

Many of the documents recommend 'technology neutrality', the assumption that with the right supporting policies, markets will choose the best options among fuel and engine technologies through competition and deliver required greenhouse gas emissions reductions with no major changes to mobility trajectories. Several documents call on government to ensure a level playing field rather than pick winners, stressing manufacturers' support for a technology neutral approach. Some government documents reaffirm allowing industry to develop the low-carbon technologies most appropriate for users. Government and industry seem to be reassuring each other through the narrative of technological neutrality, possibly a sign of a regime alliance between incumbents and policymakers (Geels, 2014). This could be interpreted simply as prudent government action in the face of technological uncertainty: hesitation by the state has been observed elsewhere in relation to EVs and is particularly likely when the free market paradigm prevails (Nykvist and Nilsson, 2015). For the industry, this could be another *delay tactic* that preserves the regime's stability as ICEV improvements can meet emission targets in the short term.

This analysis highlights the profoundly political nature of visioning, seeking to order the institutional constellation and identify and assign responsibilities to different groups of actors. Delay tactics prevent more profound changes, allowing more efficient ICEVs (or hybrids) to continue their dominance in coming years. Industry seeks close alliance with the government with public investment and assurances, while arguing for regulatory approaches that favour incumbents to minimise risk.

Finally, some factors are under-emphasised in the central vision. Behaviour changes beyond choice of vehicle for purchase are portrayed as marginal, and the heterogeneity of mobility patterns and complexity of awareness and acceptance of LCVs are simplified. Deeper changes to the automobility system are almost entirely absent. These could include infrastructural shifts, such as compact cities that reduce the need for personal motorised travel or shifting norms away from seeing high travel demand as normal and high mobility as a right. Even the observed trend of peak car is ignored. *The central vision is limited in scope, hindering genuine transformation, as the unsavoury parts of the transition are downplayed, problematised or ignored.*

Conclusions and policy recommendations

This chapter has analysed visioning documents of the future of the UK's system for personal transport, integrating lessons from transition pathways and institutional dynamics. The focus is on the role of EVs, in the context of growing landscape pressure of climate change-related policy to reduce greenhouse gas emissions from transport.

Some regime actors might consider LCVs to be inevitable, and others might be supporting EVs in order to hedge their bets, and to serve their agendas by influencing the future configuration of the system (Bakker, 2014). Either way, regime actors are working to define future trajectories and make the emerging niche as regime-symbiotic as possible, using visioning documents among other things. They work for slower change, by pressing for a continued high demand, private car-based personal transport system, which allows ICEVs to persist for years to come, and crucially – acts for LCVs to adapt to current automobility, rather than allowing the mobility system to adapt to and be shaped by these technologies. From a transitions perspective, when landscape pressures make the dynamically stable *reproduction* pathway untenable, some incumbents work towards a *transformation* pathway, which minimises disruption and allows many regime actors to maintain power.

One of the most striking features of the central vision is the focus on incremental change with an almost complete lack of discontinuities or serious disruptions to the regime or the transport system more generally. This matches other recent research that found that 'alternative visions of mobility that really might challenge the incumbent regime are rather rare' (Weyer et al., 2015, p. 20). This is unrealistic, as EVs cannot simply replace conventional cars without allowing new approaches to vehicle manufacture or design by new entrants (Kay et al., 2013), perhaps even redefining the role of the car in society (Augenstein, 2015). The fuel shift away from oil would itself be hugely disruptive to the system and is not engaged with in these trajectories. This vision is limited, and limiting, in scope, potentially preventing a deeper transition towards sustainability by locking out alternative futures and limiting EVs to the role of a techno-fix, rather than explore vast possibilities of electrical mobility. For example, EVs' role as electricity storage is a common topic in electricity futures, but its effect on travel practices should be considered in transport futures.

The main recommendation for policymakers is to engage with visions that include a larger variety of futures. These must include scenarios of disruption and shocks to the system, and possibilities of failing to meet emission reduction and other targets. This could be achieved by commissioning visioning documents from a larger variety of actors, including outsiders and niche players, who can challenge, rather than support, futures such as the central vision described here. This would offer more scope and choice for policymakers to meet policy goals and targets and leave us better prepared for foreseeable and unforeseeable changes to transport in the future.

Notes

1 This includes both battery–electric vehicles (BEVs) and plug-in hybrids.
2 In this chapter, *trajectories* describe possible unfolding transport futures, while *pathways* are theoretical, archetypal trajectories, like those detailed in the transitions literature.
3 For a detailed explanation of the methodology, see Bergman et al. (2017).

References

Anderson, B. (2010) Preemption, precaution, preparedness: Anticipatory action and future geographies. *Progress in Human Geography* 34(6): 777–798.

Augenstein, K. (2015) Analysing the potential for sustainable e-mobility – The case of Germany. *Environmental Innovation and Societal Transitions* 14: 101–115.

Bakker, S. (2014) Actor rationales in sustainability transitions – Interests and expectations regarding electric vehicle recharging. *Environmental Innovation and Societal Transitions* 13: 60–74.

Bakker, S. and Farla, J. (2015) Electrification of the car – Will the momentum last?: Introduction to the special issue'. *Environmental Innovation and Societal Transitions* 14: 1–4.

Bergman, N. (2017) Stories of the future: Personal mobility innovation in the United Kingdom. *Energy Research and Social Science* 31: 184–193.

Bergman, N., Schwanen, T. and Sovacool, B.K. (2017) Imagined people, behaviour and future mobility: Insights from visions of electric vehicles and car clubs in the United Kingdom. *Transport Policy* 59: 165–173.

BERR and DfT (2008) Investigation into the Scope for the Transport Sector to Switch to Electric Vehicles and Plugin Hybrid Vehicles. Department for Business Enterprise and Regulatory Reform and Department for Transport, London, UK.

BNEF (2017) Big Oil Just Woke Up to Threat of Rising Electric Car Demand, Bloomberg New Energy Finance. Available at: https://about.bnef.com/blog/big-oil-just-woke-up-to-the-threat-of-rising-electric-car-demand/.

van Bree, B., Verbong, G.P.J. and Kramer, G.J. (2010) A multi-level perspective on the introduction of hydrogen and battery-electric vehicles. *Technological Forecasting and Social Change* 77(4): 529–540.

Budde, B., Alkemade, F. and Weber, K.M. (2012) Expectations as a key to understanding actor strategies in the field of fuel cell and hydrogen vehicles. *Technological Forecasting and Social Change* 79(6): 1072–1083.

CCC (2010) The Fourth Carbon Budget: Reducing Emissions through the 2020s. Committee on Climate Change, London, UK.

CCC (2013) Fourth Carbon Budget Review – Technical Report: Sectoral Analysis of the Cost-effective Path to the 2050 Target. Committee on Climate Change, London, UK.

CCC (2014) Meeting Carbon Budgets – 2014 Progress Report to Parliament. Committee on Climate Change, London, UK.

Cohen, M.J. (2012) The future of automobile society: A socio-technical transitions perspective. *Technology Analysis and Strategic Management* 24(4): 377–390.

Curry, A., Hodgson, T., Kelnar, R. and Wilson, A. (2006) Intelligent Infrastructure Futures: The Scenarios – Towards 2055. Department of Trade and Industry, London. Available at: www.foresightfordevelopment.org/library/55/1337-intelligent-infrastructure-futures-scenarios-toward-2055-perspective-and-process.

DECC (2011) The Carbon Plan: Delivering our Low Carbon Future. Department of Energy and Climate Change, London, UK.

Dijk, M., Orsato, R.J. and Kemp, R. (2015) Towards a regime-based typology of market evolution. *Technological Forecasting and Social Change* 92(C): 276–289.

Dings, J. (2009) *How to Avoid an Electric Shock. Electric Cars: From Hype to Reality*. Transport and Environment, Brussels.

Doughty, K. and Murray, L. (2016) Discourses of mobility: Institutions, everyday lives and embodiment. *Mobilities* 11(2): 303–322. DOI: 10.1080/17450101.2014.941257.

Driscoll, P.A. (2014) Breaking carbon lock-in: Path dependencies in large-scale transportation infrastructure projects. *Planning Practice and Research* 29(3): 317–330.

EST (2002) *Pathways to Future Vehicles*. Energy Saving Trust, London, UK.

EST (2007) *Passenger Car Market Transformation Model*. Energy Saving Trust, London, UK.

Fuenfschilling, L. and Truffer, B. (2014) The structuration of socio-technical regimes – Conceptual foundations from institutional theory. *Research Policy* 43(4): 772–791.

Fuenfschilling, L. and Truffer, B. (2016) The interplay of institutions, actors and technologies in socio-technical systems – An analysis of transformations in the Australian urban water sector. *Technological Forecasting and Social Change* 103: 298–312.

Fujimura, J. (2003) Future Imaginaries: Genome Scientists as Socio-cultural Entrepreneurs. In: Goodman, A.H., Heath, D., Lindee, M.S. (Eds) *Genetic Nature/*

Culture: Anthropology and Science beyond the Two Culture Divide. University of California Press, Oakland, CA, USA, 176–199.

Geels, F.W. (2005) Processes and patterns in transitions and system innovations: Refining the co-evolutionary multi-level perspective. *Technological Forecasting and Social Change* 72(6): 681–696.

Geels, F.W. (2011) The multi-level perspective on sustainability transitions: Responses to seven criticisms. *Environmental Innovation and Societal Transitions* 1(1): 24–40.

Geels, F.W. (2012) A socio-technical analysis of low-carbon transitions: Introducing the multi-level perspective into transport studies. *Journal of Transport Geography* 24: 471–482.

Geels, F.W. (2014) Regime resistance against low-carbon transitions: Introducing politics and power into the multi-level perspective. *Theory, Culture and Society* 31(5): 21–40. DOI: 10.1177/0263276414531627.

Geels, F.W. and Schot, J. (2007) Typology of sociotechnical transition pathways. *Research Policy*, 36(3): 399–417.

Geels, F.W. and Verhees, B. (2011) Cultural legitimacy and framing struggles in innovation journeys: A cultural-performative perspective and a case study of Dutch nuclear energy (1945–1986). *Technological Forecasting and Social Change* 78(6): 910–930.

Goodwin, P. and Van Dender, K. (2013) 'Peak car' – themes and issues. *Transport Reviews* 33(3): 243–254.

Haxeltine, A., Whitmarsh, L., Bergman, N., Rotmans, J. Schilperoord, M. and Kohler, J. (2008) A conceptual framework for transition modelling. *International Journal of Innovation and Sustainable Development* 3(1–2): 93–114.

Hazeldine, T., Kollamthodi, S., Brannigan, C., Morris, M. and Deller, L. (2009) Market Outlook to 2022 for Battery Electric Vehicles and Plug-in Hybrid Electric Vehicles, AEA Group, commissioned by the Committee on Climate Change, Oxfordshire, England.

HMG (2017) Building our Industrial Strategy: Green Paper. HM Government, London, UK. Available at: www.gov.uk/government/uploads/system/uploads/attachment_data/file/611705/building-our-industrial-strategy-green-paper.pdf.

Kay, D., Hill, N. and Newman, D. (2013) Powering ahead: The future of low-carbon cars and fuels. Ricardo-AEA-RAC Foundation. Available at: www.racfoundation.org/wp-content/uploads/2017/11/powering_ahead-kay_et_al-apr2013.pdf.

King, J. (2007) *The King Review of Low-carbon Cars: Part I: The Potential for CO_2 Reduction.* HM Treasury, London, UK.

King, J. (2008) *The King Review of Low-carbon Cars: Part II: Recommendations for Action.* HM Treasury, London, UK.

Lane, B. (2011) *Market Delivery of Ultra-Low Carbon Vehicles in the UK: An Evidence Review.* RAC Foundation. Ecolane, London, UK.

Lawrence, T.B. and Suddaby, R. (2006) Institutions and Institutional Work. In: Clegg, S.R., Hardy, C., Lawrence, T.B. and Nord, W.R. (Eds) *The SAGE Handbook of Organization Studies*, SAGE, London, UK, Chapter 7.

Lente, H. van (1993) Promising technology: The dynamics of expectations in technological developments. PhD Thesis. Universiteit Twente The Netherland. ISBN: 90-5166-354-4.

Levidow, L. and Papaioannou, T. (2013) State imaginaries of the public good: Shaping UK innovation priorities for bioenergy. *Environmental Science and Policy* 30: 36–49.

McCormack, D.P. and Schwanen, T. (2011) Guest editorial: The space-times of decision making. *Environment and Planning A* 43(12): 2801–2818.

National Grid (2015) *Future Energy Scenarios: UK Gas and Electricity Transmission.* National Grid, Warwick, UK.

Nykvist, B. and Nilsson, M. (2015) The EV paradox – A multilevel study of why Stockholm is not a leader in electric vehicles. *Environmental Innovation and Societal Transitions* 14: 26–44.

Paterson, M. (2007) *Automobile Politics: Ecology and Cultural Political Economy.* Cambridge University Press, Cambridge, UK.

Penna, C.C.R. and Geels, F.W. (2015) Climate change and the slow reorientation of the American car industry (1979–2012): An application and extension of the Dialectic Issue LifeCycle (DILC) model. *Research Policy* 44(5): 1029–1048.

Ruef, A. and Markard, J. (2010) What happens after a hype? How changing expectations affected innovation activities in the case of stationary fuel cells. *Technology Analysis and Strategic Management* 22(3): 317–338.

Schot, J. and Geels, F.W. (2008) Strategic niche management and sustainable innovation journeys: Theory, findings, research agenda, and policy. *Technology Analysis and Strategic Management* 20(5): 537–554.

Schwanen, T. (2015) Automobility. In: J.D. Wright (Ed.) *International Encyclopedia of the Social and Behavioral Sciences.* Elsevier, Oxford, UK, 303–308.

Smink, M.M., Hekkert, M.P. and Negro, S.O. (2015) Keeping sustainable innovation on a leash? Exploring incumbents' institutional strategies. *Business Strategy and the Environment* 24(2): 86–101.

SMMT (2017) *Motor Industry Facts 2017.* SMMT, London, UK. Available at: www.smmt.co.uk/reports/smmt-facts-2017/.

SMMT (2018) *EV & AFV Registrations: December 2017 and Year-to-Date, Society of Motor Manufacturers and Traders.* SMMT, London, UK. Available at: www.smmt.co.uk/vehicle-data/evs-and-afvs-registrations/.

Straw, W. and Rowney, M. (2013) *Leading the Charge: Can Britain Develop a Global Advantage in Ultra-low Emission Vehicles?* Institute for Public Policy Research, London, UK.

Unruh, G. (2000) Understanding carbon lock-in. *Energy Policy* 28: 817–830.

Urry, J. (2004) The 'system' of automobility. *Theory, Culture and Society* 21(4–5): 25–39.

Walker, G., Cass, N., Burningham, K. and Barnett, J. (2010) Renewable energy and sociotechnical change: Imagined subjectivities of 'the public' and their implications. *Environment and Planning A* 42(4): 931–947.

Wells, P. and Nieuwenhuis, P. (2012) Transition failure: Understanding continuity in the automotive industry. *Technological Forecasting and Social Change* 79(9): 1681–1692.

Weyer, J., Hoffmann, S. and Longen, J. (2015) *Achieving Sustainable Mobility.* 44/2015. Dortmund. Available at: https://eldorado.tu-dortmund.de/bitstream/2003/34344/1/Weyer_et_al_2015_Discontinuation_Automobility_AP-44.pdf.

5 Experimentation with vehicle automation

Debbie Hopkins and Tim Schwanen

Introduction

Deriving from the Latin *experimentum* (n.) or *experiri* (v.), the verb 'to experiment' denotes 'to try', or 'to test'. Contemporary forms of experimentation are described by Kullman (2013, p. 885) as careful processes of 'tinkering with relatively limited set-ups of bodies, materials and spaces'. Such processes should have sufficient flexibility to allow for reconfigurations, but also sufficient control to hold together. Experimentation offers diverse actors – often configurations from the public, private and third sectors – a way to make sense of the present, while also forming visions of the future (Bulkeley et al., 2015). As such, experiments are seen to be crucial to both sustainability transitions and contemporary modes of governance (e.g. Bulkeley, 2018; Evans, 2011, 2016), particularly where 'capacity to govern is recognised as fragmented and where what it means to govern well – to improve the urban condition – is subject not only to uncertainty but to contestation' (Bulkeley, 2018, p. 1). There is a burgeoning literature on urban experimentation (e.g. Evans et al., 2016), which has pointed to its potentialities and, increasingly, its limitations. In this literature, the relation between experimentation and the 'real-world' setting has received some attention (e.g. Karvonen et al., 2014) but requires further exploration because the latter is not pre-existing but carefully constructed, managed and sanitised alongside and as part of the experiment.

In this chapter, we argue that the real-world character of urban experiments needs to be understood as an artefact. This is not only because, as authors like Caprotti and Cowley (2017) argue, the real world is rendered knowable to actors involved in experiments predominantly through quantitative metrics. It is especially so because most what defines a city – heterogeneity, multiplicity, surprise – is often carefully selected and sanitised by experimental actors. Analysing to what an experiment is exposed and to what it is not therefore contributes to understanding the politics and power-laden nature of urban experimentation.

These propositions are informed by research into the emergence of connected and autonomous vehicle (CAV) technologies, and the ongoing 'real-world' experimentation as part of the innovation process. We empirically examine CAV experimentation in the UK urban settings of Oxford and the London Borough of Greenwich – Greenwich hereafter. CAV innovation and experimentation in the UK has been stimulated by national government funding priorities and justified by way of the intersecting and wicked problems of congestion and air quality together with hoped-for economic benefits as part of the UK government's new Industrial Strategy (HM Government, 2017). Claims of energy demand reduction associated with automated vehicles relate to: the uptake of alternative fuels – although this is less likely for freight vehicles than passenger; automation enabling efficient driving practices (e.g. smooth acceleration and deceleration); and automation-enabled practices such platooning, which relies on vehicle-to-vehicle connectivity for potential fuel efficiency benefits.

Understanding experimentation

The function of experimentation

The distinction between *in vitro* and *in vivo* experiments offers a way to understanding the current popularity of experimentation in cities and more generally. Drawing from the life sciences, Muniesa and Callon (2007) distinguish between *in vitro* experiments that take place in laboratory settings and those run *in vivo*, in 'real-world' ('in situ') environments. The 'real-worldness' of *in vivo* experiments leads to claims that these types of experiments unfold in uncontained settings and are less controllable than *in vitro* experiments (Bellamy et al., 2017), thereby generating alternative types of knowledge and data.

The distinction between *in vitro* and *in vivo* experimentation is, however, underscored by selective interpretations of a (pre-)existing 'real-world' environment, and the objectives of the experimental outcomes. For Caprotti and Cowley (2017), the real world or real *urban environment* becomes known by a somewhat messy set of variables and parameters – traffic flows, NO_x levels, urban developments, and so on – which may include and exclude particular infrastructures, places, policies and publics, 'lend[ing] itself to a potentially normative epistemological approach to the city' (p. 1445). Importantly, such exclusion(s), management and control of experiments may limit the *exposure* of whatever innovation is at stake to unintended experiences or 'surprises' that occur in the real world. This is potentially problematic since surprises are a critical aspect of successful experimentation (Gross, 2010, 2016). Thus, selectivity in experimental actors, practices and spaces, and carefully managed degrees of openness and control restricts the potential benefits of *in vivo* experimentation.

Traditional *in vivo* experimentation fulfils a number of functions according to different literatures. For instance, for Strategic Niche Management, experimentation plays a role in reducing wide-ranging uncertainties, including: techno-economic uncertainties (e.g. competing sociotechnical configurations), finance and investment-related uncertainties (e.g. securing funding for research, development and experimentation), cognitive uncertainties (e.g. views and perceptions that differ from those of the regime), and social uncertainties (e.g. unstable networks), which can hinder the development of niche innovations, and thereby delay or prevent their emergence and subsequent diffusion (Geels et al., 2018).

From a transitions perspective, experiments occur within niches or particular environments which are protected from conventional political economy, with experimental actors managing how system elements (new technologies, institutions, and actor and network constellations) evolve and potentially align in a way that would become stable, and that could replace the current, unsustainable system (Bulkeley et al., 2016; de Wildt-Liesveld et al., 2015). However, Marvin and Silver (2016, p. 14) go further to show how 'real-world' urban experimentation can also play a role in the validation and legitimation of particular ways of thinking about and performing innovation and transitions, which often aligns with neoliberal ideas, and may fit with entrepreneurial urban governance.

From a transition perspective, therefore, experimentation is about various forms of learning and institutionalisation by different stakeholders. On one level, experimental or demonstration projects are one way to aid niche development through first-order learning by public and private sector niche actors and wider publics about innovations 'in real-life circumstances' (Geels, 2012, p. 472). On another level, experimentation is about second-order learning through changes to underlying values and norms and about getting 'outsiders' and regime actors involved and thereby increase the innovation's legitimacy, its institutional embedding, and the availability of resources for its further development (Schot and Geels, 2008). In this context, real-world experimentation can increase an innovation's legitimacy by 'publicly demonstrating proof of concept before scaling up' (Gross, 2016, p. 615) and is a critical step in the innovation pathway.

A framework for analysis: experiments and their settings

It is clear from the growing body of literature on experimentation that both the material (e.g. people, infrastructures, technologies) and the discursive have key roles to play in the recently increased interest in urban experiments with CAVs. In this chapter, we analyse the emergence of those experiments using an analytical framework developed on the basis of the recent literature on urban experimentation that highlights both the material and the discursive. The framework focuses on four key aspects: (1) the underlying framings and logics, (2) the practices of experimentation, (3) the multiple and intersecting roles of the experimental subject(s) and (4) the places and spaces for experimentation.

Underlying framings and logics

In their work on urban laboratories[1] – a term that retains the vernacular of *in vitro* experiments, while introducing the real-world situatedness of *in vivo* experiments – McCormick and Hartmann (2017) distinguish between three

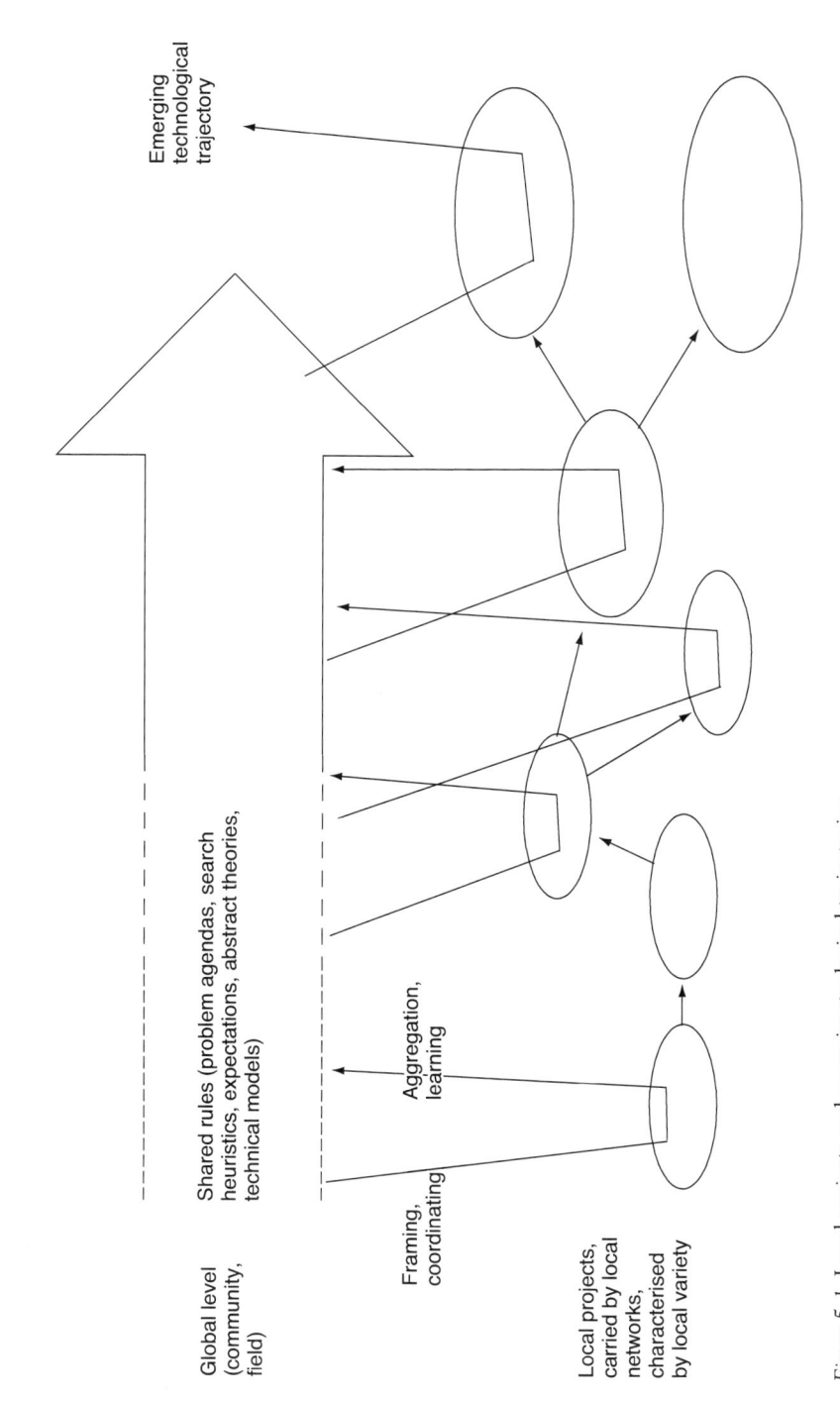

Figure 5.1 Local projects and emerging technical trajectories.
Source: Geels and Deuten (2006, p. 274), with permission.

experimental configurations: *strategic*, *civic*, and *grassroots*. Strategic laboratories are led by government or large private sector actors, and often include multiple projects. *Civic* laboratories are led by urban actors (e.g. universities, urban developers), and may involve just one project, often include co-funding. *Grassroots* laboratories, in contrast, are led by not-for-profits or urban civil society actors. They often have limited budgets and focus on broad agendas (e.g. well-being). In some circumstances, the laboratories will evolve over time, morphing along with funding regimes and actor group involvement. Building on the work of Geels and Deuten (2006, Figure 5.1), we can expect specific laboratories to inform global level actions with interconnections across various sites and domains into coordinated framings and learning.

These configurations can presumably operate according to different *logics* – a term used by Marvin and Silver (2016) to refer to the 'drivers that shape the purpose' of an experiment and frame the rationale and expectations. From their research, Marvin and Silver (2016) found four main types of logic: economic growth, education/knowledge production, techno-orientation and post-capitalist living. In practice, multiple arrangements and logics of experimentation may co-exist at any one place and time, depending on the size of the experiment, the actor groups involved, collective priorities, and the experiment's relationship to a wider innovation pathway.

Practices of experimentation

This aspect concerns the action-of-experimentation as a nexus and integration of discursive and material elements. It highlights that experimentation encompasses a variety of activities (Marvin and Silver, 2016) and that the types of work undertaken may differ across experiments. Work may include research and development (e.g. with new technologies), piloting and testing of new ideas (with less emphasis on research), and/or both formal and informal training and education (e.g. skill development, behaviour change). Such work can be focused on a particular topic (e.g. new technology), or outcome (e.g. sustainability) as previously noted, dependent on the experimental arrangement there may be one project, or multiple (e.g. McCormick and Hartmann, 2017). Funding will have a large impact on the size and duration of experiments (Marvin and Silver, 2016), with some experiments operating on long(er) timeframes and being well-funded, while others are temporary, short term, provisional and constantly on the look-out for secure future funding. Nevertheless, narrow depictions of funding rounds can overlook the connectedness between local projects, which are often part of a wider innovation pathway (Figure 5.1). Hence, while a specific project may appear short term and provisional due to limited funding, these may be continued in different guises by project consortia.

As expected, experiments take up a multitude of configurations. In reference to urban laboratories, McCormick and Hartmann (2017) distinguish between *capacities* by way of the *trial*, the *enclave*, the *demonstration*, and the *platform*. These terms are often used interchangeably but have subtle yet powerful differences in structure and performance. The *trial* points to the 'real-world' testing of products, technologies and/or processes, with the 'real-worldness' determined by key actors. The *enclave* refers to the spatial segregation of the laboratories for innovations to be protected as niches. The *demonstration* context is a managed form of showcasing or exhibiting possible urban futures. Finally, the *platform* seeks to present arenas through which new urban configurations can be fostered. The different intentions captured by these configurations may have powerful impacts on the practices of experimentation, but are likely to be intersecting, with innovation experiments being designed around multiple capacities. McCormick and Hartmann's (2017) capacities are used in this chapter to make sense of the ongoing practice of experimentation with automation in Oxford and Greenwich.

Experimental subjects

Contemporary urban experiments often include wide-ranging actor groups, with differing purposes and priorities. While involvement in experiments can be driven by the prospect of commercial benefits for some, others may have been drawn in by the feeling of involvement in 'cutting edge innovation as it unfolds on the ground' (Karvonen *et al.*, 2014, p. 105). It is tempting to adopt the triple helix approach (Leydesdorff and Etzkowitz, 1998) to analyse how CAV experiments bring together start-up businesses, incumbent actors from motor, ICT, insurance and other allied industries with local and national governments and engineers and data scientists from universities. Caprotti and Cowley (2017), however, suggest that analyses of experimental subjects should go beyond the actor groups involved and consider deeper questions about 'on whom the experimenting is carried out? And by implication, who decides what is to be an experiment?' (pp. 1445–1446). Addressing such questions recognises the power-laden and political nature of urban experimentation and helps to shed light on artificial nature of its real-world character.

Places and spaces of experimentation

Experimentation is often geographically situated and linked to a specific place and time. There often are clear temporal and spatial boundaries to experiments, imposed by limits to funding and jurisdiction. Yet, according to Turnheim *et al.* (2018), experiments are increasingly breaking free of these confines and taking on new configurations. Experiments reach across spatial scales simultaneously drawing from the national (e.g. government funding and priorities, and national organisations), regional (e.g. councils and industry operators) and local (e.g. sites of experimentation). They also stretch beyond national boundaries, to connect internationally through the transfer of know-how, expertise and technologies, connecting to global flows, networks and innovation races (e.g. Hopkins and Schwanen, 2018a; see also Figure 5.1). For research, the significance of connections in shaping experiments calls for relational analytical frameworks that are sensitive to both the circulations of discourses, practices, technologies, actors, and so forth and to the wider political economies in which those circulations and experimentation are situated.

The experiment *arena* – the site of/for experimentation – can be the city-as-a-whole through to campuses or individual buildings but will frequently draw upon expertise, technologies, policies and practices from other territories. However, where claims of whole-city experiments are made, practices of experimentation are often concentrated in or even limited to particular sites within the city – areas of (re)development, urban centres and wealthy suburbs. As such experiments are often co-opted as sources of economic development and identity formation, linked to regional provenance claims, and used to attract inward investment often by means of public–private partnerships. Thus, such experiments will rarely engage with questions of equity, equality, access and social inclusivity (e.g. Evans, 2016), even where these may be priorities of certain project partners such as local councils.

Experimentation with connected and autonomous vehicles (CAVs)

The emergence of CAVs

While framed as particularly novel and contemporary technologies, CAVs have undergone multiple waves of hype and experimentation. The current wave is perhaps the most geographically dispersed and technologically advanced, although previous attempts to develop driverless technologies saw similar dynamics in terms of high expectations, public trials and government funding in the 1920s and the 1960–1970s (Hopkins and Schwanen 2018b). Each of these periods ended with the eventual discontinuation of public funding, as technologies failed to meet heightened expectations – a fate that on the face of it seems unlikely for present-day CAV experimentation but might yet befall them, should future controversies, for instance around casualty numbers in traffic accidents, become unmanageable.

The resurgence of industry and policy interest in CAVs can be related to three factors. First, increased technological capabilities following breakthrough developments in deep machine learning, coupled with a growth in sensor technologies and smart connectivity, have provided the technological foundations for modern CAV innovation. Second, ongoing and increasingly politicised concerns over traffic congestion, road safety, air pollution and inequitable transport provision have stimulated interest in CAV innovation as a potential solution. Third, for the UK specifically, interest in domestic innovation through indigenous research and development and in a resurgence of manufacturing has been intensified because of techno-optimistic and neoliberalised governance formations and an economy struggling in the wake of the Global Financial Crisis and more structurally because of limited long-term improvements in productivity, coupled with instabilities associated with the UK's exit from the European Union ('Brexit') (Hopkins and Schwanen, 2018a).

Uncertainties in the innovation pathway and potential contributions to sustainability agendas notwithstanding, governments across the Western world are preparing for, and seek to accelerate, a transition towards automated mobility. The UK government, for instance, has been active in niche development, undertaking a range of tasks, including: future visioning, network building, harmonising domestic legislation and collaborating on international standards, developing a code of practice for experimentation and testing, and funding CAV research and development including demonstration projects and consortia (Hopkins and Schwanen, 2018b). Since 2014, over £120 million has been committed to the research, development and demonstration of (elements of) CAV technologies, from discrete sensor technologies through to smart and automated transport ecosystems (CCAV, 2018).

In vivo experimentation with CAVs is widely seen to have multiple, interrelated benefits. Not only are CAV technologies, infrastructures, practices and policies experienced directly in 'real-world' conditions, the various automated configurations are also made and remade through interactions with 'real-world' people, places and infrastructures. The resulting algorithmic, behavioural, sensory and tacit knowledges are considered indispensable to the further development and subsequent scaling of CAVs. Moreover, across the transport sector, the Volkswagen emissions scandal – a.k.a. 'Dieselgate' – has raised deep suspicion about the validity and transparency of *in vitro* laboratory testing, exposed the (intentional) human interference with putatively objective *in vitro* testing, and highlighted the divergence between *in vitro* and *in vivo* trials. Dieselgate thus seems to have reinforced a need for public, 'real-world' experiments and claims of superior knowledge. Nevertheless, the notion of a (set of) complex 'real world(s)' that will produce additional knowledges can be subject to contestation, not least due to the very specific contexts in which automation is being 'experimented' with, as elaborated further below.

UK national policy and case studies

This chapter draws empirically from 15 interviews with CAV stakeholders in both Oxford and Greenwich, across the public, private and third sectors. The interviews, part of a broader study about CAV experimentation across the UK, were conducted in 2017, audio-recorded and professionally transcribed. The interviews were then thematically coded using NVivo 10 qualitative software. The case studies presented below draw selectively from this analysis to examine the ways through which automation experimentation is taking place in Oxford and Greenwich.

National contexts

The UK government's approach to CAV innovation is broadly in line with the processes suggested by the Transition Management cycle (Loorbach, 2010): considerable effort has been expended on the *strategic* activity of future visioning, for instance through policy statements like the 'Pathway to Driverless Cars' document (Department for Transport, 2015) and the new 'Industrial Strategy' (HM Government, 2017); the *tactical* activities of network building, the creation of domestic legislation, and the alignment of international standards; and *operational* activities relating to technology trials and demonstrations (Hopkins and Schwanen, 2018b). With regard to those operational activities, national government support has actively stimulated city and regional council involvement in the development and testing of automated technologies. This activation of local government needs to be seen in the wider context of almost a decade of national-level austerity politics, which are affecting local government deeply. On the one hand, an austerity-driven push for devolution is placing increasing responsibilities onto local authorities, while budget cuts on the other hand are significantly narrowing their room to manoeuvre and forcing them into all kinds of partnerships with private sector actors.

The replacement of Regional Development Agencies with Local Enterprise Partnerships (LEPs) in 2010 by the Conservative–Liberal Democrat coalition paved the way for local innovation strategies. In 2014, the LEPs were given responsibility for delivering part of the EU Structural and Investment Funds for 2014–2020. Our interviews found LEPs defined in the following terms: 'to drive economic growth and to provide the strategic leadership that's necessary in order to plan for long term economic growth and transformation'. Oxfordshire Local Enterprise Partnership (OXLep), established in 2011, is one of 39 LEPs in the UK. OXLep focuses its work on four intersecting themes – people, place, enterprise and connectivity – which feature prominently in the strategic plans of most LEPs across the country. For Greenwich, as a borough of the City of London, the LEP works across the London boroughs to 'lead economic growth and job creation'.

Oxford, Oxfordshire

Oxford is part of the so-called 'Golden Triangle' – together with Cambridge, Milton Keynes and London – of innovation-intensive industries that are viewed as nationally important drivers of economic development. The City of Oxford's growth agenda underscores priorities for the regional and city council, and includes increases in retail space, and growth in housing and jobs, each putting renewed demands on the stretched regional transport system which is already heavily car dependent as far as connections among concentrations of people and economic activity is concerned. The City of Oxford has long been associated with research and innovation (Lawton Smith, 2003), with numerous institutions including the city's two universities, research institutes and associated spin-out companies, such as Oxbotica – a key player in UK CAV innovation.

Oxford's links to the motor industry date back to at least 1912 when William Morris moved his motor company to southeast Oxford to begin vehicle mass production, and the city now is a hub for the motorsports industry which occupies multiple sites across the Oxfordshire countryside. Building on its historical legacies and driven by the global success of Oxbotica and the strategic activities and enthusiasm of 'middling technocrats' (Larner and Laurie, 2010, p. 219), Oxford has – through a variety of means – signalled its intent to become a forerunner in the development, testing and demonstration of CAV technologies. Experimentation with vehicle automation in Oxfordshire takes many guises, with the city and its various actors hosting a number of discrete projects.

Royal Borough of Greenwich

Greenwich is a relatively diverse London borough situated in the southeast of the city. Greenwich is in a period of change and at the cusp of a range of opportunities linked to the growing economy and regional development, but also facing significant challenges, including the transportation of people, goods and waste. Greenwich has a historical town centre, and pressure from traffic travelling through Greenwich from the southeast into central London via the Blackwall Tunnel. Infrastructural responses to Greenwich's transport issues include increasing the capacity of and access to London's public transport networks (e.g. underground, overground, bus) but this has done little to address ongoing congestion of vehicles travelling through Greenwich to central London. Continuing urban development and intensification (e.g. on Greenwich Peninsula), is likely to further exacerbate travel demand through the movement of goods (e.g. construction processes, waste, produce) and people (e.g. residents and workers).

As one of the original sites of CAV demonstration, funded by the UK government, Greenwich has become 'a major centre for testing and demonstration' (Digital Greenwich, 2016). The borough launched its 'Smart City' strategy on 22 October 2015, setting out the Council's proposal to implement 'smarter' responses to the borough's challenges, and the creation of new opportunities for local business and communities. Digital Greenwich was founded as an in-house team of the Royal Borough of Greenwich Council, charged with the development and implementation of its Smart City strategy.

Framings and logics

An experimental approach to transport innovation was understood by research participants as an opportunity to find solutions to address air pollution and other contested environmental concerns in a context of fragmented governmental capacities (cf. Bulkeley, 2018). Participants from Oxford articulated clear expectations about CAV technologies, particularly relating to energy demand reductions. However, as shown below, these focus particularly on demand management and electrification of the powertrain rather than on automated technologies per se.

> (CAVs will aid energy demand reductions) in two different ways. The one is the powertrain in the vehicles, we look forward to electrification. There are areas that you cannot just substitute electric vehicles for traditional

vehicles because of the grid capacity. So, there would need to be a demand reduction or local grids or different solutions. The other part is the intelligent part of the mobility which is minimising the demands, both in terms of cars and automatically then in terms of power to move these cars around.

(O2: Oxford, Local Government, CAV Innovation)

When you look at something like connected and autonomous vehicles, the point there is that you would think there would be more vehicle sharing, think there would be more efficient driving, think there'd be better use of road space. You'd hope that you'd got these views along with it that you're going to have lower emission vehicles, you're going to have something closer and more akin to a sharing economy with cleaner vehicles that operate.

(O8: Oxford, Local Government, Economic Development)

In Greenwich, associated with congestion were concerns about transport-related pollution, 'the pollution from the diesel engines is something that's attracting a lot of attention, a lot of concern because this is already a hugely polluted part of the world' (G2: Greenwich, Community Group). As in Oxfordshire, a connection was made between current transport concerns – largely related to congestion and air pollution – and the emergence and 'real-world' testing of CAVs as an appropriate response by local government. The focus on automation was viewed to be 'quite forward-looking' and driven by a desire for 'different methods to try and address' (G3: Greenwich, Local Government, Digital Greenwich) 'wicked' transport problems (cf. Ney, 2009). However, this is not the only logic at play in Greenwich, with terms such as 'productivity', 'growth' and 'foreign investment' explicitly linked by Digital Greenwich to its focus on the testing of CAV technologies (Digital Greenwich, 2016). Both Oxford and Greenwich appear to see CAV experiments as part of the dominant urban agenda of neoliberal economic growth (Karvonen et al., 2014), and technological solutionism to environmental and social concerns.

Experimental practices

Practices of experimentation differed between Oxford and Greenwich due to the different actor groups, and capabilities. Table 5.1 outlines a series of CAV activities in Oxford and Greenwich, demonstrating clear overlap between the two sites in terms of experimental actors and thus underscoring the need for a relational analytical perspective (see above).

Two key projects that are undertaking a variety of experimental practices in Oxford are the 'Smart Oxford' lab and the DRIVEN consortium. These two projects represent diversity in terms of formality, funding and focus. For instance, while Smart Oxford was identified as an 'informal partnership' and 'quite a nebulous idea' (O8: Oxford, Local Government, Economic Development), the practices of the DRIVEN consortium focus specifically on the development of

CAV technologies. The Smart Oxford Lab has a broad remit and a focus on building partnerships, creating opportunities, and 'horizon scanning'. Thus, Smart Oxford 'has a legitimacy but it isn't a decision-making board' (O8: Oxford, Local Government, Economic Development). In this way, Smart Oxford helps to create and nurture an ecology that is conducive to, and makes possible, the experimental practices of others, while also actively constructive positive visions around Oxfords innovation capabilities.

In contrast, the DRIVEN project (www.drivenby.ai) aims to deploy a fleet of highly automated vehicles on urban roads and motorways, leading to multiple journeys between Oxford and London in 2019. The DRIVEN website states that it is 'the most complex CAV trial that's ever been attempted' with 'groundbreaking' research. The novelty appears to lie in its use of 'fleets of cars' – six autonomous vehicles working 'cooperatively' – sharing information between the six vehicles to 'help them know how to behave' (DRIVEN, 2018). The experimental practices of DRIVEN to date fit with what McCormick and Hartmann (2017) refer to as a *demonstration* as the innovation has been showcased to tightly managed, invited audiences.

CAV projects in Greenwich include partnerships between a variety of private and public sector actors and are funded by grants from the European Union, Innovate UK and other agencies. The GATEway project (**G**reenwich **A**utomated **T**ransport **E**nvironment) was one of the first CAV experiments in the UK, and thus presented one of the first opportunities for selected publics – by way of an online selection process – to travel in a CAV. Media reporting of the experimental practices may have also contributed to more distributed awareness of CAV innovation. MOVE_UK, another partially government-funded consortium, speaks specifically of processes of data collection and analytics as part of its experimental practices. This include, for instance, the real-time capture of vehicle operation and data from the automated driving system technologies, transmitted via 3G, as well as the automatic transfer of data to a secure storage which enables advanced data analytics. These practices effectively operate as an amalgamation of particular elements of both *in vivo* and *in vitro* experimentation, with the boundaries between controlled and uncontrolled, and degrees of experimental containment unsettled.

MOVE_UK and the GATEway project now both sit under the umbrella of the Transport Research Laboratory (TRL)'s Smart Mobility Living Lab in Greenwich. This is 'an 18-month project ... set up as a testbed in and around the whole Borough of Greenwich to allow real-world testing of AVs' (G3: Greenwich, Local Government, Digital Greenwich), designed to 'speed up' CAV research and development and building upon the portfolio of government grants led by TRL and its partners. As an umbrella organisation – and something of an experimental *platform* (McCormick and Hartmann, 2017) – the Living Lab offers conditions for experimentation and facilitates practice by way of on- and off-site sites.

Another public experiment taking place in Greenwich is led by a private sector start-up, Starship Technologies (www.starship.xyz/company/), which has been

Table 5.1 Examples of connected and autonomous vehicle (CAV) activities in Oxford and Greenwich

City	Experimental practices	Projects	Partners	Stakeholders
Oxford	A fleet of six fully autonomous (SAE level 4) vehicles deployed in urban areas and on motorways. The project will culminate in an end-to-end journey from Oxford to London. To date, experimental practices have included private testing for product development, and showcases of vehicle-to-vehicle communication to invited media.	DRIVEN Consortium (2017–2019) http://drivenby.ai	Oxbotica, Oxford Robotics Institute, re/insurer XL Catlin, Nominet, Telefonica O2 UK, Transport Research Laboratory, the UK Atomic Energy Authority's RACE, Oxfordshire County Council, Westbourne Communications and Transport for London	Innovate UK, UK Government
	Developing, testing and trialling of products including Selenium (autonomous control system), a vehicle agnostic operating system and Caesium, a cloud-based fleet management system. This has taken place within projects (e.g. GATEway), and as part of discrete research and development activities.	Oxbotica (2014–ongoing) www.oxbotica.ai/	Oxbotica	
	Initiation and facilitation of relationships with the goal of stimulating new partnerships in line with the goals of the Smart City. This includes identifying opportunities, matching these with local needs, connecting partners, and convening consortia. Specific projects include Mobox – a partnership that promotes urban mobility, transportation and innovation in Oxford.	Smart Oxford Laboratory (2015–ongoing) www.oxfordsmartcity.uk/oxblog/home/	Oxford Strategic Partnership (OSP): a collaboration between public sector (e.g. City and County Councils, NHS and Police), academia (e.g. University of Oxford and Oxford Brookes University), business (e.g. Unipart) and voluntary and community organisations (e.g. Oxfordshire Community and Voluntary Action)	University of Oxford, Oxford City Council, UK Government

Greenwich	A fleet of driverless pods providing a shuttle service around the Greenwich Peninsula to understand public acceptance of, and attitudes towards, driverless vehicles.	GATEway Project (2015–2018) https://gateway-project.org.uk/	TRL: The Future of Transport, Royal Borough of Greenwich, Royal College of Art, University of Greenwich, RSA, O₂, GOBOTIX Ltd, Commonplace, Westfield, Heathrow DG Cities	Innovate UK, Centre for Connected and Autonomous Vehicles, UK Government
	Trials of automated driving systems in real-world traffic conditions, whilst capturing real-time data, advanced analytics, and an evidence base for regulatory approaches.	MOVE_UK (2016–2018) www.move-uk.com	Bosch, TRL: The Future of Transport, Jaguar Land Rover, The Floow, Direct Line Group, Royal Borough of Greenwich	Innovate UK, Centre for Connected and Autonomous Vehicles, UK Government
	The creation of positive discourse around CAV innovation and Greenwich as a site for experimentation, facilitating partnerships and attracting additional funding.	Smart Mobility Living Lab (2016–ongoing) http://uklivinglab.trl.co.uk/	TRL: The Future of Transport, Innovate UK, EPSRC, Bosch, Jaguar Land Rover, UMTRI, Telefonica, Shell, CEDR, RSA, Direct Line Group, Westfield, Heathrow and Oxbotica	MOVE_UK, GATEway Project

Source: the authors.

testing their last-mile delivery technologies in Greenwich since 2016. These experiments involve 'live' deliveries using their pavement-based delivery bots, designed for 'local delivery of goods and groceries for consumers for under £1 per shipment' and drawing from sensing technologies including radar, stereo vision, ultrasonics and machine learning. The 'bots' are currently operated by human workers-at-a-distance, as the experiment gathers data and develops the technology. However, the intention is to achieve autonomous mobility with human 'oversight' in due course. Thus, Starship's experimental practices are anchored in learning-by-doing and the gathering of data to enable fully automation.

There seem to be a difference in durability and resilience of experimental practices in the two cities. A sense of continuity emerges from the capability-driven aspects of Oxford's innovation ecosystem, given that both Oxbotica and the UK Atomic Energy Authority's Remote Applications in Challenging Environments (RACE) facility – a centre conducting research on robotic and autonomous systems based in Abingdon to the immediate south of Oxford – are only to a limited extent dependent on CAV-specific government funding priorities and research grants. In the case of RACE, this is due to the broader remit on remote applications of robotics in challenging environments (e.g. nuclear, petrochemical and space exploration), which offers their experimental practices resilience to funding regimes. In Greenwich, the various activities described above are broadly dependent on government funds, albeit with industry co-funding projects. Project funding nevertheless appears diverse and ongoing, confirming a wider trend of success breeding success in local-level innovation in transport in the UK (Schwanen, 2015). However, funding regimes are just one aspect influencing the durability and resilience of experimental practices. In the long term, the stability of Oxford's innovation ecosystem could be impacted by wider politico-economic and technological trends and buy-outs of smaller firms by large transnational corporations. Both could reconfigure the local innovation ecosystem considerably.

Experimental subjects

McCormick and Hartmann's (2017) *strategic* configurations prevail in CAV experimentation in Oxford and Greenwich, with claims of civic involvement often poorly substantiated. Both government (national and local) and private sector actors lead the various experimental projects often by way of (partial) government funding. Greenwich was successful in the UK government's first round of funding for demonstration cities for CAV technologies in 2015. The early success established relationships between local government, TRL and Oxbotica that have outlasted the initial project. The Council, the Mayor and other public sector actors were seen to be pivotal in the construction of Greenwich as a laboratory for the experimentation and testing of CAV technologies.

> I think it's basically because the Council got in very early … The current Leader of the Council is Denise Hyland who is really, to a great extent,

responsible for this organisation. She used to be the Cabinet Member for Transport and she's very aware of the importance of keeping the city moving, keeping the borough moving and the need for interconnectivity. So, I think it was pushing at an open door when this was proposed, absolutely. Greenwich has always been quite forward thinking in economic development, in social mobility.

(G3: Greenwich, Local Government, Digital Greenwich)

Oxford stakeholders, including the local government actors, tell a story of emergence that is capability led, with clusters of innovation and a retrospective action to develop cohesion among these groups of actors. The Innovation team at the County Council, OXLep, Oxbotica and RACE were all noted as being important in the establishment of an innovation ecosystem. In some senses, however, Oxbotica can be identified as the most important actor – 'an extremely capable company', and 'number two after Google in terms of autonomous vehicle technology' (O2: Oxford, Local Government, CAV Innovation). This framing of Oxbotica undergirded claims in interviews with local government actors of local advantage and opportunity that sought to differentiate Oxford from other UK cities with CAV experiments, such as Greenwich, Bristol and Milton Keynes. Provenance is central to the claims made by urban stakeholders in Oxford, for whom intellectual property of CAV innovation is seen to be emerging directly from, and as a result of, Oxford's supportive ecosystem.

Beyond the experimental actor constellation, the role(s) for other experimental subjects – those on whom the experiment was taking place – are largely passive. Decisions on sites and practices of experimentation are decided upon by selected groups of actors who retain authority and are therefore able to include and exclude other experimental subjects. Thus, paying attention to these dynamics, as suggested by Caprotti and Cowley (2017), reaffirms the constructed nature of 'real-world' urban experimentation. Observation of experiments on public roads was noted by consortia partners as a way through which publics were involved in the experiments, and there have so far been few attempts to harness active engagement of publics in Oxford. Key actors in ongoing experiments rather seem to think that flawless technical performance of CAVs is what drives public acceptance, as suggested by a DRIVEN project press release:

Possibly the most important thing about these trials is … the building of our confidence in how it works, because that will be key to public acceptance of driverless vehicles both as road users and in time as potential passengers.

(Oxbotica, 2017, NP)

Similarly, in Greenwich, interviewed members of a community group reported learning about the experiments through national newspapers but had not experienced those experiments first hand. The limited and passive role of the 'general public' in both cities replicates their marginalisation in national-level discourses and activities around CAVs (see Hopkins and Schwanen, 2018a). It draws attention to the political nature of CAV experimentation, which has so far been configured around technical performance rather than exposure to the starkly differentiated and sometimes unexpected responses and actions of city residents and visitors who might one day share the road with CAVs or use them for work, care or pleasure.

Spaces and places of experimentation

Both Oxford and Greenwich make claims of geographic and traffic-related 'uniqueness' that is beneficial to the trialling of CAVs. In London this related to diversity and being 'Europe's only megacity', and in Oxford the rhetoric to the complexity of the transport system, including the historical town centre. However, in both places, current sites of experimentation are largely limited to new and/or high-income neighbourhoods, and this is particularly evident in Greenwich. For instance, the experiments linked to the 'Smart Mobility Living Lab @ Greenwich' occur on the real-world sites of the Digital Greenwich Innovation Centre on the Greenwich Peninsula – a new development on industrial land – and also 'off-site' at the TRL's research and development headquarters in Berkshire.

Claims of public engagement and interaction are, as shown above, often overstated, with a very limited and particular type of public included. Moreover, despite claims of visibility of urban experiments, even passive interactions are unlikely. This was noted by both local government and community participants in Greenwich:

> In reality, trials are occurring within a very bounded and restricted portion of the borough, which represents the population, infrastructures of Greenwich rather little. Instead, the site of demonstration is a newly developing area on Greenwich peninsula, newly gentrified from its industrial roots.
> (G3: Greenwich, Local Government, Digital Greenwich)

> The difficulty for those of us in Greenwich to actually exercise an opinion about it is that it seems to have taken place on the peninsula which, of course, is new build, new roads and on the residential streets of the peninsula there's far less traffic than you would actually get on the Blackwall Tunnel approach road or down this road at all. So, we haven't seen much of it. I've seen the press saying it's been happening but haven't actually seen much sign of it.
> (G2: Greenwich, Community Organisation)

Some of the CAV experimentation in Greenwich, such as the GATEway project, thus resembles what McCormick and Hartmann (2017) call *enclaves* as a result of deliberate attempts to shape to what encounters the vehicles are exposed. This spatial reality stands in marked contrast to prevailing discourse mobilised by Smart Mobility Living Lab @ Greenwich who frame the experiment as designed 'to assist with research and development, concept testing and validation, launching new technology or services, and understanding how new technology is perceived in a real-world environment' (Digital Greenwich, 2016, NP).

Categorisation as enclave is less appropriate for Oxbotica's experimental activities in Oxford, which were described in interviews as

> Running in and amongst cars, buses, pedestrians, cyclists, motorbikes, people cutting across the front of them, round the back of them, down the side of them, reversing out in front of them, in real time and reacting to whatever's in front of them.
> (O6: Oxford, Local Government, Local Enterprise Partnership)

Yet, these too appeared to initially take place in a limited portion of the city, i.e. the wealthy suburbs of north and central Oxford near to Oxbotica's Summertown offices where, across diurnal and weekly cycles, traffic levels and composition will be somewhat less challenging than other parts of the city, including demographically and socioeconomically more diverse neighbourhoods. A logic of light segregation appears therefore to be at play.

Conclusions and policy recommendations: experimenting with automation

This chapter has examined processes of experimentation with automation currently underway in Oxford and Greenwich. These two cities are neither alone nor unique in their experimentation with CAVs but part of a broader movement across the UK stimulated to some degree by national priorities, and the international race to automation. From our relational approach to CAV experimentation, we now draw several conclusions.

From our case study, it is evident that what constitutes the 'real world' in urban experiments is not pre-given but artificial. It is constructed through the actions and decisions of key experimental actors that are typically networked across multiple experimental projects and sites. Those actors include some groups (e.g. selected innovators, investors, selected publics) while excluding others (community groups) and seek to condition CAVs' – and presumably other innovations' – exposure to the heterogeneity and unpredictability that is integral to the city. The (discursive) publicity of 'real-world' experiments and CAV technology may in the short run heighten the already-hyped expectations around the innovation and its purported benefits, but in the long term frustrate its diffusion if it turns out that earlier *in vivo* experimentation has insufficiently exposed CAVs to the vagaries, surprises and complexities of urban traffic.

Similarly, the real-worldness of *in vivo* experiments – at least those with CAVs – is more artificial and political than the discourse mobilised by many influential experimental actors suggests. That real-worldness reflects what those actors find important and relevant, which for CAVs in the UK at this point in time is more about technical performance, (future) profitability and economic development than about interaction with the inevitably diverse general public as potential users, fellow road users, citizens, sceptics, and so forth. Besides, CAV experimentation is only partly about engagement with the real world. The data gathered from such engagement is used for machine learning and digital experimentation using simulation techniques that thus hybridise in vivo and in vitro elements. This use of experimentation also suggests that data generation should be added to trial, enclave, demonstration and platform as a configuration in McCormick and Hartmann's (2017) heuristic scheme.

Another insight in relation to McCormick and Hartmann's (2017) classification of urban laboratories, and presumably experiments, is that trial, enclave, demonstration and platform cannot be seen as mutually exclusive categories. The urban experiments with CAVs in Oxford and Greenwich tend to be configured as trials and demonstrations but this capacity is made possible by their simultaneous character of being enclaves, with the degree of segregation varying across sites and projects. Similarly, by highlighting how *in vivo* experiments are is used to generate data for subsequent (quasi-)*in vitro* experiments using digital technologies, our empirical research also begins to highlight how *in vivo* and *in vitro* experimentation differ not so much in kind as in degree. *In vitro* experimentation using digital techniques is not as isolated and free from real-world processes as it might appear.

Schot and Geels (2008) point to the need for second-order learning, the development of underlying norms and values through the niche development phase. Achieving this requires the involvement of 'outsiders' in the experimentation process. Experiments in Oxford and Greenwich could offer an opportunity for learning about the innovation in 'real-life circumstances', as previously suggested by Geels (2012), but the limited and managed actor constellations at the current stage of experimentation are significantly restricting learning about (potential) user engagement with CAV technologies. The need for exposure to surprise and opportunities for second-order learning suggests that wide-ranging experimental subjects need to be better understood and incorporated into processes of urban experimentation with automation.

The benefits of current forms of real-world experiments are two-fold. First, the technology developers are gaining new insights into interactions between the innovation and its (carefully constructed, highly selective) environments. Second, for the hosts of the experiments – likely to include public-sector actors – the experiments are often linked to economic development, identity formation, and regional provenance claims, and used to attract inward investment often by way of public–private partnerships. Proclamations of 'real-world experimentation' have important implications for perceptions of CAVs and their capabilities. For some audiences, real-world conditions may represent particular connotations of vehicle capabilities, including the ability to travel on busy inner city public roads. Unless the artificial and highly selective character of the real-worldness of CAV experimentation is recognised and carefully communicated, ongoing *in vivo* experiments may further entrench already-hyped expectations about the capabilities of vehicle automation and the time scales for widespread diffusion. This is a risk perhaps best to avoided considering previous failures of CAV innovation.

Based on these conclusions, we have three policy recommendations.

Policy recommendation 1: Innovation literatures highlight the importance of second-order learning around niche innovations, such as CAVs. The experimentation process needs to offer opportunities for inclusion of a broader range of publics and experimental subjects which offer not only the opportunity to learn about the innovation but also opportunities for surprises and unintended outcomes. The real-world experiment needs to be acknowledged as highly subjective and managed.

Policy recommendation 2: Caution in the communication of real-world experimentation and CAV capabilities is encouraged, as real-world experiments can contribute to elevated expectations and hype, which may ultimately be unmet.

Policy recommendation 3: As a technique deployed in policymaking, *in vivo* experimentation is not as neutral and innocent as it may seem. Given that real-worldness in experimentation is constructed, policymakers – and others – should be reflexive about what forms of exposure are privileged and marginalised in actual (urban) experiments, and what the consequences of this might be for wider innovation pathways and broader attempts to address wicked problems such as road congestion, air pollution, social inequality and excessive energy consumption of fossil fuels.

Note

1 Karvonen and van Heur (2014, p. 380) explain that the urban laboratory 'is part of a wider discursive field that includes ideas of Mode 2 science, triple helix formations, engaged research, service learning, transdisciplinarity, living laboratories, applied innovation and the co-production of knowledge'.

References

Bellamy, R., Lezaun, J. and Palmer, J. (2017) Public perceptions of geoengineering research governance: An experimental deliberative approach. *Global Environmental Change* 45: 194–202.

Bulkeley, H.A. (2018) Urban living laboratories: Conducting the experimental city?, *European Urban and Regional Studies*. [In Press].

Bulkeley, H., Castan-Broto, V. and Edwards, G. (2015) *An Urban Politics of Climate Change: Experimentation and the Governing of Socio-technical Transitions*. Routledge, London, UK.

Bulkeley, H., Coenen, L., Frantzeskaki, N., Hartmann, C., Kronsell, A., Mai, L., Marvin, S., McCormick, K., van Steenbergen, F. and Palgan, Y.V. (2016) Urban living labs: Governing urban sustainability transitions. *Current Opinion in Environmental Sustainability* 22: 13–17.

Caprotti, F. and Cowley, R. (2017) Interrogating urban experiments. *Urban Geography* 38(9): 1441–1450.

Centre on Connected and Autonomous Vehicles (CCAV) (2018) UK Connected and Autonomous Vehicle Research and Development Projects 2018. UK Government, London, UK. Available at: www.gov.uk/government/publications/connected-and-autonomous-vehicle-research-and-development-projects.

de Wildt-Liesveld, R., Bunders, J.F.G. and Regeer, B.J. (2015) Governance strategies to enhance the adaptive capacity of nice experiments. *Environmental Innovation and Societal Transitions* 16: 154–172.

Department for Transport (2015) Pathway to Driverless Cars: Summary Report and Action Plan. UK Government, London, UK. February 2015. Available at: www.gov.uk/dft.

Digital Greenwich (2016) Introducing the UK Smart Mobility Living Lab @ Greenwich, 18 February 2016. Available at: www.digitalgreenwich.com/introducing-the-uk-smart-mobility-lab-greenwich/.

DRIVEN (2018) Ground Breaking Trials. Project Website. Available at: http://drivenby.ai/2017/04/24/groundbreakingtrials/.

Evans, J.P. (2011) Resilience, ecology and adaptation in the experimental city. *Transactions of the Institute of British Geographers* 36(2): 223–237.

Evans, J. (2016) Trials and tribulations: Conceptualizing the city through/as urban experimentation. *Geography Compass* 10(10): 429–443.

Evans, J., Karvonen, A. and Raven, R. (2016) *The Experimental City*. Routledge, Abingdon, UK.

Geels, F.W. (2012) A socio-technical analysis of low-carbon transitions: introducing the multi-level perspective into transport studies. *Journal of Transport Geography* 24: 471–482.

Geels, F.W. and Deuten, J.J. (2006) Local and global dynamics in technological development: a socio-cognitive perspective on knowledge flows and lessons from reinforced concrete. *Science and Public Policy* 33(4): 265–275.

Geels, F.W., Schwanen, T., Sorrell, S., Jenkins, K. and Sovacool, B. (2018) Reducing energy demand through low carbon innovation: A sociotechnical transitions perspective and thirteen research debates. *Energy Research and Social Science* 40: 23–35.

Gross, M. (2010) *Ignorance and Surprise*. MIT Press, Cambridge, USA.

Gross, M. (2016) Give me an experiment and I will raise a laboratory. *Science, Technology and Human Values* 41(4): 613–634.

HM Government (2017) Building our Industrial Strategy. Green Paper. January 2017. HM Government, London, UK. Available at: https://beisgovuk.citizenspace.com/strategy/industrial-strategy/supporting_documents/buildingourindustrialstrategygreenpaper.pdf.

Hopkins, D. and Schwanen, T. (2018a) Automated mobility transitions: Governing processes in the UK. *Sustainability* 10(956): 1–19, DOI: 10.3390/su10040956.

Hopkins, D. and Schwanen, T. (2018b) Governing the Race to Automation. In: Marsden, G. and Reardon, L. (Eds) *Governance of Smart Mobility*. Emerald, Bingley, UK.

Karvonen, A. and van Heur, B. (2014) Urban laboratories: Experiments in reworking cities. *International Journal of Urban and Regional Research* 38(2): 379–392.

Karvonen, A., Evans, J. and van Heur, B. (2014) The Politics of Urban Experiments: Radical Change or Business-as-Usual? In: Marvin, S. and Hodson, M. (Eds) *After Sustainable Cities*. Routledge, London, 104–115.

Kullman, K. (2013) Geographies of experiment/experimental geographies: A rough guide. *Geography Compass* 7(12): 879–894.

Larner, W. and Laurie, N. (2010) Travelling technocrats, embodied knowledges: Globalising privatisation in telecoms and water. *Geoforum* 41(2): 218–226.

Lawton Smith, H. (2003) Local innovation assemblages and institutional capacity in local high-tech economic development: The case of Oxfordshire. *Urban Studies* 40(7): 1353–1369.

Leydesdorff, L. and Etzkowitz, H. (1998) The Triple Helix as a model for innovation studies. *Science and Public Policy* 25(3): 195–203.

Loorbach, D. (2010) Transition management for sustainable development: A prescriptive, complexity-based governance framework. *Governance: An International Journal of Policy, Administration, and Institutions* 23(1): 161–183.

Marvin, S. and Silver, J. (2016) The Urban Laboratory and Emerging Sites of Experimentation. In: Evans, J., Karvonen, A. and Raven, R. (Eds) *The Experimental City*. Routledge, London.

McCormick, K. and Hartmann, C. (2017) The emerging landscape of urban living Labs: characteristics, practices and examples. *GUST: Governance of Urban Sustainability Transitions*. Available at: http://lup.lub.lu.se/search/ws/files/27224276/Urban_Living_Labs_Handbook.pdf.

Muniesa, F. and Callon, M. (2007) Economic Experiments and the Construction of Markets. In: MacKenzie, D., Muniesa, F. and Siu, L. (Eds) *Do Economists Make Markets? On the Performativity of Economics*. Princeton University Press, Princeton, USA, 163–189.

Ney, S. (2009) *Resolving Messy Policy Problems. Handling Conflict in Environmental, Transport, Health and Ageing Policy*. Earthscan, London, UK.

Schot, J. and Geels, F.W. (2008) Strategic niche management and sustainable innovation journeys: Theory, findings, research agenda, and policy. *Technology Analysis and Strategic Management* 20(5): 537–554. DOI: 10.1080/09537320802292651.

Schwanen, T. (2015) The bumpy road towards low-energy urban mobility: Case studies from two UK cities. *Sustainability* 7: 7086–7111. DOI: 10.3390/su7067086.

Turnheim, B., Kivimaa, P. and Berkhout, B. (2018) Experiments and Beyond: An Emerging Agenda for Climate Governance Innovation. In: Turnheim, B., Kivimaa, P. and Berkhout, B. (Eds) *Innovating Climate Governance: Moving Beyond Experiments*. Cambridge University Press, Cambridge, UK, Chapter 12.

6 The United Kingdom smart meter rollout through an energy justice lens

Kirsten E.H. Jenkins, Benjamin K. Sovacool and Sabine Hielscher

Introduction

The United Kingdom's (UK) Smart Meter Implementation Programme (SMIP) creates the legal framework so that an in-home display unit (IHD) and a smart gas and electricity meter can be installed in every household by the end of 2020. Intended to reduce household energy consumption, the SMIP is one of the world's most complex smart meter rollouts. It is also proving to be a challenging one as a series of obstacles has characterised and potentially restricted implementation. This chapter first gives background to the most recent smart meter roll-out developments in the UK and second, uses energy justice criteria to explore the emergent challenges under the headings of distributional justice, procedural justice and justice as recognition.

We delve into pertinent technical and non-technical issues arising through the lens of energy justice, – a now commonly used structure first introduced by Schlosberg (2004, 2007) and Walker (2009) and used in a policy setting by McCauley *et al.* (2013). As introduced in full in Chapter 3, this includes a focus on three core tenets, which in this chapter manifest as: (1) an analysis of potential distributional benefits and ills, (2) illustrations of the overlaps between smart meters and other social vulnerabilities including poverty, ill health, social integration or rural marginalisation, which lead to the need to recognise particular sections of society and (3) explorations of the role of procedural engagement during the roll-out.

While the energy justice literature is gaining ever-increasing popularity – including in the field of energy 'end-use' – no energy justice research has investigated the development of smart meter technologies to date. This chapter makes an early step towards this goal. The success and necessity of doing so is clear as applying these three concepts to an analysis of the UK SMIP provides opportunities to accurately record, present and expose potential forthcoming injustices, with potential policy implications. Thus, in our conclusions and policy recommendations section we present a series of take-home recommendations for policymakers and advocates. We note from the outset that the material presented here builds directly on Sovacool *et al.* (2017).

Smart meters in the United Kingdom: an introduction

To begin, it is necessary to provide background context. This section of our chapter presents the technologies being deployed in the SMIP before describing the anticipated benefits of the scheme and the rollout timeline.

Smart meters have been implemented in several European countries, with Italy and Sweden being the first to complete their rollout processes. Other countries have prepared or started their rollout over the last 5–10 years. The UK is among the European Union (EU) countries anticipating a positive business case, with the Department of Business, Energy and Industrial Strategy (BEIS) presenting that case that as of 2016, 'the net present value (NPV) for the domestic rollout of smart meter in GB is now estimated to be £3.8bn' (BEIS, 2016, p. 13).

The goal of the SMIP is to develop and implement technologies that can collect, distribute, and analyse electricity and gas use and production data in order to assist current energy demand and supply management (Pullinger et al., 2014). In this regard, a 'smart meter' in the UK setting has been defined by the Office of Gas and Electricity Markets (Ofgem, 2011) as, 'a gas and electricity meter that is capable of two-way communication'. It measures energy consumption in the same way as a traditional meter but has a communication capability that allows data to be read remotely and displayed on a device within the home or transmitted securely externally. The meter can also receive information

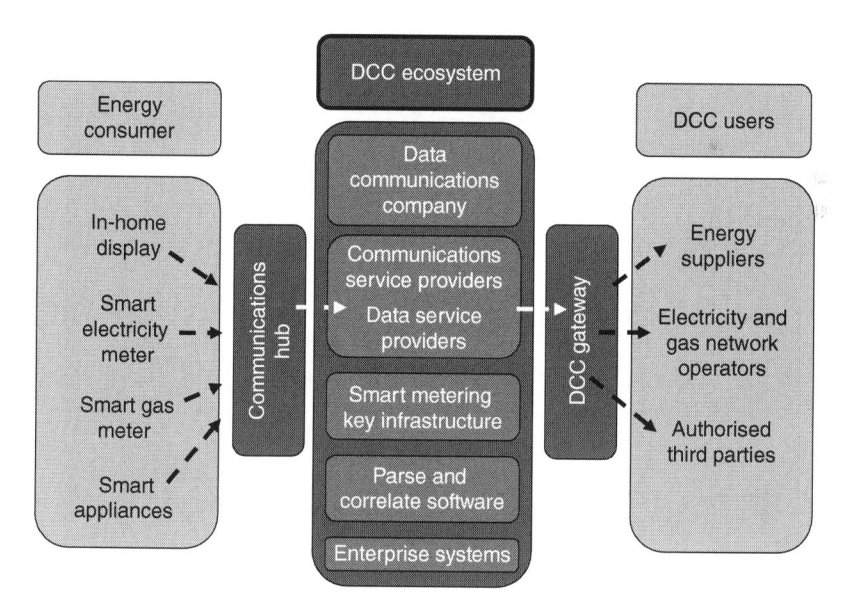

Figure 6.1 Key components of the smart meter communication service and service providers.

Source: the authors.

remotely, e.g. to update tariff information or switch from credit to prepayment mode. This is, in short, an elaborate way of describing automated meter reading or remote meter diagnostics. One component of the system – the 'in-home display' or IHD – is a monitor that, through a connection to the smart meter, delivers consumers with information about their energy consumption and costs. The UK is peculiar in that it pushes both separate gas *and* electricity meters and promotes the use of IHDs, which is not the norm. Beyond the provision of these core units, the SMIP also involves the changeover of wireless networks, data and communications companies (DCCs), service users, and electricity and gas suppliers – a configuration of social and technical interactions shown in Figure 6.1.

The government aims to install smart meters – or smart meter ecosystem – in every home in Scotland, Wales and England/(Smart Energy GB) by December 2020 (BEIS, 2016). Whether this is compulsory has remained confusing. Some energy companies have stated that consumers would only be 'offered' smart meter units, whereas others have, as of July 2017, implied that the rollout would be mandatory (Meadows and Brodbeck, 2017).

Proposed benefits

An advanced metering energy infrastructure enabled by smart meters is said to create the building blocks for a smart grid, including smarter energy management services. This is one of several expected benefits that have been linked to the UK rollout, many of which are sustainability oriented. As a summary, this includes the fact that:

- Energy suppliers are expected to reduce their operational overheads associated with high level of consumer service enquires and manual meter reading and as a result increase customer loyalty. Moreover, they are able to introduce wider range of tariffs and services.
- Network operators are said to benefit through improving the efficiency and responsiveness of the network and enable a greater penetration of renewable technologies into the network.
- Consumers are argued to benefit through empowerment to more appropriately control their energy flows, reducing their bills as they become more aware of their energy consumption. Smart meters are also said to allow easier transferring between energy suppliers and increased billing accuracy (see the summary of benefits in, for instance, McKenna *et al.*, 2012).

Indeed, these are a few of 67 short- and longer-term benefits identified by Sovacool *et al.* (2017) (and summarised in Table 6.1).

According to the BEIS (previously the Department of Energy and Climate Change (or DECC)) the total costs of the SMIP were estimated to be approximately £8 to £11 billion, whereas the financial benefits could be as high as £17.1 billion when factoring in improvements in air quality and the savings anticipated for both consumers and suppliers.

Table 6.1 Sixty-seven anticipated short- and long-term benefits to smart meters in the UK

No.	Short-term benefits
1	Offer an alternative to pre-payment meters or bring down costs of pre-payment meters
2	Help consumers to budget
3	Increase energy efficiency awareness
4	Feedback on energy use
5	Carbon savings
6	Provide real-time information on energy costs
7	Provide information to make informed choices/greater understanding
8	Remote reading, avoid home calls
9	Energy bills accurate
10	Saving energy/reduce consumption
11	Manage their energy use, avoid waste
12	Customers install microgeneration
13	Remote switching credit and prepayment
14	Smoother switching between suppliers
15	Wide range of tariffs and incentive packages from suppliers
16	Suppliers to reduce costs
17	Customers save money/reduce costs
18	Better services from energy companies
19	Energy network planning
20	Drive uptake of renewable electricity
21	Reduce demand for heat
22	Billions in net benefits to the economy
23	Future innovation
24	Jobs
25	Drive a more vibrant and competitive market
26	Offset price rises
27	Access a full range of energy management tools
28	Changing the way we think about energy
29	Help vulnerable customers
30	Pre-payment replaces by smart meter
31	Promote community energy
32	Consumers more active in the energy system
33	Suppliers offer more cost effective tariffs
34	Record how much consumed 1/2 h period
35	Promote distributed generation or distributed energy resources
36	One day switching
	Long-term benefits
37	Demand-side management
38	Reduce peak loads via time of day tariffs
39	Network reinforcement and peak generation avoided
40	Advanced management techniques/automated demand-side response
41	Reduced energy consumption
42	Consumers more flexible and responsive to market signals
43	Smart grid
44	Electric vehicle promotion
45	Automated responses to changes in network
46	Enhanced monitoring flow across the network

continued

Table 6.1 Continued

No. Long-term benefits

47 Deal with intermittence
48 New products and services/innovation
49 Vibrant, competitive market in energy supply and energy management
50 Improved network efficiencies
51 Update microgeneration
52 Turning off non-essential electrical appliances
53 Energy network management
54 Smart Energy Services supported
55 Smart energy market
56 Network operators understand loads on infrastructure
57 Network operators plan investments
58 Network operators respond faster to supply loss
59 Avoid the need to invest in additional network/and generation capacity
60 Generate capacity to meet peak demand
61 Support smart apps and automated appliances
62 Enhance resilience
63 New opportunities for storage
64 Consumers take advantage of lower price periods
65 Peak shaving
66 Develop a domestic smart appliance industry
67 Large industrial consumers and small-scale generators capacity market

Source: Sovacool *et al.* (2017), with permission.

Rollout programme

The rollout of the UK's SMIP policy, technical and regulatory apparatus began in 2010 and is composed of several phases: (1) the policy design stage, (2) the foundation stage and (3) the main installation stage. Within the policy design stage (July 2010–March 2011) the technical specifications and implementations model were negotiated. It was decided that energy suppliers should take responsibility for the rollout and installation of smart meters. The Smart Meter Central Delivery Body (SMCDB) was established with the aim to increase awareness about smart meters, alongside the currently operative Smart Energy GB marketing campaign ('the national campaign for the smart meter rollout').

In addition to setting up the implementation programme, a large-scale trial project, Energy Demand Research Project (ERDF) was conducted from 2007 onwards and trials ended in autumn 2010. As argued by Darby (2011, p. 6), 'they took place at a time when policy of smart metering was already developing, so the full findings and lessons have not been available to inform policy development during this period'.

Skipping forward in the timeline, the household installation phase during which most will have their smart meter installed, started at the end of 2016. This start date was pushed back twice from 2014 until November 2016 because of several technical difficulties. This included the technical difficulties of enabling broader interoperability so that consumers can switch more easily between suppliers. Currently, energy supplier still install the SMETS 1 (Smart Metering Equipment Technical Specification) smart meter version that go 'dumb' once the householder switches suppliers and the roll out of SMETS 2 devices that enable switching has been repeatedly delayed.

Despite positive potential (Table 6.2), however, capturing these benefits has been elusive, and the implementation of the SMIP has been replete with obstacles. The programme is substantially delayed. Up until September 2017, 8.6 million smart meters have been installed (BEIS, 2017b), corresponding to 15.38 per cent of the target number of 56 million. Now, to make sure that the SMIP meets targets for the remainder of the programme, suppliers must complete installations at a rate of about 40,000 per *day*. The Institute of Directors stated in 2017 that

> The programme has already failed to deliver interoperable meters for switching, is behind schedule, is over-budget and wedded to out of date technology. Not only that, the legal obligation on suppliers to install potentially incompatible meters by the deadline of December 2020 or else pay large fines is already pushing up inflationary costs in wages and advertising.
>
> (IoD, 2017)

Alongside the above issues, concerns have also arisen around potential customer resistance to smart meters (see National Audit Office, 2014).

Table 6.2 Estimated benefits to the smart meter implementation programme in the UK

	Domestic (£m)	Non-domestic (£m)	Total (£m)
Consumer benefits (from energy saving and microgeneration)	4,295	1,437	5,732
Supplier benefits (including avoided site visits, reduced inquiries etc.)	7,970	295	8,265
Network benefits (reduced losses, reduced outage notification calls, fault fixing, avoided investment from ToU (distribution/transmission) etc.)	877	112	947
Generation benefits (avoided investment in generation from peak shifting through ToU)	803	49	852
UK-wide benefits (including CO_2 reduction, air quality)	867	440	1,307
Total	**14,812**	**2,333**	**17,103**

Source: Sovacool *et al.* (2017), with permission.

Note
ToU = Time of Use. UK = United Kingdom. CO_2 = Carbon dioxide.

Conceptualising smart meters through an energy justice lens

In the following sections we present the results of our restructured analysis according to each tenet of the energy justice framework, distributional justice, justice as recognition and procedural justice (see also Chapter 3). We do so in order to demonstrate the *social* challenges associated with the rollout process, as well as to position a series of policy recommendations that may smooth its execution. We do not provide an in-depth overview of each of these terms as to do so would duplicate material elsewhere in this book. As an abbreviated summary, however, the notion is as follows: distributional justice refers to the unequal distribution of benefits and ills resulting in this case, as a result of an energy project (Jenkins *et al.*, 2017); procedural justice considers access to decision-making processes that govern the distributions outlined previously and manifests as a call for equitable procedures that engage all stakeholders in a non-discriminatory way; whereas justice as recognition expresses concern for the fair engagement of all individuals in conditions where they are free from physical threat and are offered complete and equal political right (Jenkins *et al.*, 2016).

Distributional justice

Our analysis revealed numerous technical and non-technical distributional challenges resulting from the SMIP. The most frequently discussed relate to the technology and unequal access to it. In the work published by Sovacool *et al.* (2017), across a systematic review of 47 articles, 60 per cent discussed variations of 'technological', 'technical' or 'engineering' impediments.

As one illustration, the early meters distributed in 2014 would not function properly in one-third of UK homes, including in particular, high-rise flats, basements and remote areas. This leads to unequal access and the marginalisation of particular locations and social groups. Part of the cause of this difficulty was the chose to pursue a less-customary Wi-Fi or Bluetooth mechanism called Zigbee. Within high-rise blocks, where meters tended to be located in basements, the signals failed to make it through thick walls. The result was the need to test more expensive hardwired cables as well as area network radio systems. Lewis and Kerr (2014) draw the conclusion that the SMIP forsake plans to implement the rollout to inaccessible flats and tower blocks, removing approximately 7 million homes from participation.

Technical disbenefits also appeared for people that *did* have access. Gosden (2015) reports that as of early 2015, one-tenth of the 1.3 million smart meters installed in the UK did not function as 'smart' meters, but as traditional ones requiring manual readings. Moreover, OVO Energy, who supply both gas and electricity, reported that 6 per cent of their customers were unable to view or pay their bills on their installed meters. EDF also reported technical issues with 0.5 per cent of their installations (Palmer 2015a). More recently, *Utility Week's* (2017, p. 4) market research implies that over one-tenth of households will require multiple visits to complete smart meter instillations – a figure that could add around £1 billion to the total cost of the SMIP programme. In this scenario, limitations with the technology led to the unequal distribution of systems burdens.

Building from the above theme, numerous media outlets reported that large numbers of households were 'trapped' with malfunctioning smart meters, as well as reporting large swings in usage rates displayed on the IHDs, and many consumer complaints (Shannon 2015). These issues were compounded by apparent meter incompatibility where, if a household wanted to change energy suppliers, they had to wait (occasionally over a year) for a new meter. In some instances, smart meters were converted back to 'dumb' types that depended on manual estimates or readings (Meadows and Brodbeck, 2017; Palmer, 2015b). Other consumers complained that their units no long worked even if they were only switching between tariffs even with the same energy company (Brignal, 2016). Concerns have also arisen around the ability of hackers and cyber-terrorists to interfere with systems, where there is some fear that individuals may be able to break in, disrupt grid reliability or intercept bills and personal data, thereby enabling theft or fraud.

Beyond device siting and accessibility, distributional injustices also related to the uneven impacts of the cost of the SMIP programme and its components. Lewis and Kerr (2014) outline that the IHD requirement will cost approximately £800 million in total, but that there are fears that users fail to engage with the units long term. They also state that the IHD requirement could be removed and replaced by a phone, tablet or personal computer app that could connect to the network with no additional need for hardware installation or cost (although we must acknowledge that these cheaper alternatives may also present distributional challenges as not everyone has ready access to smart phones, tablets or an internet connection (with an impact on the vulnerable groups we later go on to discuss)). Here then, the burdens and benefits of the rollout are distributed unevenly. What is more, even in 2012 there were several devices capable of identifying and displaying consumption data for electric appliances in a similar way and at a much cheaper cost (Thomas, 2012). This includes the OWL meter, an approximately £40 plug-in device that displays energy use over time, giving consumers 'a clear, accurate picture' of their usage (Thomas, 2012, p. 1061). In this regard, there appeared to be cheaper alternatives that would not present such a high burden to customers through their energy tariffs.

Through their cost–benefit analysis, the National Audit Office (2011) outlined that despite anticipated savings for energy suppliers, the empowerment benefits for consumers were more uncertain, especially as costs will be passed directly to them through higher tariffs and bills. Furthermore, the overall gain for households appears dependant on the extent to which energy suppliers pass savings onto consumers – something that is not guaranteed. The Public Accounts Committee (2012) of the House of Commons noted similar concerns, outlining that (1) most benefits will be distributed to the suppliers and not the consumers (who have to pay for the cost of the smart meter), (2) that the benefits of smart meters will only be realised if uptake is widespread and they are used appropriately, both of which are again, not guaranteed and (3) that still, benefits were unlikely to reach vulnerable customers and/or those on prepayment meters. This is an even greater risk when you consider that the £430 cost for a gas and electricity meter will be related to consumers directly. The fact that most benefits appear to go to producers while the consumers pay the costs for the rollout may be considered one of the biggest distributional inequities. In critical accounts, some academics (e.g. Strengers, 2013 and Balta-Okzan et al., 2014) question the promises associated with smart meters, smart grids, smart homes and demand-side benefits, for instance, arguing that they are promoted by powerful interests. Verbong et al. (2013, p. 121) suggest that: 'Although smart grids are claimed to be in the interest of the end-user, there is some ambiguity about that. (…) Despite all promises, it is not so clear which interests are primarily served by smart grid developments.'

Procedural justice

Alongside more technical barriers preventing equal access were a series of procedural justice concerns relating to how particular social groups were or were not engaged in decision-making processes about the SMIP rollout (recognising that not everyone can be), and whether the way they would use the system had been considered.

Pullinger et al. (2014, p. 1158) give the opinion that the SMETS standards (the technical standards behind the SMIP) have been developed in a top-down, industry-led way 'with little input from, or attention to the householder'. This lack of consumer engagement implies disconnect between the technical operationalisation of the SMIP and the narrative of consumer benefit being at the heart of it. This may be one reason for what appears to be growing resistance to the programme (typified by disengagement with the smart meter, or never becoming a smart meter user, for example). Chilvers and Longhurst (2016, p. 596) note, in this regard, that during one of the SMIP trials (the Visible Energy Trial) individuals 'resisted' the smart meter by intentionally failing to use it appropriately. This delayed the collection of data and results and ultimately, convinced others not to engage in the trial. They go on to suggest at least two potential reasons for this resistance: (1) that the IHD was perceived as giving only incremental, inconsequential reconfigurations of consumption and behaviour change that did not meaningfully save energy and (2) users felt that their meters placed unfair burden on them to take responsibility for carbon reductions over industry and government groups, for example. Here, there are clear breaches in what is deemed to be necessary procedural justice.

Compounding concerns over manifestations of consumer resistance, Balta-Ozkan et al.'s (2014) comparison of smart meter perceptions in Europe illustrated resistance frames in terms of accountability and responsibility. Their UK-wide focus groups revealed the expectation that the government was

The smart meter rollout and energy justice 103

responsible for addressing climate change, not individuals. IHD and smart meter users also opposed because of control and privacy concerns – the belief that they represented an extension of power companies, which were taking over their private lives and homes. Indeed, this manifested as recurring concern for the potential of smart meters to 'compromise security' and 'invade privacy' (Balta-Ozkan *et al.*, 2014, p. 1185).

Yet more consumers expressed frustration caused by the technical problems by intentionally overriding the system or practicing inefficient behaviour. One implication is that users may opt out of the SMIP. A second is that those coerced into participating could disconnect their IHD, sabotage their meter or could be unwilling or reluctant to share their data with apparently 'devious' companies. In short, all reasons culminate in mounting resistance where Rose and Thed (2014) report reservations from one supplier that 'up to 20 per cent of customers will refuse to have smart meters installed' and two firms have documented additional costs from dealing with 'reluctant customers'. A recent Smart Energy GB report outlines that 16 per cent of the UK adult population are 'indifferent' to smart meters, with a further 18 per cent in the category of 'rejector' (Smart Energy GB, 2017). Furthermore, smart meters can even *increase* consumption – an early (2004) EDF trial showed that gas consumption rose by 'almost 50%' as users became aware of significant under-heating (Mott MacDonald, 2007).

Furthermore, although an August 2017 'Customer Experience study' from BEIS involving 2,015 households found mostly positive reactions to smart meters, it did note that 7 per cent of customers reported being very dissatisfied or somewhat dissatisfied with their smart meter; that 73 per cent of customers were not given any information on data storage and privacy, or how data would be shared; and that 18 per cent of customers indicated they did not even look at the information given on their IHDs (BEIS, 2017a).

In the event observations reported on by Sovacool *et al.* (2017), which took place between 2015 and 2016, there was also very little focus on what consumers wanted. The only exception was a national grid representative who questioned 'can they manage to deliver what customers and consumers want at the end of the day?' (October 2015). This neglect reinforced the idea that the SMIP is about increasing the provision of information in order to change behaviour (one-way influence), as opposed to creating an energy system that encourages user participation and more 'active' consumerism where participants influence the future of the energy system. This could explain the lack of interest or appearance of resistance.

Where terms such as 'protection', 'engagement', 'consumer benefit', 'enabling' and 'empowerment' appeared in the events observed by Sovacool *et al.* (2017) there was no real reference to what occurs *after* consumers were enabled or empowered. Very few considered acceptance, trust and experience. Indeed, in the last event the team observed in November 2016, (the lack of) 'consumer acceptance' was outlined as one of the three key challenges remaining for the rollout. Smart Energy GB were quoted as saying:

There is no mandate on the part of the consumer although there is mandate on the part of the energy supplier. And that is a real challenge I think for a consumer engagement campaign; how do we make sure every consumer is empowered to say 'yes'.

It is a somewhat bleak representation to suggest that resistance always or frequently occurs, of course. A 2017 commercial survey of more than 1,000 UK consumers advised that 64 per cent with installed meters 'enjoyed better visibility of their energy costs'; 36 per cent reported achieving savings and 76 per cent were impressed by the expertise of the people completing their installation (*Utility Week*, 2017, pp. 2–3). Nevertheless, improved procedures may be necessary for engaging consumers in both smart meter implementation and use.

Justice as recognition

It is reasonable to acknowledge that particular groups will encounter problems with any new innovation that required technological or behavioural change, but without careful implementation, the SMIP programme raises the potential *exacerbation* of pre-existing vulnerabilities among some consumer classes – an issue of justice as recognition or the unequal burdens (or benefits) presented to particular groups. Notably, early evidence shows that the SMIP may unduly burden the elderly, ill, less educated, those living in social housing and/or those in the rural periphery at the expense of a preference for supplier and company-oriented cost savings and economic competition. If the SMIP can lessen the expenditure needed for meter readings, network operation, grid reinforcement and electricity generation, among other areas, then consumers should benefit from lower energy practice whether or not they make behavioural changed. In reality, however, many forms of social inequity and injustice emerge.

A synthesis report compiled by DECC (2015a, 2015b) presented the results of a 4,016 consumer survey, 169 in-depth household interviews with users on both credit and prepayment meters, 12 focus groups, and consumption data for over 10,000 households and revealed, ultimately, that consumers from so-called 'vulnerable groups' 'are likely to need more help if they are to obtain the full benefits of smart metering' (DECC, 2015a, p. 22). The report outlined in particular that

> Older smart meter customers, those from lower social grades, those with the lowest total annual household incomes (below £16,000), those with no formal qualification and those who lived with someone who had a long-term health condition or disability were less likely to say the IHD was easy to use or to say they knew how to operate its different functions.
> (DECC, 2015a, pp. 22–23)

Barnicoat and Danson (2015) ran an experimental project for elderly tenants in rural Scotland where they used sensors and IHDs to measure and display household energy costs over a period of seven months. The aim was to investigate how households interacted with IHDs, or what they term 'smart energy monitors'. The elderly were soon a particularly relevant sample group given that they were often in their homes for large amounts of time, are high users of domestic energy, may receive fixed incomes that necessitate fuel rationing, require higher temperatures due to old age and may also suffer physical limitations that affect their interactions with technological equipment. Barnicoat and Danson's research revealed that despite greater price feedback, little 'awareness' occurred. Specifically, households did not understand the link between IHDs and electrical appliance use. Further, the study suggested that the primary benefit of the IHDs was not perceived as being for households, but for the energy supply company engineers as it gave them information about household use. This was at odds with expectations. Citizens Advice (2017) rehearsed similar concerns in a report critiquing the SMIP for its potentially negative impacts on low-income and elderly households, including in particular, those without a formal education, who did not speak English fluently, or who suffered from a long-term illness. According to their report, some consumers were distressed, confused about, or unable to use smart meter information.

On the grounds of the events the authors observed, Sovacool *et al.* (2017) discuss a mix of responsibilities necessary to make sure vulnerable groups are accounted for and benefit from, smart meters. Smart Energy GB discussed 'shared responsibility with suppliers around behaviour change' (Smart Energy GB, 2016) but did not go on to outline how this responsibility was shared. The idea of 'partnering up' between different organisations to support vulnerable users was also discussed with the aim of 'mobilising' energy champions, volunteers, and a community fund in order to support people so they could 'make use of the benefits'. Again, however, Sovacool *et al.* reported that there was little discussion of who would guide these activities or what role they would play in the SMIP rollout. This may be the cause of DECC's (2015b) move to ask Smart Energy GB to advance stronger advisory and supporting materials, including increasing its role as a facilitator of knowledge exchange and enabling local-level networks and partnerships.

Alongside elderly populations, which are the most frequently discussed in connection to the SMIP, there are several less frequently documented increased vulnerabilities. Echoing Sovacool *et al.* (2017), we briefly discuss two: (1) increased rural peripheralisation and (2) externalities and lifecycle impacts. The SMIP may worsen the urban/rural divide, or unwittingly increase the preference for smart meter instillation in urban areas but not rural, countryside homes. For example, access to the dedicated network connection services required for a functioning smart meter system varies widely across the UK, with notably lower accessibility outside of urban centres (Ofcom, 2016). The housing stock is also more challenging to access, meaning that installation often requires more travel mileage and person hours. This leads to an inequitable, albeit understandable, focus on 'easy to manage' areas with well-established delivery and logistic networks and larger installation volumes.

Conclusions and policy recommendations

As an outcome of our energy justice evaluation, the main contribution of this chapter is to inform current policies and practices concerning the SMIP and national energy policy attempts to decarbonise electricity and heat in the UK. Most especially, we highlight insufficient consumer engagement with both the rollout and the physical infrastructure of smart meters and IHDs and failure to fully engage with potential burdens and benefits for vulnerable households. We position this finding in the light of critiques coming from a diverse set of actors such as academics, consumer bodies, parliamentary committees and newspaper articles.

We suggest that it is necessary to better account for, and manage, potential vulnerabilities as well as produce a broader range of outreach and communication materials that are easier to understand, especially among the elderly or the poor (Citizens Advice, 2017). While we acknowledge that Smart Energy GB have created a marketing partnership programme to reach out to vulnerable customers and increasingly emphasise community engagement (Smart Energy GB, 2018), this process must focus not only on outreach and communication, but also on the depth of engagement and alternative models for it, such as community-led rollout promotions. We encourage the creation of meaningful feedback mechanisms to engage consumers, which requires, in part (more) time to trial different mechanisms in diverse settings. Doing so can help overcome social barriers, perhaps increasing smart meter implementation and ultimately, long-term affectivity. Otherwise, the route to a smarter energy system will be littered with social obstacles.

Finally, for several expected benefits to be realised, consumers need to play an active part with regards to how and when they consume (and sometimes produce) energy (Buchanan et al., 2016). Particularly in the UK, ambitious consumer-oriented aims have been set out in the SMIP (Pullinger et al., 2014). Smart meters are argued to 'putting consumers in control, ending estimated bills and helping people save energy and money' (BEIS, 2017b, p. 1). In addition to rolling out smart meters, the UK government has decided to provide every home with 'real-time' feedback with their energy use through an IHD in order to facilitate the expected benefits. The inclusion of IHD has been highly debated. Energy suppliers have argued that cheaper digital options would be more applicable, as they would reduce the cost of the overall rollout whereas academics have argued over the effectiveness of feedback devices to help people manage their energy consumption (such as Darby, 2006). In 2010, Darby (2010, p. 449) argued that 'it is risky for utilities and their regulators to adapt a 'fit and forget' attitude to any new technology in the belief that it will, unsupported, achieve their goals and be acceptable to consumers'.

While we have used the UK as a case study, our material contributes to debates beyond this one case study. Some €51 billion will be spent on smart meter initiatives in the near future across the EU (Darby, 2010). In 2013, only about 10 per cent of households in the EU had a smart meter, but the

European Commission has mandated that this number rise dramatically to 80 per cent by 2020 (Viitanen et al., 2015). The European Commission (2017) reports that Member States have committed to rolling out close to 200 million smart meters for electricity and 45 million for gas by 2020 at a total potential investment of €45 billion. This study elucidates some of the technical and social elements befuddling attempts to rapidly diffuse smart meters across homes and cities – findings that have relevance for those wishing to better understand the temporality and complexity of both national and household energy transitions.

References

Balta-Ozkan, N., Amerighi, O. and Boteler, O. (2014) A comparison of consumer perceptions towards smart homes in the UK, Germany and Italy: Reflections for policy and future research. *Technology Analysis and Strategic Management* 26(10): 1176–1195.

Barnicoat, G. and Danson, M. (2015) The ageing population and smart metering: A field study of householders' attitudes and behaviours towards energy use in Scotland. *Energy Research and Social Science* 9: 107–115.

Brignal, M. (2016) Smart meters: an energy-saving revolution or just plain dumb? *Guardian*. 1 October 2016. Available at: www.theguardian.com/money/2016/oct/01/smart-meter-energy-saving-revolution-cut-bills-gas-electricity.

BEIS (2016) Smart Meter Roll-out Cost-benefit Analysis. HM Government, London, UK. Available at: www.gov.uk/government/publications/smart-meter-roll-out-gb-cost-benefit-analysis.

BEIS (2017a) Smart Meter Customer Experience Study: Post-Installation Survey Report. August 2017. Available at: www.gov.uk/government/uploads/system/uploads/attachment_data/file/641076/Post-install_key_findings_report_FINAL_24082017_PUBLICATION.pdf.

BEIS (2017b) Smart Meters Implementation Programme 2017 Progress Update. HM Government, London, UK. Available at: https://assets.publishing.service.gov.uk/government/uploads/system/uploads/attachment_data/file/671930/Smart_Meters_2017_update.pdf.

Buchanan, K., Banks, N., Preston, I. and Russo, R. (2016) The British public's perception of the UK smart metering initative. Threats and opportunities. *Energy Policy* 91: 87–97.

Chilvers, J. and Longhurst, N. (2016) Participation in transition(s): Reconceiving public engagements in energy transitions as co-produced, emergent and diverse. *Journal of Environmental Policy and Planning* 18(5): 585–607.

Citizens Advice (2017) Smart support: Support for vulnerable consumers in the smart meter roll-out. 9 March 2017. Available at: www.citizensadvice.org.uk/Global/CitizensAdvice/Energy/Smart%20Support.pdf.

Darby, S. (2006) The effectiveness of feedback one energy consumption – A review for DEFRA of the literature on metering, billing and direct displays. *Environmental Change Institute*. Oxford University, UK.

Darby, S. (2010) Smart metering: what potential for householder engagement? *Building Research and Information* 38(5): 442–457.

DECC (2015a) Smart Metering Implementation Programme: DECC's Policy Conclusions – Early Learning Project and Small-scale Behaviour Trials. Department of Energy and Climate Change, London, UK. Available at: www.gov.uk/government/publications/smart-metering-early-learning-project-and-small-scale-behaviour-trials.

DECC (2015b) Smart Metering Early Learning Project: Synthesis Report. Department of Energy and Climate Change, London, UK. Available at: www.gov.uk/government/publications/smart-metering-early-learning-project-and-small-scale-behaviour-trials.

European Commission (2017) Smart Metering Deployment in the European Union Joint Research Centre. Available at: http://ses.jrc.ec.europa.eu/smart-metering-deployment-european-union.

Gosden, E. (2015) Energy smart meter roll-out may be 'costly failure', MPs warn. *Telegraph*, 5 March 2015. Available at: www.telegraph.co.uk/news/earth/energy/11456193/Energy-smart-meter-roll-out-may-be-costly-failure-MPs-warn.html.

IoD (2017) Future Proofing Energy. Available at: www.iod.com/Portals/0/PDFs/Campaigns%20and%20Reports/Infrastructure/Future-proofing-Energy.pdf?ver=2017-06-01-183745-103.

Jenkins, K., McCauley, D., Heffron, R., Stephan, H. and Rehner, R. (2016) Energy justice: A conceptual review. *Energy Research and Social Science* 11: 174–182.

Jenkins, K., McCauley, D. and Warren, C.R. (2017) Attributing responsibility for energy justice: A case study of the Hinkley Point Nuclear Complex. *Energy Policy* 108: 836–843.

Lewis, D. and Kerr, J. (2014) Not Too Clever: Will Smart Meters be the Next Government IT Disaster? Institute of Directors, UK. Available at: www.emraware.com/Documents/smart_meters_not_too_clever.pdf.

Meadows, S. and Brodbeck, S. (2017) Smart meter roll-out: Is getting one still compulsory? *Telegraph*, 8 July 2017. Available at: www.telegraph.co.uk/bills-and-utilities/gas-electric/smart-meter-roll-out-getting-one-still-compulsory/.

McCauley, D., Heffron, R., Stephan, H. and Jenkins, K. (2013) Advancing energy justice: The triumvirate of tenets. *International Energy Law Review* 32(3): 107–110.

McKenna, E., Richardson, I. and Thomson, M. (2011) Smart meter data: Balancing consumer privacy concerns with legitimate applications, *Energy Policy* 41: 807–814.

Mott MacDonald (2007) Appraisal of Costs and Benefits of Smart Meter Roll Out Options. Final Report. Mott MacDonald, Brighton, UK.

National Audit Office (2011) Preparations for the Roll-out of Smart Meters, HC 1091, Session 2010–2012, 30 June 2011, London. Available at: www.nao.org.uk/idoc.ashx?docId=6854152F-711C-4B5A-AB79-C829BA80A504&version=-1.

National Audit Office (2014) Update on Preparations for Smart Metering. Available at: www.nao.org.uk/report/update-on-preparations-for-smart-metering/.

Ofcom (2016) Communications Market Report: Scotland. 4 August 2016. Available at: www.ofcom.org.uk/__data/assets/pdf_file/0024/43476/CMR_Scotland_2016.pdf.

Ofgem (2011) Factsheet 101. Smart metering – What it means for Britain's homes. 31 March 2011. Available at: www.ofgem.gov.uk/sites/default/files/docs/2011/03/consumersmartmeteringfs_0.pdf.

Palmer, K. (2015a) Billing glitches for first customers of £11bn smart meter roll-out. *Telegraph*, 23 May 2015. Available at: www.telegraph.co.uk/finance/personalfinance/energy-bills/11622798/Billing-glitches-for-first-customers-of-11bn-smart-meter-roll-out.html.

Palmer, K. (2015b) 1.5 million smart meters won't work when you switch energy supplier. *Telegraph*, 7 June 2015. Available at: www.telegraph.co.uk/finance/personalfinance/energy-bills/11643750/1.5-million-smart-meters-wont-work-when-you-switch-energy-supplier.html.

Public Accounts Committee (2012) Preparation for the Roll-out of Smart Meters, Sixty-third Report of Session 2010–2012, HC637, The Stationery Office, London, UK. 17 January 2012. Available at: www.publications.parliament.uk/pa/cm201012/cmselect/cmpubacc/1617/1617.pdf.

Pullinger, M., Lovell, H. and Webb, J. (2014) Influencing household energy practices: A critical review of the UK smart metering standards and commercial feedback devices. *Technology Analysis and Strategic Management* 26(10): 1144–1162.

Rose, D. and Thed, M. (2014) Unveiled: New £200 'smart' meters every household must pay for (but may not work), *The Mail on Sunday*, 7 July 2014. Available at: www.thisismoney.co.uk/money/bills/article-2681954/Unveiled-New-200-smart-meters-household-pay-not-work.html.

Schlosberg, D. (2004) Reconceiving environmental justice: Global movements and political theories. *Environmental Politics* 13(3): 517–540.

Schlosberg, D. (2007) *Defining Environmental Justice: Theories, Movements and Nature*. Oxford University Press, Oxford, UK.

Shannon, L. (2015) Just how smart are these energy meters? *Financial Mail on Sunday*, 17 May 2015. Available at: www.thisismoney.co.uk/money/bills/article-3084432/ERROR-Smart-energy-meters-leave-hundreds-thousands-households-billing-limbo.html.

Smart Energy GB (2016) Citizens Advice Partners with Smart Energy GB to Spread the Word about Smart Meters. 18 July 2016. Available at: www.smartenergygb.org/en/resources/press-centre/press-releases-folder/citizens-advice-partnership.

Smart Energy GB (2017) Smart Energy GB Consumer Engagement Plan and Budget 2018. 15 December 2017. Smart Energy GB, London, UK. Available at: www.smartenergygb.org/en/about-us/essential-documents.

Smart Energy GB (2018) Annual Report and Accounts: Year Ended 31st December 2017. Available at: www.smartenergygb.org/en/about-us/essential-documents.

Sovacool, B.K., Kivimaa, P., Hielscher, S. and Jenkins, K. (2017) Vulnerability and resistance in the United Kingdom's smart meter transition. *Energy Policy* 109: 767–781.

Strengers, Y. (2013) *Smart Energy Technologies in Everyday Life*. Palgrave Macmillan, Basingstoke, UK.

Thomas, S. (2012) Not too smart an innovation: UK plans to switch consumers to smart gas and electricity meters. *Energy and Environment* 23(6–7): 1057–1074.

Utility Week (2017) Smart metering: Challenging times lead to strange bedfellows, 10 May 2017. Available at: http://utilityweek.co.uk/news/smart-metering-challenging-times-lead-to-strange-bedfellows/1302322#.WU_6OWjyvIV.

Verbong, G.P.J., Beemsterboer, S. and Sengers, F. (2013) Smart grids or smart users? Involving users in developing a low carbon electricity economy. *Energy Policy* 52: 117–125.

Viitanen, J., Connell, P. and Tommis, M. (2015) Creating smart neighborhoods: Insights from two low-carbon communities in Sheffield and Leeds, United Kingdom. *Journal of Urban Technology* 22(2): 19–41.

Walker, G. (2009) Globalizing environmental justice, *Global Social Policy* 9(3): 355–382.

7 Overcoming the systemic challenges of retrofitting residential buildings in the United Kingdom
A Herculean task?

Donal Brown, Paula Kivimaa, Jan Rosenow and Mari Martiskainen

Introduction

In Greek mythology, the Hydra was a giant serpent with many heads. The second of the 12 labours of Hercules was to kill the Hydra. However, when one of the Hydra's heads was cut off, two more grew in its place. In many ways, overcoming the 'multi-headed-challenges' of achieving widespread energy efficiency (EE) retrofit is an equally Herculean task. Policy initiatives in the UK, such as the Energy Companies Obligation (ECO) and the Green Deal, have sought and failed to achieve the mass uptake of residential retrofit. This chapter will argue that such policies have failed to address four systemic challenges that constrain uptake for whole house retrofits, and that a more comprehensive and wide-reaching policy approach will be needed to overcome each of these challenges. The chapter is therefore focused on some of the solutions to these challenges from the perspective of three key elements of a retrofit: the business model, financing and intermediaries. It also discusses the ways in which policy could support these outcomes.

Retrofit of buildings involves the 'construction approach involving the action of introducing [retrofitting] new materials, products and equipment into an existing building with the aim of reducing the use of energy of the building' (Baeli, 2013, p. 17). This is different from renovating or refurbishing – which refers to work undertaken to repair homes or make them more aesthetically pleasing (Baeli, 2013). Retrofits of residential buildings have significant potential to reduce carbon dioxide emissions (CCC, 2016), fuel poverty (Sovacool, 2015), and improve occupant health and wellbeing (Willand et al., 2015). However, in the UK, much of this potential is yet to be realised. Residential buildings account for almost a quarter of the UK's carbon emissions (CCC, 2016), and for every £1 spent on retrofitting fuel poor homes an estimated £0.42 is saved in National Health Service spending (UKGBC, 2017). The Committee on Climate Change (CCC, 2015b) estimates that there is cost-effective[1] potential to reduce direct emissions[2] from all buildings by a third by 2030, with the need to achieve near-zero emissions from the sector by 2050 (CCC, 2016). It is estimated that this level of retrofit activity would create a Gross Domestic Product (GDP) effect of approximately £25.3 billion in gross value added (Guertler and Rosenow, 2016). The UK government has therefore announced a target for all UK homes to achieve an Energy Performance Certificate (EPC) rating of C or above by 2035 (HM Government, 2017).

To achieve these targets, an increasingly comprehensive whole house approach to residential retrofit will be needed (Hansford, 2015). Such an approach involves multiple measures with strategies for insulation, draught proofing, ventilation, heating systems and low-carbon microgeneration (ibid.). However, the traditional policy approach to residential retrofit has tended to incentivise single measures and piecemeal interventions, which may cause damaging unintended consequences;[3] such as mould growth, poor air quality and in some cases structural damage (Davies and Oreszczyn, 2012). Thus, a comprehensive whole house retrofit; where the entire building is treated as a system rather than as individual elements or measures, can mitigate such issues and achieve greater reductions in emissions (Hansford, 2015). Much literature in this area has focused on the key 'barriers' to uptake (Fylan et al., 2016; Sorrell et al., 2004; Kangas et al., 2018). However, this focus on barriers has tended to characterise retrofit decision-making in terms of rational choices while ignoring broader social and contextual factors (Walker et al., 2014). This framing also carries the assumption that there is a latent *demand* for retrofit (Wilson et al., 2015).

The UK is an interesting case study – although achieving major progress in power sector decarbonisation, it still has one of the least efficient housing stocks in Europe. This is despite recent policy initiatives for residential EE. This chapter starts with a brief overview of recent UK policy on residential retrofit. It then moves onto characterising four challenges that constrain demand for retrofits, then proposes solutions centred around three key elements of successful whole house retrofits: business models (Brown, 2018); financing (Borgeson et al., 2013) and intermediaries (Kivimaa and Martiskainen, 2018). Drawing on recent empirical work at the CIED, we then argue that achieving these ambitions will require a comprehensive mix of policies (Kern et al., 2017; Kivimaa et al., 2018).

UK policy on residential retrofit

Improved EE has played a pivotal role in reducing the UK's energy use and carbon emissions. On a temperature corrected basis, total UK household energy use decreased by 19 per cent between 2002 and 2016, despite a 12 per cent increase in the number of households and a 10 per cent increase in population (BEIS, 2016). Per-household energy consumption fell by 37 per cent between 1970 and 2015, with most of this decrease (29 per cent) occurring since 2004 (ibid.). EE improvements in individual households have offset the 46 per cent increase in the number of households, the 5.6°C increase in average internal temperatures and the rapid growth in appliance ownership over this period, with the result that total household energy consumption has increased by only 7 per cent in 45 years.

Although rising energy prices and the 2008 recession contributed to recent trends, the bulk of the reduction in per-household energy consumption can be attributed to public policies to improve EE (CCC, 2017; CEBR, 2011; DECC, 2015; Odyssee, 2017). Of particular importance have been the major home insulation programmes funded by successive 'supplier obligations' (SOs) such as the Carbon Emissions Reduction Target (CERT – 2008 to 2012) (Rosenow, 2012) and ECO – 2013 onwards. Since 1994, energy and carbon-saving targets imposed on electricity and gas suppliers have allowed them to recover the costs through a levy on household energy bills. Also important were the requirement for condensing boilers within the UK Building Regulations and the progressive tightening of EU standards on the EE of electrical appliances (CEBR, 2011). Evaluations of these policies have shown them to be highly cost-effective, both in terms of the cost savings to participating households and in terms of broader societal welfare (Lees, 2006, 2008; Rosenow and Galvin, 2013). This experience supports the argument that market forces alone cannot deliver all cost-effective investments in residential buildings, owing to multiple and overlapping market failures. Instead, policy intervention can be used to increase the uptake of residential retrofit through a mix of regulation, public engagement and incentives.

Despite dozens of instruments in the broader EE policy mix targeting residential buildings (Kern *et al.*, 2017) and the apparent success in reducing energy demand through policy, in more recent years there has been a marked shift in the policy landscape. Previously, SOs supported relatively low-cost EE measures, and dedicated grant programmes funded through general taxation provided support for low-income households to invest in EE measures. The last version of such grant programmes – Warm Front, was terminated in 2011 and the government decided to radically change the way EE was delivered in the UK. Through the introduction of the Green Deal in 2013, an on-bill-repayment loan scheme, the government intended to trigger substantial investment in EE retrofits while the SO would fund only the costlier EE measures. It is now widely recognised that this approach failed – the Green Deal was effectively terminated in 2015 and funding provided through SOs has been significantly reduced (Rosenow and Eyre, 2016). As a result, the uptake rate of EE improvements has stalled since 2012.

There are, however, signs of a change to the approach taken. The Clean Growth Strategy, launched by the UK government on 12 October 2017, sets out ambitious long-term targets for EE – especially for buildings and would require a significant increase of the current EE improvement delivery rate. The targets specify that all homes as far as possible should reach EPC band C by 2035 and all fuel poor homes by 2030. This requires both adjusting the ambition levels of existing policies and the implementation of new instruments. At the time of writing, government is consulting on several new policy measures, and has recently introduced minimum energy efficiency standards (MEES) for the private rented sector.

Key challenges for residential retrofit

The limited uptake of cost-effective EE measures, characterised as the 'energy efficiency gap' (Jaffe and Stavins, 1994), remains the focus of much academic and policy research. This is especially the case with residential buildings, where the benefits of retrofitting go beyond emissions reductions, including improvements to health and wellbeing, social welfare and economic development (UKGBC, 2017).

Previous literature on retrofit has adopted key 'barriers' to uptake as the theoretical basis for understanding this gap (Fylan et al., 2016; Sorrell et al., 2004). Yet the original focus of much of this barriers literature, such as Sorrell et al. (2004), was on firm level decision-making, rather than on households. As such, the focus on barriers has tended to characterise retrofit decision-making in terms of rational economic choices, while downplaying social and contextual factors (Walker et al., 2014). This framing also carries the inherent assumption of latent *demand* for retrofit once these barriers are removed (Wilson et al., 2015). This framing has come to dominate the design of recent policy initiatives such as the Green Deal and ECO, which were predicated on households saving money on their energy bills (Rosenow and Eyre, 2016).

We argue that this framing is problematic, primarily because it misrepresents how and why home renovation decisions are made, and by whom. This chapter instead frames the problem in terms of four interrelated challenges that continue to contribute to low household uptake of residential retrofits.

Information, engagement and trust

A lack of knowledge of the specific options and benefits of retrofit remains widespread among households in the UK (Marchand et al., 2015). While many of the technologies and tools exist to retrofit existing buildings, their uptake is not widespread, largely due to a lack of household interest (Bonfield, 2016). Public engagement and marketing schemes have tried to generate demand but tended to be top-down (Rosenow and Eyre, 2016), short term and focus on specific subsidy schemes (UKGBC, 2017). This has also created a supply chain largely reliant on short-term policy incentives (CCC, 2015a). Complicated government programmes such as the Green Deal have often been difficult for households to grasp (Marchand et al., 2015). Households who do decide to retrofit often have to interact with multiple tradesmen and installers, who influence decisions on technology choices and subsequent use (Maby and Owen, 2015). These challenges of gaining appropriate advice, concerns over post-retrofit performance, combined with poor-quality workmanship, has undermined trust with the wider public (Pettifor et al., 2015).

Uncertain benefits and quality

Predicted energy and cost savings from retrofits are based on modelled energy performance. There is consistently a 'performance gap' between these models and actual energy performance outcomes (Fylan et al., 2016). This is characteristic of an industry with a reputation for low quality, with few contractual penalties for under-performance (Bonfield, 2016). Equally, retrofit interventions may alter a building's existing features, affecting a household's routines and practices in ways that may make them reticent to change (Wilson et al., 2015). By only focusing on financial savings, policies have also failed to recognise that retrofits could be framed and promoted in terms of aesthetics, comfort and wellbeing (Rosenow and Eyre, 2016). Much evidence now suggests that those who undertake energy retrofits do so because of these non-economic *sources of value*, such as environmental concerns, desire for improved comfort and living standards, property longevity and aesthetics (Fawcett and Killip, 2017; Kivimaa and Martiskainen, 2018).

Complexity, disruption and timing

Whole house retrofits involve multiple activities carried out by multiple contractors and consultants. Management of this process is complex and time consuming for the household (Pettifor et al., 2015). Alongside the significant disruption of extensive works, this can be a major deterrent to uptake (Snape et al., 2015). Thus, households may prefer to retrofit gradually, when it is less disruptive to do so, despite the higher costs and longer duration (Fawcett, 2014). Consequently, energy retrofit may only be considered during wider renovations (Wilson et al., 2015). Identifying such 'trigger points' could therefore promote retrofit in certain circumstances, such as moving into a new home (Maby and Owen, 2015).

Capital cost and split incentives

While retrofits result in long-term energy savings, whole house retrofits typically require long periods before the capital cost can be recovered in energy savings (Gouldson et al., 2015). Thus, many households lack access to up-front capital, with the benefits of the investment not being realised when moving house or in a landlord-tenant situation – termed 'split incentives' (Sorrell et al., 2004). While the up-front cost barrier has largely been the focus of recent policy initiatives in the UK, the economics of long-term financing is extremely sensitive to interest rates (Gouldson et al., 2015), particularly if energy bill neutrality[4] is required (Rosenow and Eyre, 2016). Further, while households may value funding for wider non-energy measures, such as general repairs, the majority of policies fund EE measures alone (Borgeson et al., 2013). These four related challenges are shown in Figure 7.1.

Typically, policy interventions in this area have targeted one or at most two of these issues. However, to overcome these 'multi-headed challenges' and deliver on the promise of residential retrofit, a systemic approach across multiple sectors and involving multiple government departments will be necessary (see the Conclusions and Policy Recommendations section). This chapter draws on

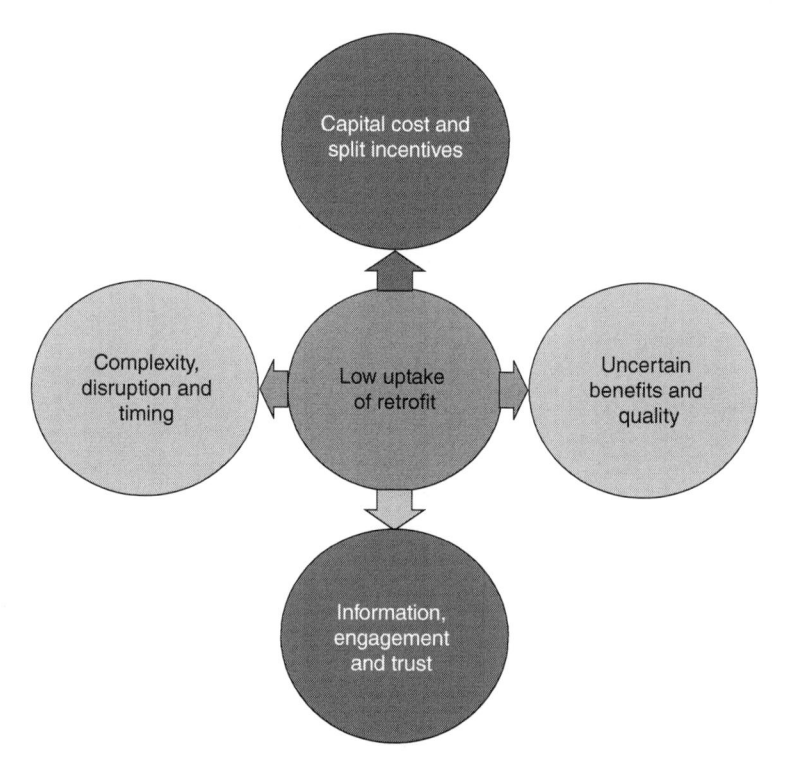

Figure 7.1 Key systemic challenges for driving retrofit uptake.
Source: the authors.

three emerging research themes: business models, financing and intermediaries. Building on these insights we then propose policy solutions to overcome the challenges for the widespread diffusion of whole house residential retrofit.

Overcoming the challenges for residential retrofit

In the following section we explore how best practice approaches to retrofit – business models, financing and intermediaries – can overcome many of the challenges that constrain uptake identified in the previous section.

Retrofit business models

A business model is defined as the nature of the products or services delivered to customers, the activities involved in delivering these and the means of capturing revenue from these activities (Boons *et al.*, 2013). Many radical innovations only became widespread once a complementary business model enabled their diffusion (Teece, 2010). Examples such as the MP3 player, low-cost air travel and smartphones owe their success to the effective pairing of the technology

116 *Donal Brown* et al.

with an appropriate business model and in many cases financing package. Emerging 'service-based' business models provide the useful end service rather than the technology or commodities themselves, shifting incentives for resource efficiency onto suppliers (Bocken et al., 2014). Consequently, energy service business models are promoted as a means of reducing energy demand (Labanca et al., 2014). Innovations such as distributed energy[5] and whole house retrofit may therefore require novel, complementary business models before they are viable on a large scale (Hall and Roelich, 2016). Drawing on recent research at the CIED (Brown, 2018), we argue that despite significant policy action in this area, a major reason for the lack of uptake of whole house retrofit is the limitations of the traditional business model.

The dominant business model for residential retrofit (Figure 7.2) is characterised by a piecemeal offering; with a fragmented supply chain, a focus on single (rather than multiple, complementary) measures, and no guarantees on performance. This is typically marketed on estimated energy cost and carbon savings and involves measures and technologies installed by separate contractors. Customers procure the individual measures, energy audits and finance separately, with the result that multiple interfaces are required for a comprehensive residential retrofit. The offer of energy savings is based on modelled impacts of measures, and no performance guarantees are provided. Therefore, any finance package is based on estimated rather than guaranteed savings. Such an approach has typified the delivery of the EE through UK policies such as ECO and the Green Deal.

This approach introduces significant *complexity* for customers in managing multiple interfaces with sub-contractors, energy auditors and finance providers, also tending to result in major *disruption* for a whole house retrofit. Equally, the narrow emphasis on estimated cost savings, without performance or ongoing

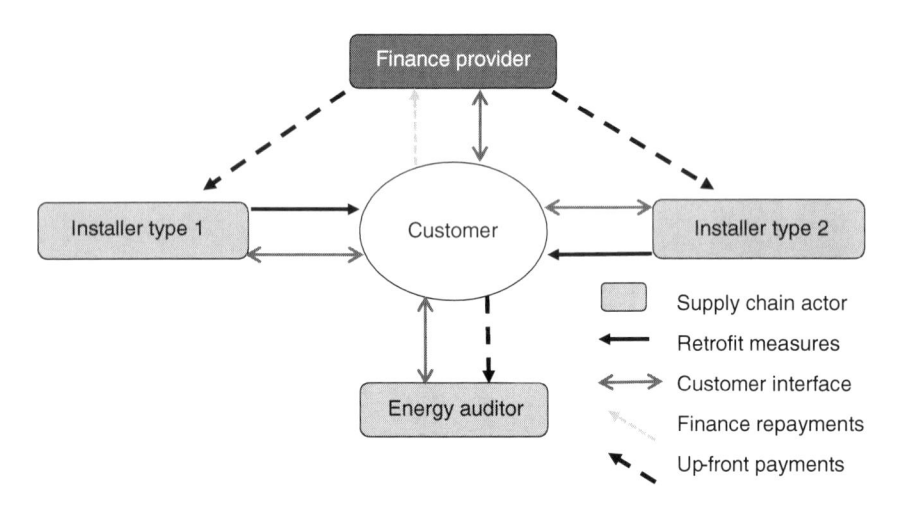

Figure 7.2 The incumbent 'atomised market model' for residential retrofit.
Source: Brown (2018), with permission (and without changes).

Retrofitting buildings in the UK 117

maintenance guarantees, means *uncertain benefits* for the customer and provides limited trust on installation *quality*. Unsurprisingly, this approach has resulted in low demand for comprehensive residential retrofits.

Recently, novel, integrated business models for residential retrofit have begun to emerge. These approaches emphasise a broader source of value for a whole house retrofit; focused upon aesthetics, increased property value, comfort, health and wellbeing alongside energy and carbon savings. Such approaches involve integrated and increasingly industrialised supply chains providing comprehensive whole house retrofits, through a single contractor or well-integrated network of sub-contractors. These approaches are characterised by a simplified customer interface with a single expert point of contact to coordinate the entire project. Some examples also offer integrated financing packages, and in some cases energy performance guarantees.

The Energiesprong initiative originated in the Netherlands and has expanded into the UK (Energiesprong, 2017). Customers are offered a comprehensive residential retrofit, based upon net-zero energy consumption. Typically, an Energiesprong retrofit involves the rapid delivery and installation of off-site manufactured, insulated wall facades, integrated with renewable heat systems and photovoltaic panels as well as ventilation and controls. The provider offers a 30-year energy performance guarantee (based on set internal temperature) for annual net-zero energy consumption, with specified energy usage limits, alongside an upstream financing package. An energy service contractor (ESCO) also takes on responsibility for the payment of the energy bill of the customer to provide 'total energy management'. This represents a holistic energy services offering to the household, commonly termed a Managed Energy Services Agreement (MESA) (Kim et al., 2012; Figure 7.3). This approach is currently being trialled in multi-family buildings and across large social housing estates.

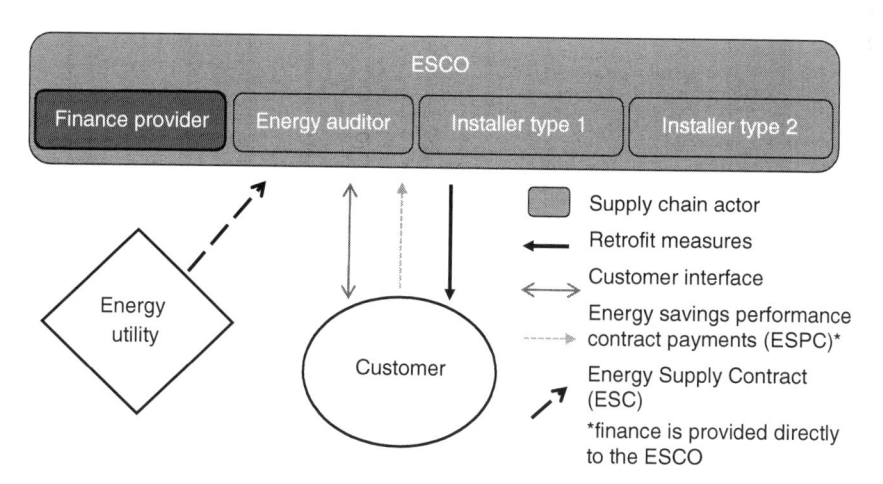

Figure 7.3 The Energiesprong Managed Energy Services Agreement (MESA).
Source: Brown (2018), with permission (and without changes).

Integrated business models such as the MESA have significant potential to drive demand for residential retrofit. By emphasising broader sources of value and including additional renovation measures as part of the offering, suppliers can attract customers by appealing to the *wider benefits* of improved aesthetics, increased property value, comfort, health and wellbeing alongside energy and carbon savings. Creating a simplified customer journey through an integrated supply chain, project co-ordination and a financing offer reduces *complexity* and minimises *disruption* for households. Further, the offer of energy performance guarantees provides *certainty* surrounding the ongoing performance benefits of the retrofit and the *quality* of the installation. While this may be the optimal solution, it is worth noting that integrated business models also face barriers and their uptake has been slow in the residential sector (Kangas *et al.*, 2018).

Business model innovation involves novel approaches and relationships for the delivery of products and services (Chesbrough, 2010). However, incumbent business models may be heavily embedded with existing industry practices, technological artefacts and regulatory regimes (Hannon, 2012). Therefore, adopting integrated energy service business models remains a challenge for an industry dominated by small-scale small and medium-sized enterprises (SMEs).

Retrofit finance

The up-front *capital cost* of retrofit measures and the *split incentives* faced by tenants and landlords remain a key challenge for the scaling up of comprehensive residential retrofits. Many UK households are also still in fuel poverty – defined as the necessity to spend more than 10 per cent of household income on energy bills (Sovacool, 2015).

As noted above, the UK's market-based SOs have funded significant loft and cavity wall insulation, low-energy lightbulbs and other low-cost measures (Rosenow and Eyre, 2014). ECO, the latest evolution of the SO policies, was initially designed to fund more expensive retrofit measures, such as solid wall insulation. It has since been criticised for its focus on single measures (Brown, 2018), dis-incentivising comprehensive installations, with no funding for complementary work such as ventilation and damp prevention (Hansford, 2015). SO policies require a levy on all households' energy bills, and thus increase the energy bills of households that do not benefit from programmes such as ECO (Rosenow *et al.*, 2013). The ECO has now been redesigned to focus on the 'fuel poor'. Although, having added approximately £50 a year to average household bills – a total of £1.3 billion annually (DECC, 2013), policies like ECO are arguably a poor tool for addressing fuel poverty (Rosenow *et al.*, 2013).

Meeting the UK's retrofit targets will require an estimated £85.2 billion of net investment to 2035 (Rosenow *et al.*, 2017). Achieving this level of investment through an SO like ECO could introduce politically unacceptable bill rises (Kern *et al.*, 2017) and be particularly regressive for the fuel poor who do not adopt retrofit measures (Rosenow *et al.*, 2013). Previous fuel poverty policies such as Warm Front did not raise wider energy bills as they were funded by general taxation (Sovacool, 2015). A fuel poverty policy funded by general taxation is also more consistent with targeting the co-benefits of social welfare (Rosenow et al., 2013) and improved health and wellbeing (UKGBC, 2017).

Alongside fuel poverty grants, there is a likely need for repayable retrofit financing for the 'able-to-pay' segment (Freehling and Stickles, 2016). The UK's Green Deal policy involved a novel *finance mechanism*, intended to deliver approximately 2 million retrofit installations per year and leverage billions of pounds of private sector investment. The scheme was based on private sector lending to households, paid back through energy bills – known as on-bill-repayment. However, the scheme achieved a fraction of its targets, and resulted in a significant loss to the UK taxpayer before its premature scrappage in 2015 (Rosenow and Eyre, 2016).

A range of other retrofit finance mechanisms have been developed in the UK, wider EU and USA, including several that have been markedly more successful than the Green Deal (EEFIG, 2015). Examples include: property assessed clean energy finance (PACE) in the USA, repaid through property taxes (Kim et al., 2012); low-cost public loans (such as the German KfW scheme) (Schröder et al., 2011); utility funded on-bill-financing (State and Local Energy Efficiency Action Network, 2014); retrofit mortgages (EEFIG, 2015); state-backed guarantee funds (Borgeson et al., 2013); and energy service agreements (ESA) – where finance for measures is procured upstream by an ESCO as part of an energy performance contract (Kim et al., 2012).

Examples of successful retrofit financing programmes, including Germany's KfW programme and California's PACE scheme, share some common features. These schemes typically include: a *cost of capital* that is low enough not to deter households and enable deeper retrofit measures to remain cost-effective (Rosenow and Eyre, 2016); a simplified *customer journey* – with finance often arranged by the contractor or project manager (Brown, 2018), use of an existing repayment channel (such as property taxes), attaching the debt to the property not the householder (resolving the spilt incentive issue); and funding for *broader sources of value*, such as wider renovation work or essential home improvements, that are often more highly valued by households (Fawcett and Killip, 2017).

By contrast the Green Deal involved a complex vetting and application process, that required a separate interface with a third-party provider, with no funding available for wider improvements. Introducing significant complexity that was likely to be offputting for most households. The Green Deal also had relatively high interest rates of 7–11 per cent (Marchand et al., 2015). Indeed, the total cost of capital amounted to at least 49 per cent of total Green Deal Plan costs over 15 years (UKGBC, 2014). Programmes such as the KfW scheme offer finance at extremely low or zero interest rates (>2 per cent) (Schröder et al., 2011). Such offers are likely to be more appealing to households (Marchand et al., 2015) and drastically improve the economics of whole house retrofits with longer payback periods (UKGBC, 2014).

Several approaches exist to reduce the cost of capital for retrofit finance. Privately funded schemes such as PACE and retrofit mortgages are *secured* against the property and can be securitised and sold to secondary markets – reducing risk and transaction costs for investors (Borgeson et al., 2013). State actors may also assist in reducing the cost of capital, particularly where customers face difficulties or high costs in accessing finance. Policy options include interest subsidies (Gouldson et al., 2015), state provision of subordinated (high risk) capital (State and Local Energy Efficiency Action Network, 2014), investor guarantee funds (Borgeson et al., 2013) or the direct provision of low-cost loans, as has been the case in Germany's KfW programme (Schröder et al., 2011).

However, there are limits to what financing alone can achieve. In most cases financing is likely to be an *enabler* of retrofit projects rather than a *driver* of demand (Borgeson et al., 2014). Consequently, policymakers can introduce a range of incentives to promote demand for retrofit. These include fiscal or energy supplier incentives, such as variable property taxes (i.e. stamp duty or council tax), income tax rebates, VAT reductions or EE feed-in tariffs (Rosenow and Cowart, 2017; UKGBC, 2013). Some can be made fiscally neutral through penalising properties that do not meet a certain performance level (UKGBC, 2013). Incentives are likely to be particularly effective when they are available at key junctures when broader renovation decisions are being made. Thus, approaches that target key *trigger points* such as when properties change hands, during extensive renovations or heating replacements, are likely to be most successful (Maby and Owen, 2015).

Retrofit intermediaries

Intermediaries – that can be individuals, organisations or platforms – facilitate innovation processes (and broader transition processes) by educating, collecting and allocating financial and human resources, assessing new technologies and practices, creating partnerships, and influencing changes in regulations and rules (Stewart and Hyysalo, 2008). They may also shape how innovation occurs when it faces the user and negotiate on behalf of other actors (ibid.). Intermediaries may act as a single point of contact between households and retrofit contractors. In this section, we focus on how intermediaries can (1) stimulate, guide and manage different whole house retrofit projects, and (2) aid the creation of a market for new retrofit business models and financing solutions, supporting a transition towards a low-energy housing stock.

To address the challenges of *information, engagement and trust* as well as the complexity of whole house retrofits, intermediary actors are needed both at project level (e.g. specific retrofits) and the broader market level. In the former, intermediaries interconnect different technological, human and financial solutions. In the latter, they can have a crucial role in building trust and aggregating and disseminating clear and reliable information on retrofit techniques, suppliers and contractors.

A review of European case studies (Kivimaa and Martiskainen, 2018) shows that two types of intermediaries are specifically important in driving the market for retrofit. First, *innovation funders* such as Innovate UK are important in supporting successful piloting of complex architectural or systemic innovation (i.e. interconnecting innovative and standard solutions to deliver whole house retrofits). Second, *social housing providers and local community actors* are crucial in market creation and advancing retrofits in practice. Yet, the role of social housing providers has lately diminished through policies introducing rent caps and 'right to buy' schemes, as well as local authority budget cuts – leaving less resources for housing providers to carry out retrofits in their building stock. In addition, *business networks*, such as the Passive House Platform in Belgium (Mlecnik, 2013) are important in pooling together different types of companies and solutions to create new business models and promote retrofitting. In the UK, the Green Deal Pioneering Places also stimulated cooperatives to deliver retrofits. What still seems to be largely lacking in the UK are intermediaries that can effectively stimulate the market for whole house retrofitting by owner–occupiers and private landlords, at the community level.

At the project level, intermediaries are needed to stimulate interest in whole house retrofits, share experiences among home owners, and provide necessary expertise during planning and implementation. Platforms, such as Eco Open Houses in the City of Brighton and Hove, organised in 2008 and between 2010–2015, enabled people to see and visit sustainable homes. These cases demonstrate that such events have been extremely useful in providing information, stimulating engagement and sharing knowledge on whole house retrofits, as well as providing details of trusted local tradespeople and installers. When planning and executing whole house retrofits, individual actors taking up intermediary roles – for example, *architects, building cooperatives* or *local authority officers* taking actions beyond their usual roles – are valuable in helping households make choices over technologies and materials. Previous research has shown the importance of local authority energy managers, planners (Lovell, 2008) and sustainability officers (Martiskainen and Kivimaa, 2018) as important intermediaries in project planning and implementation.

Recent CIED research involved a case study of a three-bedroom terraced home built in 1860 in Southampton Street, Brighton. The house was part of a local project obtaining funding from the 'Green Deal Pioneer Places' Programme (a national government-funded programme that sought to demonstrate the benefits of EE). The house has undergone an extensive retrofit, motivated by the owners' interest in climate change and sustainable living, though the owners had no specific knowledge or interest to carry out a retrofit themselves. This was coordinated by the Green Building Partnership, which was formed initially to take part in the programme. The owners therefore did not have to acquire knowledge on the technical or policy aspects of the retrofit. The retrofit measures included external solid wall insulation, loft insulation, improved windows, new boiler and heating controls – involving multiple partners. While the Green Building Partnership led the process, from the perspective of the owner, there was no one key intermediary communicating between the broader scheme and the owners, leading to some confusion. Southampton Street later became part of local Eco Open Houses event, acting as an example to others.

Without these intermediary roles, projects may become much more complicated. Intermediaries provide information on the retrofit options available for the building projects and help to create a plan that meets regulations. More support is, however, needed from dedicated intermediaries, to facilitate 'one-stop-shops' for retrofitting (Brown, 2018), through which households can access trustworthy advice on technological and financing options, as well as tradesmen, contractors and installers. In this way intermediaries are often the key actors in providing *information* for households on the options and benefits of undertaking comprehensive retrofits; *engaging* communities and supply chains to promote retrofit at a local level; and are also likely to be more *trusted* than actors with a financial stake in promoting certain services or products.

Overall, some factors for successful intermediaries can be depicted. On a broader scale, most impact occurs over a longer timeframe. For example, the Centre on Alternative Technology established in the 1970s still influences the expectation and visions behind home retrofits. While the Eco Open Houses events have been popular in Brighton, they were not organised in 2016–2017, creating uncertainty about future knowledge exchange and example setting locally. Another important determinant in market formation is the positioning between ambitious sustainability aims and connections to business and supply chains.

Innovative business models, such as the Energiesprong approach, owe much of their success to dedicated intermediaries, often initiated by government policy. Energiesprong was brought into being through a €50m grant from the Dutch government, and the setting up of a *market development team* (Energiesprong, 2017). These market development teams brought together stakeholders including the construction industry, housing providers, policymakers and financiers to radically re-think the business model through which EE retrofit is delivered. While these approaches still face challenges, they could represent a template for how the UK could deliver on its ambitious retrofit targets.

Conclusions and policy recommendations

In the ancient Greek myth, the Hydra was invulnerable only if it retained at least one head. Heracles, realising that he could not defeat the Hydra alone, worked with Iolaus, and through a combination of decapitating the beasts' multiple heads and burning the stumps with a firebrand, stopped them growing back. The Hydra's remaining immortal head was cut off with a golden sword given to Heracles by Athena. Heracles placed the head – still alive and writhing – under a great rock and shot it with an arrow dipped in the Hydra's poisonous blood. Thus, his second task was complete.

The previous sections outline how tackling the 'multi-headed-challenges' of whole house residential retrofit will require a similarly sophisticated and multifaceted approach. Promoting business model innovation, delivering a range of financing options and incentives along with the establishment of strategic intermediaries, at both local community and national levels, will require a wide-reaching and systemic policy strategy. This strategy should incorporate a mix of regulations,

Table 7.1 Policy mix for achieving widespread comprehensive residential retrofit in the UK

Policy type	Policy	Challenge addressed	Government department
Regulation (sticks)	EE as an Infrastructure Priority (Frontier Economics Ltd, 2015)	All	HMT, NIC
	Minimum EE Standards (MEES) moving to EPC C in 2035 (Sustainable Energy Association, 2017)	All	DCLG
	New retrofit quality assurance standard such as home Quality Mark (Bonfield, 2016)	Uncertain energy savings and quality	BEIS, DCLG
Financial (carrots)	Financial Incentives at trigger points, options could include: (UKGBC, 2013) • Variable Stamp duty • Variable Council tax • 0% VAT on renovation work that includes retrofit • Income tax rebates • EE Feed-in Tariff	Capital cost and split incentives Complexity, disruption and timing	BEIS, HMT, HMRC
	Government backed low-interest financing mechanism secured to property and available at point of sale of retrofit (Borgeson et al., 2013)	Capital cost and split incentives Complexity, disruption and timing	BEIS, HMT, NIC
	Fuel poverty obligation funded by general taxation (Rosenow et al., 2013)	Capital cost and split incentives	BEIS, HMT, DH, DWP
New institutions and intermediaries (tambourines)	National Retrofit Taskforce/Agency (Rosenow et al., 2017) with central Information Hub and a Data Warehouse	All	BEIS, DCLG, HMT, DfE, DH, NIC
	Area-based intermediaries based on Community Social Enterprise or Local Authority Arm's Length Management Organization (ALMO) delivery models (UKGBC, 2017). Market facing intermediaries and standardised procurement frameworks (Nolden et al., 2016)	Information, engagement and trust Complexity, disruption and timing	BEIS, DCLG, HMT, DfE, DH, NIC

Source: the authors.

financing and incentives along with the establishment of new institutions and the recognition of EE as a strategic infrastructure priority. Equally, different solutions will be required for socially rented, privately rented and owner–occupier sectors. This will require joined up action across multiple government departments including but not limited to: Business Energy and Industrial Strategy (BEIS), Department for Communities and Local Government (DCLG), Ministry of Housing, Communities and Local Government's (MHCLG), the Treasury (HMT), Education (DfE) and Health (DH), the Department for Work and Pensions (DWP), Her Majesty's Revenue and Customs (HMRC) and the National Infrastructure Commission (NIC). The following section provides an outline of the range of policies (Table 7.1) that could contribute to achieving the enormous potential for the comprehensive retrofit of residential buildings.

Standards and regulations

EE retrofits create economic benefits that are often several multiples of the initial investment (Guertler and Rosenow, 2016). Cost-effective investments in residential EE to 2035 have a current net present value of £7.5 billion. With wider benefits such as gross domestic product (GDP) effects and health improvements that could be up to £47 billion (Rosenow et al., 2017). Thus, EE investments share the characteristics of other forms of infrastructure as identified in HM Treasury's valuation guidance (Frontier Economics Ltd, 2015). Therefore, we argue that *EE should be re-framed as an infrastructure priority* by the UK government and given the level of strategic support and status as other forms of infrastructure; such as road, rail and supply side energy infrastructure and be included within the remit of the NIC (Rosenow and Cowart, 2017).

The UK Clean Growth Plan set an aspirational goal for all domestic buildings to achieve and EPC level C or higher by 2035. We support these aims, but argue the government could go further, mandating MEES for the owner–occupier sector in the 2020s. This could take the form of a gradual step change through to a *minimum EPC level of C by 2035* at the point of sale, with potential for ever-tightening standards moving into the 2040s and beyond (Sustainable Energy Association, 2017).

There remain concerns surrounding the standard and quality of many installations currently funded under ECO, particularly solid wall insulation, which is to be a key part of the UK's targets (Hansford, 2015). Therefore, we support the findings of the recent Each Home Counts – 'The Bonfield Review', that the government should establish a *new quality assurance standard* such as a home Quality Mark (Bonfield, 2016). Such a policy should be designed not to introduce a further regulatory and cost burden on SMEs and could build on existing standards of good practice along the lines of the Investor Confidence Project in the commercial sector (Investor Confidence Project, 2015).

Taken together these three high-level regulatory policies would set the strategic direction for UK residential retrofit policy and would send market signals for the development of integrated business models, novel financing solutions and market intermediaries. However, on their own, top-down regulations are unlikely to build a sufficient market for whole house retrofit.

Financial measures

Overcoming the up-front capital cost of EE retrofit remains a challenge for many households. Current fuel poverty schemes such as ECO are limited in size and have inherent design flaws (Rosenow et al., 2013). For those in fuel poverty we instead propose that these costs should largely be met through government grants in the form of *a fuel poverty obligation paid for by general taxation*. This would allow the government to better spread the costs of such a scheme, and if properly designed could reduce spending in areas such as health, social care and welfare (Rosenow et al., 2017).

For the 'able-to-pay' segment a range of market-led financing mechanisms may eventually emerge, including mortgage-based approaches and other private sector offerings. Yet, we argue that the government should learn the lessons of the failed Green Deal and promote a *new low-cost financing mechanism tied to the property*, perhaps retaining the on-bill repayment channel. Successful financing schemes such as Germany's KfW programme have used government funds to provide a *low cost of capital*, involved a simplified *customer journey* and funded *broader sources of value* such as wider renovation works, which are likely to be perceived as higher value by households (Schröder et al., 2011).

Although providing sources of lending for EE measures is key to *enabling* retrofit projects, it is unlikely that low-cost financing alone will be *drive* demand for retrofit (Borgeson et al., 2014). Therefore, government can introduce a range of *fiscal incentives at key trigger points* to promote uptake. These might include: variable VAT, stamp duty land tax, council tax, income tax rebates or an EE feed-in tariff for households who have undertaken measures – with increasing benefits for deeper retrofits (UKGBC, 2013). Such approaches will be most effective when they are targeted at key trigger points such as moving home or when undertaking major renovations (Maby and Owen, 2015).

New institutions and intermediaries

A key challenge for residential retrofit remains the paucity of information, engagement and trust within communities. Recent work at the UK Green Building Council (2017) has highlighted a *new role for intermediaries* to catalyse retrofit and regeneration activity in *local areas*. These actors would engage local communities on the benefits of retrofit and regeneration and be the key point of contact for: information, marketing, financing and project delivery, through dedicated project managers/coordinators – drawing on the pre-existing networks of diffuse intermediaries already operating in many communities (Martiskainen and Kivimaa, 2018). These intermediaries could be based on Community Social Enterprise or Local Authority Arm's Length Management Organization

(ALMO) delivery models, and funded through a combination of local authority budgets, central government grants and community shares (UKGBC, 2017).

Intermediaries also play a role in promoting business model innovation for the delivery of comprehensive residential retrofit. Examples such as the Dutch Energiesprong scheme (Brown, 2018) and the RE:FIT programme in London (Nolden et al., 2016) demonstrate how public bodies can promote business model innovation, through the *creation of new market facing intermediaries* and standardised procurement frameworks. These initiatives help reduce transaction costs and bring together stakeholders to foster learning, new funding approaches and supply chain integration.

Achieving the promise of residential retrofit and tackling the 'multi-headed-challenges' that stand in the way, will require a joined up and coordinated strategy – as outlined in this chapter. To deliver this vision, we argue that the UK government should set up a *National Retrofit Taskforce*. This body would be responsible for the planning and delivery of the MEES targets through an overarching strategy, monitoring and verification process that brings together key stakeholders, including, Government, Third sector, Industry and Consumer groups (Rosenow and Cowart, 2017). This new high-level intermediary would also be responsible for the management of a central *Information Hub* (to act as a collection point for best practice advice and guidance) and a *Data Warehouse* (to act as a store for property-level data and information) (Bonfield, 2016). Advising multiple government departments, this body could monitor progress towards the UK's targets for the sector and propose polices to keep this progress on track.

Climate change is perhaps the biggest challenge facing humanity in the twenty-first century. Buildings are perhaps the biggest single contributor to carbon emissions, with the existing residential buildings by far the largest component (CCC, 2016). Such a Herculean challenge will require an equally Herculean effort. We argue that the considerable rewards are more than worth rising to this challenge, and that the proposals presented here could go a long way towards achieving this.

Notes

1 The CCC define the cost-effective path as comprising measures that cost less than the projected carbon price across their lifetimes together with measures that may cost more than the projected carbon price but are necessary in order to manage costs and risks of meeting the 2050 target (CCC 2013).
2 Those that result from heating, ventilation and cooling systems as well as and hot water. This term excludes emissions from electricity consumption.
3 Such as mould growth, poor air quality and interstitial condensation due to poor detailing, and insufficient consideration of building physics, airtightness and ventilation.
4 Energy bill neutrality may include requirements that modelled savings are 'cash-flow positive' meaning that finance repayments are equal to, or result in, net energy cost savings (Borgeson et al., 2013).
5 Defined as electricity generation feeding into the local distribution network (operating from 132 kV down to 230 V), as opposed to the regional or national transmission grid (which operates from 400 kV and 275 kV).

References

Baeli, M. (2013) *Residential Retrofit: 20 Case Studies*. RIBA Publishing, London, UK.

BEIS (2016) Energy Consumption in the UK 2016. HM Government, London, UK. Available at: www.gov.uk/government/statistics/energy-consumption-in-the-uk.

Bocken, N.M.P., Short, S.W. and Evans, R.S. (2014) A literature and practice review to develop sustainable business model archetypes. *Journal of Cleaner Production* 65: 42–56. DOI: 10.1016/j.jclepro.2013.11.039.

Boons, F., Montalvo, C., Quist, J. and Wagner, M. (2013) Sustainable innovation, business models and economic performance: An overview. *Journal of Cleaner Production* 45: 1–8.

Bonfield, P. (2016) Each Home Counts – An Independent Review of Consumer Advice, Protection, Standards and Enforcement for Energy Efficiency and Renewable Energy, Department for Business, Energy and Industrial Strategy (BEIS) and Department for Communities and Local Government (DCLG), London, UK.

Borgeson, M., Todd, A. and Goldman, C. (2013) Getting the Biggest Bang for the Buck Exploring the Rationales and Design Options for Energy Efficiency Financing Programs. Available at: https://eta.lbl.gov/sites/default/files/publications/lbnl-6524e.pdf.

Borgeson, M., Zimring, M. and Goldman, C. (2014) The Limits of Financing for Energy Efficiency. Available at: https://eta.lbl.gov/sites/default/files/publications/lbnl-limits-of-financing-aceee-ss2012-final2.pdf.

Brown, D. (2018) Business models for residential retrofit in the United Kingdom; A critical assessment of 5 key archetypes. *Energy Efficiency* 11(6): 1497–1517.

CCC (2013) Committee on Climate Change. Fourth Carbon Budget Review – Part 2. HM Government, London, UK. Available at: www.theccc.org.uk/wp-content/uploads/2013/12/1785b-CCC_TechRep_Singles_Book_1.pdf.

CCC (2015a) Sectoral Scenarios for the Fifth Carbon Budget. Available at: www.theccc.org.uk/wp-content/uploads/2015/11/Sectoral-scenarios-for-the-fifth-carbon-budget-Committee-on-Climate-Change.pdf.

CCC (2015b) The Fifth Carbon Budget – The next step towards a low-carbon economy – November 2015. Available at: www.theccc.org.uk/wp-content/uploads/2015/11/Committee-on-Climate-Change-Fifth-Carbon-Budget-Report.pdf.

CCC (2016) Meeting Carbon Budgets – 2016 Progress Report to Parliament. Available at: www.theccc.org.uk/publications/.

CCC (2017) Energy Prices and Bills – Impacts of Meeting Carbon Budgets. HM Government, London, UK. Available at: www.theccc.org.uk/publication/energy-prices-and-bills-report-2017/.

CEBR (2011) British Gas Home Energy Report 2011: An Assessment of the Drivers of Domestic Natural Gas Consumption. Available at: www.centrica.com/sites/default/files/bg_home_energy_report_110202_0.pdf.

Chesbrough, H. (2010) Business model innovation: Opportunities and barriers. *Long Range Planning* 43(2–3): 354–363. DOI: 10.1016/j.lrp. 2009.07.010.

Davies, M. and Oreszczyn, T. (2012) The unintended consequences of decarbonising the built environment: A UK case study. *Energy and Buildings* 46: 80–85.

DECC (2013) Energy Company Obligation (ECO) Delivery Costs. Available at: www.gov.uk/decc.

DECC (2015) Energy Consumption in the UK (2015). Chapter 3: Domestic Energy Consumption in the UK between 1970 and 2014. Available at: www.connaissancedesenergies.org/sites/default/files/pdf-actualites/ecuk_chapter_3_-_domestic_factsheet.pdf.

EEFIG (2015) Energy Efficiency – The First Fuel for the EU Economy. Available at: www.unepfi.org/fileadmin/documents/EnergyEfficiencyInvestment.pdf.

Energiesprong (2017) United Kingdom – Energiesprong. Available at: http://energiesprong.eu/country/united-kingdom/.

Fawcett, T. (2014) Exploring the time dimension of low carbon retrofit: Owner-occupied housing. *Building Research and Information* 42(4): 477–488.

Fawcett, T. and Killip, G. (2017) Anatomy of low carbon retrofits: Evidence from owner-occupied Superhomes. *Building Research and Information* 42(4): 434–445. DOI: 10.1080/09613218.2014.893162.

Freehling, J. and Stickles, B. (2016) Energy Efficiency Finance: A Market Reassessment. ACEE White Paper. Available at: http://aceee.org/sites/default/files/market-reassessment-021716.pdf.

Frontier Economics Ltd. (2015) Energy Efficiency: An Infrastructure Priority. Available at: www.energysavingtrust.org.uk/sites/default/files/reports.

Fylan, F., Glew, D., Smith, M., Johnston, D., Brooke-Peat, M., Miles-Shenton, D., Fletcher, M., Aloise-Young, P. and Gorse, C. (2016) Reflections on retrofits: Overcoming barriers to energy efficiency among the fuel poor in the United Kingdom. *Energy Research and Social Science* 21: 190–198. DOI: 10.1016/j.erss.2016.08.002.

Gouldson, A., Kerr, N., Millward-Hopkins, J., Freeman, M.C., Topi, C. and Sullivan, R. (2015) Innovative financing models for low carbon transitions: Exploring the case for revolving funds for domestic energy efficiency programmes. *Energy Policy* 86: 739–748. DOI: 10.1016/j.enpol.2015.08.012.

Guertler, P. and Rosenow, J. (2016) Buildings and the 5th Carbon Budget. ACE, London, UK. Available at: www.ukace.org/wp-content/uploads/2016/09/ACE-RAP-report-2016-10-Buildings-and-the-5th-Carbon-Budget.pdf.

Hall, S. and Roelich, K. (2016) Business model innovation in electricity supply markets: The role of complex value in the United Kingdom. *Energy Policy* 92: 286–298.

Hannon, M. (2012) Co-evolution of innovative business models and sustainability transitions : the case of the Energy Service Company (ESCo) model and the UK energy system. PhD Thesis, University of Leeds. October 2012, 1–348.

Hansford, P. (2015) Solid Wall Insulations: Unlocking Demand and Driving Up Standards. A Report to the Green Construction Board and Government by the Chief Construction Adviser. HM Government, London, UK. Available at: https://assets.publishing.service.gov.uk/government/uploads/system/uploads/attachment_data/file/476977/BIS-15-562-solid-wall-insulation-report.pdf.

HM Government (2017) The Clean Growth Strategy: Leading the Way to a Low Carbon Future. HM Government, London, UK. Available at: www.gov.uk/government/.

Investor Confidence Project (2015) Fueling Investment in Energy Efficiency. Available at: http://blogs.edf.org/energyexchange/files/2015/07/ICP-Fact-Sheet-070115.pdf.

Jaffe, A.B. and Stavins, R.N. (1994) The energy-efficiency gap. What does it mean? *Energy Policy* 22(10): 804–810.

Kangas, H.-L., Lazarevic, D. and Kivimaa, P. (2018) Technical skills, disinterest and non-functional regulation: Barriers to building energy efficiency in Finland viewed by energy service companies. *Energy Policy* 114: 63–76.

Kern, F., Kivimaa, P. and Martiskainen, M. (2017) Policy packaging or policy patching? The development of complex energy efficiency policy mixes. *Energy Research and Social Science* 23: 11–25.

Kim, C., O'Connor, R., Bodden, K., Hochman, S., Liang, W., Pauker, S. and Zimmermann S. (2012) *Innovations and Opportunities in Energy Efficiency Finance*. Wilson Sonsini Goodrich and Rosati, New York, USA.

Kivimaa, P. and Martiskainen, M. (2018) Innovation, low energy buildings and intermediaries in Europe: Systematic case study review. *Energy Efficiency* 11(1): 31–51. DOI: 10.1007/s12053-017-9547-y.

Kivimaa, P., Boon, W., Hyysalo, S., Klerkx, L. and Group, I., (2018) Towards a typology of intermediaries in sustainability transitions: A systematic review and a research agenda. *Research Policy*: 1–31. Available at: https://doi.org/10.1016/j.respol.2018.10.006.

Labanca, N., Suerkemper, F. Bertoldi, P., Irrek, W. and Duplessis, B. (2014) Energy efficiency services for residential buildings: Market situation and existing potentials in the European Union. *Journal of Cleaner Production* 109: 284–295. DOI: 10.1016/j.jclepro. 2015.02.077.

Lees, E. (2006) Evaluation of the Energy Efficiency Commitment 2002–05. Report to Defra. Eoin Lees Energy, Wantage, UK.

Lees, E. (2008) Evaluation of the Energy Efficiency Commitment 2005–08. Report to Defra. Eoin Lees Energy, Wantage, UK.

Lovell, H. (2008) Discourse and innovation journeys: The case of low energy housing in the UK. *Technology Analysis and Strategic Management* 20(5): 613–632.

Maby, C. and Owen, A. (2015) Installer Power – The Key to Unlocking Low Carbon Retrofit in Private Housing. Available at: http://ukace.org/wp-content/uploads/2015/12/Installer-Power-report-2015.pdf.

Marchand, R.D., Koh, S.C.L.L. and Morris, J.C. (2015) Delivering energy efficiency and carbon reduction schemes in England: Lessons from Green Deal Pioneer Places. *Energy Policy* 84: 96–106.

Martiskainen, M. and Kivimaa, P. (2018) Creating innovative zero carbon homes in the United Kingdom – Intermediaries and champions in building projects. *Environmental Innovation and Societal Transitions* 26: 15–31. DOI: 10.1016/j.eist.2017.08.002.

Mlecnik, E. (2013) Opportunities for supplier-led systemic innovation in highly energy-efficient housing. *Journal of Cleaner Production* 56: 103–111.

Nolden, C., Sorrell, S. and Polzin, F. (2016) Catalysing the energy service market: The role of intermediaries. *Energy Policy* 98: 420–430.

Odyssee (2017) Odyssee Decomposition Facility. Available at: www.indicators.odyssee-mure.eu/decomposition.html.

Pettifor, H., Wilson, C. and Chryssochoidis, G. (2015) The appeal of the green deal: Empirical evidence for the influence of energy efficiency policy on renovating homeowners. *Energy Policy* 79: 161–176.

Rosenow, J. (2012) Energy savings obligations in the UK – A history of change. *Energy Policy* 49: 373–382.

Rosenow, J. and Cowart, R. (2017) Efficiency First: Reinventing the UK's Energy System, Brussels, Regulatory Assistance Project (RAP). Available at: www.raponline. org/wp-content/uploads/2017/07/rap-efficiency-first-reinventing-UK-energy-system-2017-october.pdf.

Rosenow, J. and Eyre, N. (2014) Re-energising the UK's Approach to Domestic Energy Efficiency. *ECEEE Summer Study Proceedings*: 281–289.

Rosenow, J. and Eyre, N. (2016) A post mortem of the Green Deal: Austerity, energy efficiency, and failure in British energy policy. *Energy Research and Social Science* 21: 141–144.

Rosenow, J. and Galvin, R. (2013) Evaluating the evaluations: Evidence from energy efficiency programmes in Germany and the UK. *Energy and Buildings* 62: 450–458.

Rosenow, J., Platt, R. and Flanagan, B. (2013) Fuel poverty and energy efficiency obligations – A critical assessment of the supplier obligation in the UK. *Energy Policy* 62: 1194–1203.

Rosenow, J., Sorrell, S., Eyre, N. and Guertler, P. (2017) Unlocking Britain's First Fuel: The Potential for Energy Savings in UK Housing. UKERC. Available at: www.ukerc.ac.uk/publications/unlocking-britains-first-fuel-energy-savings-in-uk-housing.html.

Schröder, M., Ekins, P., Power, A., Zulauf, M. and Lowe, R. (2011) The KfW Experience in the Reduction of Energy Use in and CO_2 Emissions from Buildings: Operation, Impacts and Lessons for the UK. LSE Housing and Communities, London, UK, 1–80.

Snape, J.R.R., Boait, P.J.J. and Rylatt, R.M.M. (2015) Will domestic consumers take up the renewable heat incentive? An analysis of the barriers to heat pump adoption using agent-based modelling. *Energy Policy* 85: 32–38.

Sorrell, S., O'Malley, E., Schleich, J. and Scott, S. (2004) *The Economics of Energy Efficiency: Barriers to Cost-Effective Investment*. Edward Elgar, Cheltenham, UK.

Sovacool, B.K. (2015) Fuel poverty, affordability, and energy justice in England: Policy insights from the Warm Front Program. *Energy* 93(1): 361–371.

State and Local Energy Efficiency Action Network (2014) Financing Energy Improvements on Utility Bills: Market Updates and Key Program Design Considerations for Policymakers and Administrators. Available at: https://emp.lbl.gov/publications/financing-energy-improvements-utility.

Stewart, J. and Hyysalo, S. (2008) Intermediaries, users and social learning in technological innovation. *International Journal of Innovation Management* 12(3): 295–325.

Sustainable Energy Association (2017) Energy Efficiency – A Policy Pathway Addressing the Able to Pay Sector. Available at: www.sustainableenergyassociation.com/energy-efficiency-a-policy-pathway/articles/.

Teece, D.J. (2010) Business models, business strategy and innovation. *Long Range Planning* 43(2–3): 172–194.

UKGBC (2013) Retrofit Incentives. Task Group Report. UK Green Building Council. 1–76. Available at: www.ukgbc.org/sites/default/files/130705%2520Retrofit%2520Incentives%2520Task%2520Group%2520-%2520Report%2520FINAL_1.pdf.

UKGBC (2014) Green Deal Finance: Examining the Green Deal Interest Rate as a Barrier to Take-up. Available at: www.ukgbc.org/ukgbc-work/green-deal-finance/.

UKGBC (2017) Regeneration and Retrofit Task Group Report. Available at: www.ukgbc.org/wp-content/uploads/2017/09/08498-Regen-Retrofit-Report-WEB-Spreads.pdf.

Walker, S.L., Lowery, D. and Theobald, K. (2014) Low-carbon retrofits in social housing: Interaction with occupant behaviour. *Energy Research and Social Science* 2: 102–114.

Willand, N., Ridley, I. and Maller, C. (2015) Towards explaining the health impacts of residential energy efficiency interventions – A realist review. Part 1: Pathways. *Social Science and Medicine* 133: 191–201.

Wilson, C., Crane, L. and Chryssochoidis, G. (2015) Why do homeowners renovate energy efficiently? Contrasting perspectives and implications for policy. *Energy Research and Social Science* 7: 12–22.

Part III
Societal impacts and co-benefits

8 Exergy economics

New insights into energy consumption and economic growth

*Paul Brockway, Steve Sorrell,
Tim Foxon and Jack Miller*

Introduction

To be effective in mitigating climate change, technical and policy initiatives to reduce energy demand must have significant impacts at the aggregate level. This means they must contribute to the decoupling of primary energy consumption from economic output for both national economies and the world as a whole. However, the feasibility, difficulty and cost of decoupling are disputed.

The rate of growth of global primary energy consumption has been remarkably stable since 1850 (2.4 per cent/year ± 0.08 per cent) and shows little sign of slowing down (Jarvis et al., 2012). However, since primary energy consumption (E) has grown more slowly than gross domestic product (GDP) (Y), there has been a steady decline in global primary energy intensity (E/Y) – termed *relative* decoupling. Globally, primary energy intensity fell by ~1.3 per cent/year between 1980 and 2000, but progress has subsequently slowed to only 0.3 per cent/year. To achieve the goals of the Paris Agreement, global primary energy intensity must fall at least six times faster than this – a much faster rate than has previously been achieved (Loftus et al., 2014).

The historical decline in global energy intensity appears largely due to countries getting richer rather than from finding ways to produce particular levels of wealth with less primary energy consumption (Csereklyei et al., 2014). This in turn suggests that the technological changes that reduce energy intensity are strongly correlated with those that increase wealth – so the energy required to produce a unit of output has fallen, but the 'energy savings' have been partly offset by increased output. It is possible that there is a causal relationship between these trends (i.e. lower energy intensity leads to increased economic output), but this is difficult to establish empirically. Moreover, despite wide differences in energy intensity, very few countries have achieved *absolute* decoupling of primary energy consumption from GDP (i.e. GDP rising while energy consumption is falling) for more than short periods of time. Also, when absolute decoupling has been achieved (such as in the UK) it has partly been driven by the 'outsourcing' of energy-intensive manufacturing to other regions (Hardt et al., 2018). Such outsourcing is clearly not feasible at the global level.

This apparently strong link between energy consumption and economic activity raises important questions for both theory and policy. The orthodox view is that both increased energy consumption and improved energy efficiency provide a relatively small contribution to the growth in economic output. Consistent with this view, orthodox economists argue that technological change can reduce energy consumption with relatively little effect on economic growth. In contrast, some ecological economists claim that over the last century economic growth has largely been achieved by providing workers with increasing quantities of energy, both directly and indirectly, as embodied in capital equipment (Cleveland et al., 1984). Ecological economists therefore view energy as contributing more to economic growth than is suggested by its small share of total costs (5–10 per cent). They are correspondingly more sceptical about the feasibility of decoupling.

The success of climate policy depends in part on which of these views is correct – or more precisely, which more accurately describes the situation for different regions and stages of economic development (Foxon, 2017; Stern and Kander, 2012) But debates on this topic involve a host of theoretical and methodological issues that are both highly technical and difficult to resolve. For example, there have been several hundred studies that use sophisticated econometric techniques to explore the 'causal' relationships between GDP and energy consumption, but these have failed to reach a consistent conclusion (Kalimeris et al., 2014; Omri, 2014).

Recently, however, a new field of research has emerged which has the potential to throw new light on these long-standing questions. This approach hinges upon the thermodynamic concept of *exergy* (the portion of an energy flow that can be used to perform physical work), and the use of physical measures of energy efficiency that are based upon the second-law of thermodynamics rather than the first. The argument is that exergy is the preferred way to measure energy flows since it captures both the quantity and quality of energy, while second-law efficiency measures are preferred to first-law since they capture the distance from the theoretical maximum efficiency.

Underlying this new approach is the estimation of the *useful exergy inputs* into national economies – where useful exergy is defined as the exergy outputs of end-use conversion devices, such as the mechanical drive from an engine, the high-temperature heat from a furnace or the visible light from a lightbulb. Useful exergy, in turn, is the product of the exergy inputs to those conversion devices (which can be estimated from data on final energy consumption) and their second-law conversion efficiencies. Researchers in this field are beginning to construct consistent time series of the total useful exergy consumption of individual countries and regions (Brockway et al., 2014; Serrenho et al., 2016; Warr et al., 2010). These databases provide a measure that can be used alongside the more traditional measures of primary and final energy consumption to gain deeper insights into the role of energy in the economy.

The core claim of these researchers is that *useful exergy is a key driver of economic growth* – and that this contribution is not recognised by orthodox economics (Warr and Ayres, 2012). Increases in economic output have historically depended upon increased supplies of useful exergy, achieved through a combination of increasing use of primary energy, shifting towards higher-quality energy carriers (e.g. from oil to electricity) and improving second-law efficiencies at all stages of the energy conversion chain. Warr and Ayres (2012) go so far to suggest that improvements in second-law conversion efficiency provide a quantifiable surrogate for the majority of technical change that contributes to economic growth. Hence, *far from being a minor contributor to economic growth, the combination of increased energy inputs and improved energy efficiency becomes a key driver*. One implication of this work is that energy efficiency improvements by producers can significantly boost economic output – thereby partly or wholly offsetting the energy savings per unit of output that result from the improved efficiency. In other words, rebound effects for producers could be large.

This chapter provides an overview of this emerging field. The following section summarises the orthodox view of the relationship between energy consumption and economic growth, including the assumptions upon which this rests and the limitations of those assumptions. Next, the concept of useful exergy is is introduced to show how this may help to improve our understanding of this relationship. The following section summarises some recent research that estimates the useful exergy inputs into national economies, explores the trends in these over time, includes useful exergy within economic models and uses those models to identify the drivers of economic growth. We highlight two claims: first, that energy efficiency improvement by UK producers have provided one-quarter of UK economic growth since 1971; and second, that corresponding improvements by Chinese producers have increased global energy consumption. The chapter concludes with future directions for this line of research.

The role of energy in the economy

In the models used by orthodox economists, firms combine primary inputs (capital and labour) and intermediate inputs (energy and materials) to produce goods and services. Primary inputs facilitate production but do not form part of the product and are not used up during production (although capital may depreciate). In contrast, intermediate inputs are 'created' by production and are either embodied in products or used up during production. Subtracting the purchases of intermediates from the value of output leads to a measure of value added, which is the income received by capital and labour.

Orthodox models attribute increases in economic output to increases in primary and intermediate inputs and improvements in total factor productivity (TFP) – where the latter is the portion of growth not explained by increases in inputs (OECD, 2001; Solow, 1956). Increases in value added (including, at the aggregate level, GDP) are attributed to increases in primary inputs and TFP – with the latter accounting for a significant proportion of the total. TFP can be estimated as a residual in growth accounting studies or as a parameter in econometric studies, but it has traditionally been treated as exogenous and equivalent to technical change.[1] However, more recent models make technical change endogenous and simulate the positive externalities from education and research and development (R&D) (Aghion et al., 1998; Romer, 1994). These models also attribute a portion of economic growth to improvements in the quality of capital and labour inputs – such as better-educated workers.

Central to orthodox economics is the specification of *production functions* for firms, sectors or the economy as a whole, indicating the maximum output that can be produced from different quantities of primary and intermediate inputs (OECD, 2001). Production functions can either be specified for gross output and include all inputs or specified for value added and only include primary inputs. Specifications typically include a TFP multiplier that 'shifts' the production function over time, thereby capturing technical and other changes that allow more output to be produced from the same quantity of inputs. Production functions can be defined for different levels of aggregation using different functional forms and with differing rates of productivity improvement for each input. But it is generally assumed that production exhibits constant returns to scale, that markets are competitive, that firms maximise profits and that inputs can be substituted for one another following a change in relative prices. Using these assumptions, it can be shown that the rate of growth of output over time is a weighted average of the rate of growth of each input and the rate of growth of TFP. The weight on each input is the 'partial output elasticity' for that input, or the percentage change in output following a percentage change in that input, holding other variables constant. With these assumptions, it can be shown that the partial output elasticity is equal to the share of that input in total costs. This result has been labelled the *cost share theorem* (Kümmel et al., 2010).

The cost share theorem, together with the assumption of input substitutability, has important implications for the role of energy in economic production. Since energy represents a small share of total costs for most producers (<5 per cent), the theory implies that increases in energy inputs and improvements in the productivity of those inputs should make only a minor contribution to economic growth. Similarly, constraints on energy supplies are unlikely to have a major impact on economic growth since it should be possible to substitute away from energy. Taken together, these assumptions imply that energy consumption can be substantially decoupled from economic output.

This approach has been criticised by ecological economists, who challenge the core assumption (derived from the national accounts) that capital and labour should be treated as primary inputs, and that energy and materials should be treated as secondary inputs that make no contribution to value added. This makes little sense from a physical perspective, since all physical, biological and economic activity depends upon flows of high-quality energy that are then returned to the environment in the form of low-temperature heat. Like the biosphere, the economy is driven by solar energy, both directly and as embodied in biomass and fossil fuels. Labour and capital are not productive on their own – they only add value by harnessing the 'free' energy flows provided by nature. The productivity of capital and labour therefore depends entirely upon the associated energy flows.

Linked to this, ecological economists question the treatment of energy as a 'produced' input that can be substituted by capital and labour. Technically, the scope for substitution will be constrained at the level of the economy as a whole since producing more capital requires more of the thing that it is substituting for (Stern, 1997). In addition, many production functions violate the laws of thermodynamics, since they allow output to be produced with little or no energy. More realistic constraints on the relative magnitude of different inputs could mean that economies are less flexible in adjusting to rising energy prices than is traditionally assumed (Berndt and Wood, 1979; Lindenberger and Kümmel, 2011). Such constraints may undermine the cost share theorem, meaning that the dependence of capital and labour on energy flows could magnify the economic impact of changes in those flows (Giraud and Kahraman, 2014).

The common treatment of energy as an undifferentiated input is also problematic. Energy carriers differ in quality on multiple dimensions, including their flexibility of use, amenability to storage, energy density, economic productivity and capacity to do work (Cleveland et al., 2000). Since high-quality energy carriers are more productive, they should be given more weight in aggregate measures of energy consumption. When this is done, aggregate energy intensity is found to be declining more slowly than is commonly assumed (Cleveland et al., 2000; Berndt, 1978). Studies that neglect changes in energy quality may therefore overlook an important contributor to economic growth (Gentvilaite et al., 2015; Stern, 2010).

In contrast to the neglect of energy by orthodox economists, economic historians attribute a central role to energy in explaining previous long-term surges in economic growth – and in particular the nineteenth-century industrial revolution (Allen, 2009; Kander et al., 2014; Pomeranz, 2009; Wrigley, 2013). The continuing importance of energy is also suggested by the large impact of energy price shocks on economic output (Kilian, 2008), and by the limited decoupling that has been achieved to date. For example, Csereklyei et al. (2014) analysed 99 countries over the period 1971–2010 and found that, on average, every 1 per cent increase in per capita wealth was associated with a 0.7 per cent increase in per capita energy consumption. But it is not clear whether this strong correlation is due to economic growth causing increased energy consumption (the orthodox view), increased energy consumption causing economic growth (the ecological view), or a combination of the two. While it is possible to test this econometrically, the results are ambiguous and sensitive to the method, data and specification employed (Kalimeris et al., 2014; Omri, 2014).

These various strands of theory and evidence raise concerns about the validity of orthodox models, the accuracy of the cost share theorem and the feasibility of absolute decoupling. If economic growth depends upon increased energy consumption, it may be difficult to reduce global carbon emissions while at the same time increasing global GDP. However, the studies arguing for the importance of energy are limited in number, variable in quality and inconsistent in approach – and have largely been ignored by mainstream economists. The approach described below – termed 'exergy economics' – represents an attempt

138 *Paul Brockway et al.*

to build a bridge between these two communities. The distinctive features of this approach are: the use of exergy as a thermodynamic measure of energy quality; the focus on the 'useful' stage of the energy conversion chain, rather than the primary or final stages; and the willingness to challenge key assumptions of orthodox economics, such as the cost share theorem. These are summarised next.

Foundations of exergy economics

The concept of useful exergy

Exergy is a measure of the portion of an energy flow that can be used to perform work, i.e. the portion that is 'available' or 'useful'. As with energy, exergy is measured in joules, takes a variety of forms (e.g. kinetic, electrical or chemical) and can be converted from one form to another. But while energy is solely a measure of quantity, exergy is a measure of both quantity and quality – where quality is defined as the capacity to perform work. So, for example, 1 kWh of electricity has the same energy as 5 kg of water at 20°C, but the electricity has more exergy (i.e. is of higher quality) owing to its greater potential to be converted to physical work. The biggest difference between energy and exergy arises when considering thermal energy (heat). For any given quantity of heat within a particular environment, a temperature-dependent portion constitutes low-grade heat that has little or no ability to perform work. This represents a portion of an exergy flow that has been dissipated, or 'destroyed'.

The concept of exergy derives from the second-law of thermodynamics, which (in one form) states that every energy conversion process involves the loss of some measure of energy quality – which means that some exergy is necessarily destroyed. Energy, on the other hand, is always conserved and cannot be destroyed, as per the first-law of thermodynamics.

The term 'exergy' was first introduced by Zoran Rant (1956), although the principles on which it is based date back to the nineteenth century (Anderson, 1887). Exergy may be formally defined as the maximum physical work that can be extracted from a system as it reversibly comes into equilibrium with its environment. The exergy of a system depends upon the *differences* between that system and its environment, which may be in terms of kinetic energy, potential energy, temperature, pressure or chemical potential (Baierlein, 2001; Romero and Linares, 2014). Exergy can be defined for materials as well as for energy flows,[2] and has been proposed as a global sustainability indicator (Romero and Linares, 2014) and a universal measure of resource availability (Valero and Valero, 2011).

Here we focus upon the use of exergy as an alternative measure of the chain of energy flows within a national economy. At the top of this chain is *primary exergy*, derived from fossil fuels, nuclear fission and renewables. The exergy of fossil fuels differs from their energy content by a so-called 'exergy factor' (Szargut *et al.*, 1987), while the exergy of nuclear and renewables depends upon how they are measured.[3] Sources of primary exergy are processed and converted into commercial energy carriers (*secondary exergy*) such as electricity, gasoline and diesel which are ultimately delivered to consumers (*final exergy*) with some losses along the way (e.g. resistive losses in electricity grids). The last stage of exergy conversion takes place within end-use devices such as engines, boilers, furnaces, motors and light bulbs which convert final exergy into *useful exergy*, such as low- and high-temperature heat, mechanical power and electromagnetic radiation.

Following end-use conversion, useful exergy is preserved or trapped within *passive systems* for a period of time to produce *energy services* (Cullen and Allwood, 2010; Cullen et al., 2011). So, for example, the heat delivered from a boiler (conversion device) is held within a building (passive system) for a period of time to provide thermal comfort (energy service). Unlike useful exergy, energy services cannot be measured in common units and hence cannot be aggregated. Useful exergy is eventually dissipated as low-temperature heat, but improvements in second-law conversion efficiency (e.g. more efficient boilers) or the ability of passive systems to trap exergy (e.g. more insulation) will allow more energy services to be provided per unit of useful exergy.

Useful exergy and economic production

The main reason for thinking about useful exergy in this context is that it provides a more relevant measure of the contribution of energy to economic production and human welfare. The exergy that is destroyed at each stage of the conversion chain contributes no economic value and no energy services, but the useful exergy at the final end of the chain contributes to the production of marketable goods (e.g. the heat used to manufacture steel) and to the energy services required by consumers (e.g. the light energy used for illumination). Improvements in second-law conversion efficiency at all stages of the energy chain allow more useful exergy to be delivered from the same amount of primary energy – and is this, rather than the total energy outputs of conversion devices, that has economic value. Hence, it is the productive part of energy flows – useful exergy – that should be the focus of attention within economic models.

In a series of papers, Ayres and Warr estimated the useful exergy flows within national economies over the past century and included these within simple models of economic growth (Ayres and Warr, 2010; Ayres et al., 2003; Warr and Ayres, 2012; Warr et al., 2008). Their results suggested that the output elasticity of useful exergy was at least ten times greater than the cost share of energy, and much larger than the output elasticity of labour. Moreover, when energy was replaced with useful exergy, the estimated contribution of TFP to economic growth largely disappeared – at least for the period prior to 1970. Their explanation for these results was that energy (cf. exergy) is more productive than orthodox economists assume, and that the productivity of primary exergy has increased over time owing to continuing improvements in second-law conversion efficiency.

140 *Paul Brockway* et al.

The studies by Ayres and Warr embody a number of innovations, including: estimating aggregate time series of useful exergy; using this data within economic models (despite the fact that useful exergy is not a traded commodity); and employing an unconventional 'linear exponential' (LINEX) production function. The latter has not been accepted by orthodox economists because it violates some of the standard assumptions of neoclassical production theory.[4] However, the theoretical arguments in favour of employing useful exergy are persuasive and the concept is not wholly unfamiliar to economists since it amounts to 'quality-weighting' energy inputs.

Comparable quality-weighting of labour and capital inputs is now an standard feature of orthodox growth accounting (OECD, 2001). As an illustration, Figure 8.1 shows time series of both standard and quality-weighted capital, labour and energy inputs for Portugal over the period 1960–2010, where the latter is defined as useful exergy. Since quality-weighted inputs grow faster than the standard input measures, they can explain a larger proportion of economic growth, leaving less to be attributed to a residual (TFP). The next section summarises the contribution of researchers who building upon Ayres and Warr's work.

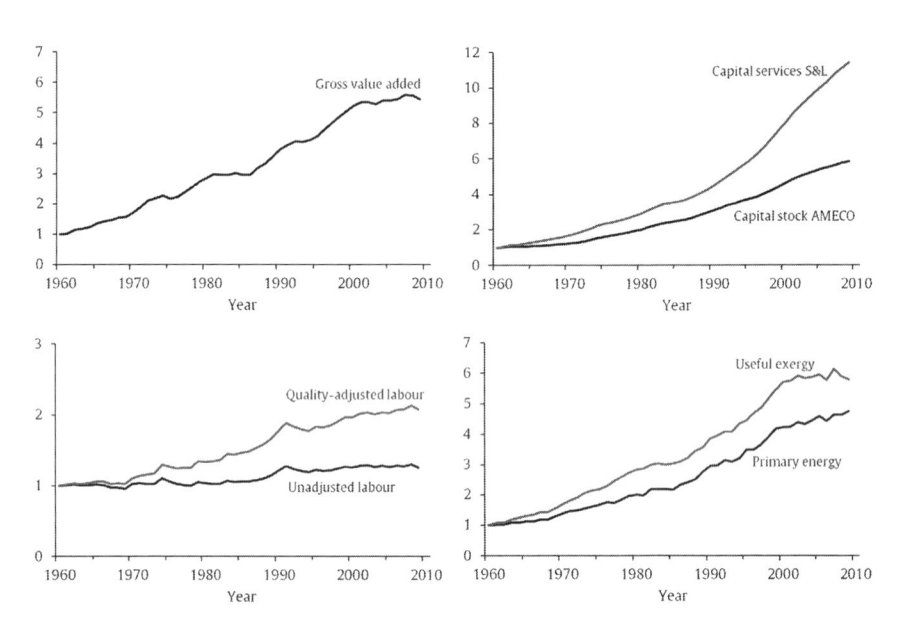

Figure 8.1 Normalised time series of inputs and outputs to production in Portugal.

Source: Santos *et al.* (2018), with permission.

Note
Capital inputs measured as a stock of assets and a flow of services. Labour inputs measured as total hours worked and total hours worked adjusted with a human capital index. Energy inputs measured as primary energy supply and useful exergy consumed.

Research findings and insights

Exergy accounting

A first step in understanding the contribution of useful exergy to economic growth is to map the exergy flows through a national economy. An estimate of these flows for a single year allows the locations and magnitudes of exergy losses to be identified, the relative efficiencies of different processes to be compared and the technical (but not necessarily economic) potential for improvements to be highlighted. Similarly, an estimate for several years allows the trends in exergy use and efficiencies to be identified, while estimates for several countries allow their relative efficiencies to be compared. Estimating useful exergy involves: (a) collecting data on primary and final energy consumption and converting these to an exergy basis; (b) mapping final exergy flows onto different

Table 8.1 Breakdown of end-uses, by useful exergy category and energy carrier group

Energy carrier group	Category of useful exergy	End-use
Combustible fuels	Heating	High-temp. industrial heat
		Med-temp. industrial heat
		Water heating
		Space heating/cooking
	Mechanical drive	Gasoline road vehicles
		Diesel automobiles
		Diesel goods vehicles
		Aviation
		Diesel marine transport
		Diesel rail
		Coal rail
		Industrial static motors
	Lighting	Non-electric lighting
Electricity	Heating	High-temp. industrial heat
		Med-temp./industrial heat
		Water heating
		Space heating
		'Wet' appliances (e.g. dishwashing)
		Cooking
	Cooling	Space cooling (AC)
		Refrigeration
	Mechanical drive	Electric motors
		Electric rail
		Electric road vehicles
	Lighting	Electric lighting
	Electronics	Consumer electronics
		Computing
Food and feed	Muscle work	Draught animal work
		Human work

Source: Miller *et al.* (2016), with permission (and without changes).

categories of useful exergy (such as mechanical power and heat) and different *end-uses* within each category (e.g. cars trucks); and (c) estimating average exergy efficiencies for those end-uses (see Table 8.1).

The first national exergy accounts were compiled by Reistad (1975), whose study coincided with rising oil prices and increasing concern about energy efficiency. Interest declined following the oil price collapse of 1981, but was reinvigorated after 2000 by Ayres and Warr (2005; Warr and Ayres, 2012). Other authors have since refined the methodology, both in terms of the level of disaggregation and the accuracy of efficiency estimates (Brockway et al., 2014; Miller et al., 2016; Serrenho, 2013). There are a growing number of single and multicountry studies within the OECD (Ayres et al., 2003; Brockway et al., 2014; Hammond and Stapleton, 2001; Serrenho, 2013; Serrenho et al., 2016; Warr et al., 2010; Williams et al., 2008) and the methodology has now being extended to Mexico (Guevara et al., 2016), India (Magerl, 2017) and China (Brockway et al., 2015). Lack of data remains a serious obstacle, but there have been significant improvements over the last few years with increasing efforts towards standardisation (Sousa et al., 2017).

Exergy efficiency

A key outcome of useful exergy accounts are time series estimates of primary to useful exergy efficiency – that is, the ratio of useful to primary exergy consumption ($\varepsilon_{PU} = B_U/B_P$) – as well as final to useful exergy efficiency ($\varepsilon_{FU} = B_U/B_F$). These estimates reflect both improvements in conversion efficiency and structural change within the economy. As an example, Serrenho et al. (2016) estimate that ε_{FU} in Portugal increased from 6 per cent in 1850 to 20 per cent in 2010, with most of this improvement occurring during the post-war period of electrification and industrialisation. Similarly, Brockway et al. (2014) estimate that ε_{PU} in the UK increased from 9 per cent to 15 per cent between 1960 and 2000, but has since remained stable (Figure 8.2). Closer examination reveals efficiency improvements in all end-use categories in the UK, with primary to useful heating efficiency increasing from 8 to 12 per cent, electricity efficiency from 8 to 14 per cent and mechanical drive from 11 to 21 per cent.

While ε_{PU} in Portugal and the UK stabilised only recently, it has remained around 11 per cent in the US for half a century (Figure 8.2). The tendency for the rate of improvement in exergy efficiency to slow in advanced economies was first observed by Williams et al. (2008), who termed it 'efficiency dilution'. The reason is the increasing proportion of exergy being used in less efficient end-uses (e.g. air-conditioning, car travel, space heating), combined with a slowdown in the rate of efficiency improvement for individual end-uses. Most advanced economies are 'outsourcing' heavy industry to emerging economies and since heavy industry is relatively exergy efficient (although exergy intensive), this reduces the aggregate exergy efficiency of those economies. In contrast, the exergy efficiency of emerging economies is improving rapidly (Figure 8.2).

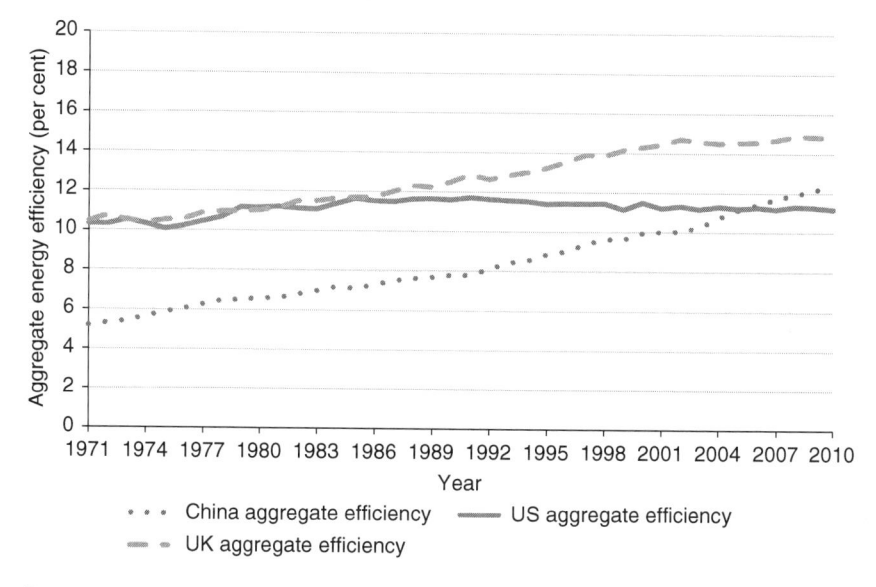

Figure 8.2 Trends in primary to useful exergy efficiency in the UK, US and China (1971–2010).

Source: Brockway et al. (2014, 2015), with permission (two separate figures merged).

Improvements in exergy efficiency mean that the useful exergy inputs to national economies are growing faster than either primary or final exergy inputs (Figure 8.3). For example, the 10-fold growth in useful exergy use in China between 1971 and 2010 was supplied by a 4-fold increase in primary exergy combined with a 2.5-fold improvement in primary to useful exergy efficiency.

Table 8.2 decomposes the trends in useful exergy consumption in the US, UK and China to identify the relative contribution of increases in primary exergy, changes in the relative importance of different end-uses (structure) and changes in the efficiency of those end-uses. If emerging economies follow the same pattern as the Organisation for Economic Co-operation and Development

Table 8.2 Decomposing the drivers of useful exergy consumption in the UK, US and China over the period 1971–2010

Country	Primary exergy (D_B)	Structure (D_S)	Efficiency (D_ε)	Useful Exergy $(D_T = D_B D_S D_\varepsilon)$
China	3.96	1.66	1.48	9.76
US	1.32	0.90	1.29	1.53
UK	1.01	0.90	1.58	1.43

Source: Brockway et al. (2015), with permission (and without changes).

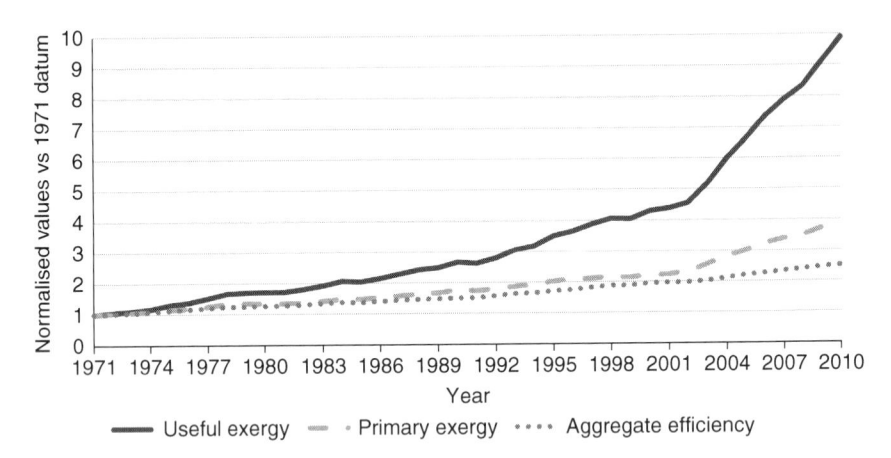

Figure 8.3 Normalised trends in primary exergy, useful exergy and primary to useful exergy efficiency in China between 1971 and 2010.

Source: Brockway *et al.* (2015), with permission (and without changes).

(OECD), their rate of efficiency improvements will decline in the future. One possible implication is that orthodox economic models may underestimate the future growth of energy consumption in those countries (Brockway et al., 2015).

Economic insights

Exergy intensity of national economies

Once estimates of primary, final and useful exergy are available, their relationship to GDP (Y) can be examined. The primary exergy intensity of a national economy ($\theta_P = B_P/Y$) can be decomposed as:

$$\theta_P = \frac{B_P}{Y} = \frac{B_P}{B_F} \frac{B_F}{B_U} \frac{B_U}{Y} \quad (1)$$

Or:

$$\theta_P = \varepsilon_{PF} \varepsilon_{FU} \theta_U \quad (2)$$

Hence, reductions in primary exergy intensity may result from either improvement in primary to useful exergy efficiency ($\varepsilon_{PU} = \varepsilon_{PF}\varepsilon_{FU}$) or reductions in useful exergy intensity (θ_U). As an illustration, Figure 8.4 illustrates the long-term trends in primary exergy and useful exergy intensity in five countries (Serrenho et al., 2016). This period includes transitions from agricultural to industrial societies (which occurred somewhat later in Portugal), together with two world wars. Although the picture is complex and there are important differences

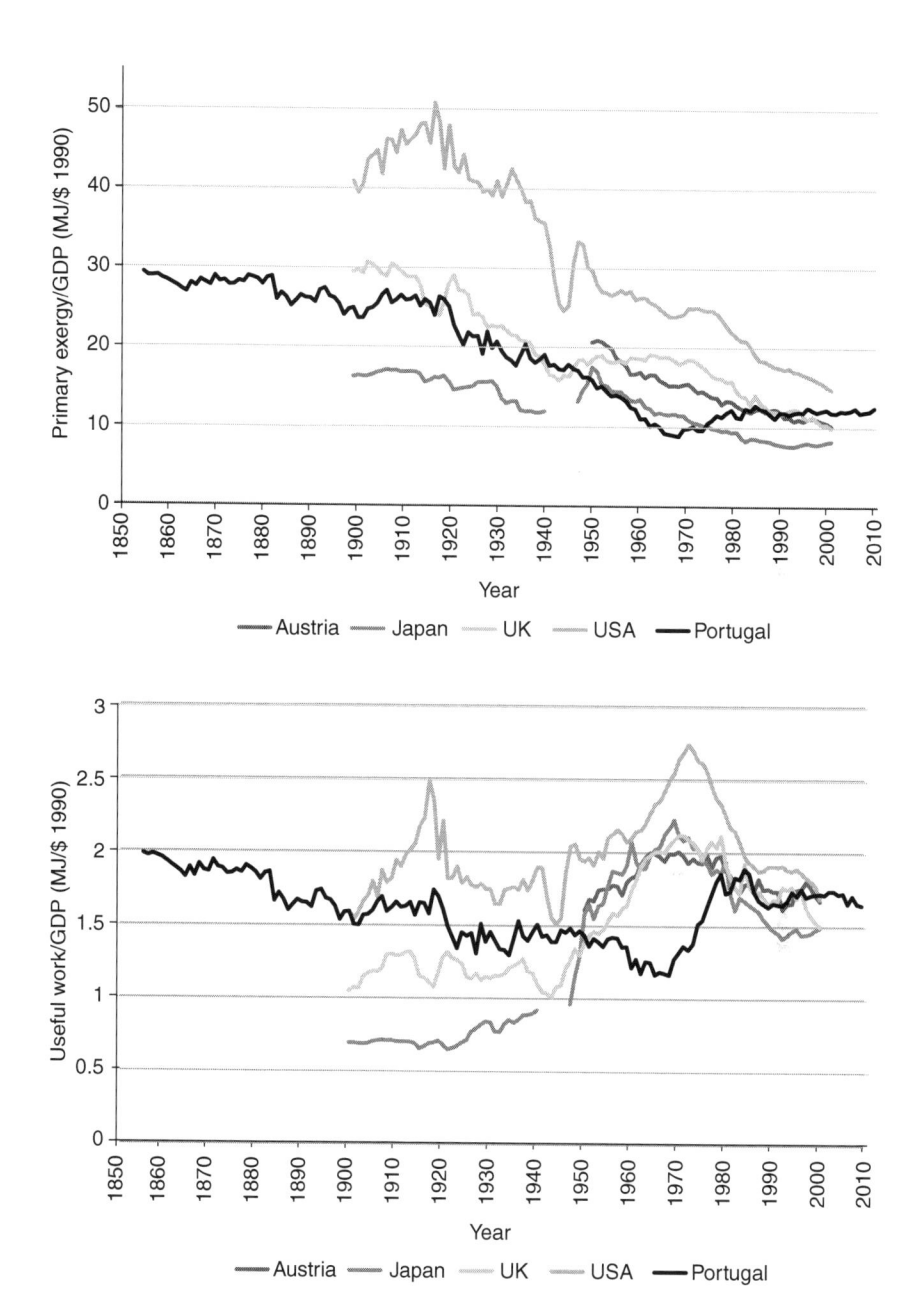

Figure 8.4 Primary exergy (top) and useful exergy (bottom) intensity of Portugal, UK, US, Austria and Japan, 1850–2010.

Source: Serrenho *et al.* (2016), with permission.

between countries, one notable feature stands out: primary exergy intensity has declined over time while useful exergy intensity has remained relatively stable.[5] This implies that most of the reduction in primary exergy intensity has derived from improvements in primary to useful exergy efficiency. While all five countries have experienced a relative decoupling of primary exergy from GDP there

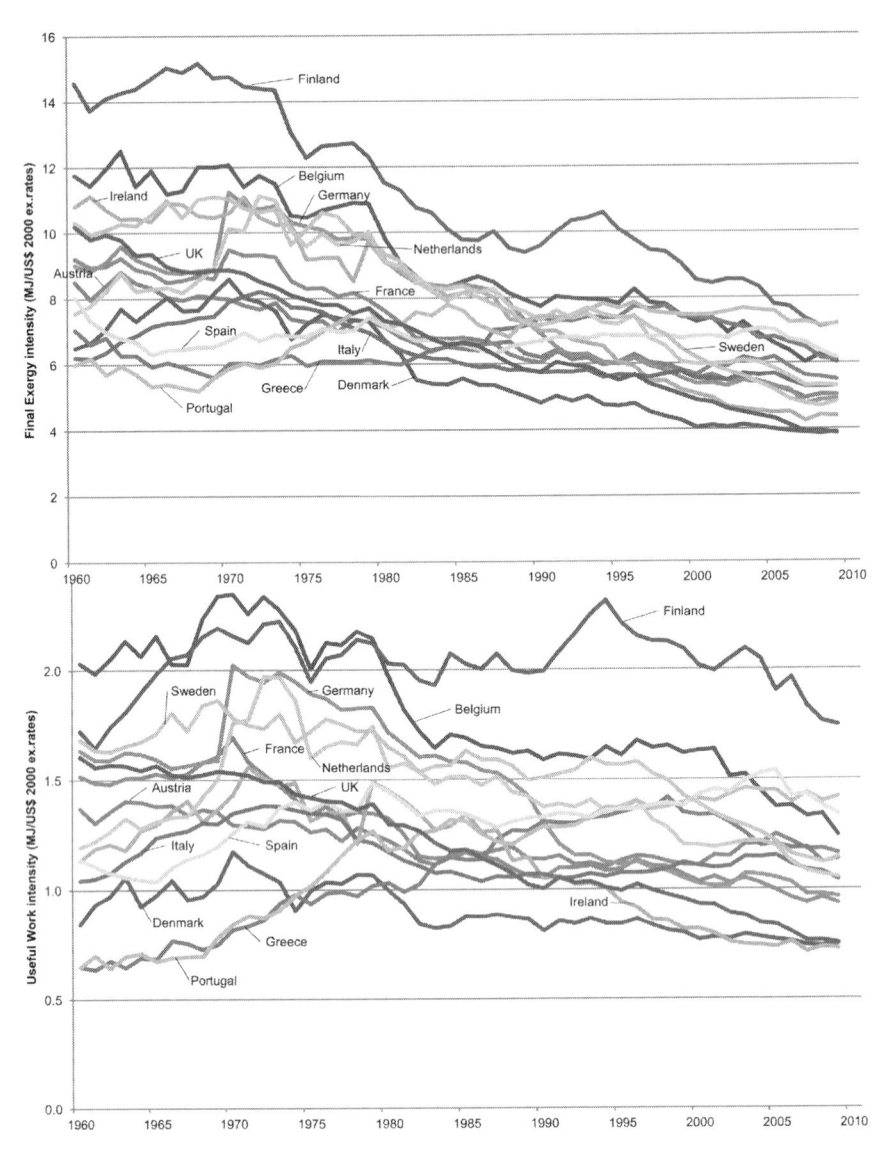

Figure 8.5 Final exergy (top) and useful exergy (bottom) intensities in the EU-15, 1960 to 2010.

Source: Serrenho *et al.* (2014), with permission.

has been little or no decoupling of useful exergy. Indeed, the useful exergy intensity of the modern Portuguese economy is comparable to what it was in 1860.

Figure 8.5 provides more recent estimates of final and useful exergy intensities for the EU 15. Again, the picture is complex, but there is little evidence of a long-term trend towards lower useful exergy intensity. Serrenho et al. (2014) show that, once differences in residential energy use (linked to average temperatures) and high-temperature heat (linked to heavy industry) are taken into account, useful exergy is statistically constant across time and equal for the EU 15. Taken together, this evidence is *suggestive* of a strong link between useful exergy and economic output. However, the trend is not universal (e.g. useful exergy intensity is rising in Mexico (Guevara et al., 2016) and falling in China (Brockway et al., 2015)) and closer investigation is required to understand the underlying determinants.

Useful exergy as a factor of production

A second insight is gained by replacing primary energy with useful exergy within models of economic production. As noted, this was first done by Ayres and Warr (2005) who employed a LINEX production function and were able to explain US economic growth over the period 1900–1973 without the need for a TFP multiplier.[6] However, there seems little prospect of this function being accepted by mainstream economists, so researchers are seeking more conventional ways to include useful exergy.

One approach is to estimate standard, three-input (capital, labour and energy – *KLE*) 'constant elasticity of substitution' (CES) production functions, with primary energy (*E*) being replaced by useful exergy ($B_U = \varepsilon_{PU} B_P$).[7] CES functions form the foundation of many macroeconomic models, but (remarkably) the parameter values tend to be assumed rather than estimated (Sorrell, 2014). Traditional approaches to estimating substitution elasticities are problematic since they rely upon the cost share theorem, but estimating the CES function requires non-linear techniques that can be unreliable (Henningsen and Henningsen, 2012; Prywes, 1986). Further, it is necessary to impose assumptions about the 'separability' of inputs that typically lack empirical justification (Sorrell, 2014). Heun et al. (2017) estimate aggregate CES functions for the UK and Portugal using data on useful exergy and other inputs over the period 1960–2009. They find that the partial output elasticity of each input varies over time and differs from the cost share – thereby questioning the validity of the cost share theorem. However, they also find the results are sensitive to the specification used and the estimated output elasticities change rapidly over short periods of time.

Santos et al. (2018) take an alternative approach, using 'co-integration' techniques to test for the existence of an aggregate production function[8] for Portugal over the period 1960 to 2009.[9] If time series of labour, capital, energy (or useful exergy) and GDP are found to be co-integrated, this suggests there is a stable, long-term relationship between them (Dickey et al., 1994). This relationship may in turn be interpreted as an aggregate production function, and the estimated parameter values can be used to derive the partial output elasticities. Santos et al. test 32 different specifications that vary in terms of whether a TFP multiplier is included, whether capital and labour inputs are quality-weighted, whether energy inputs are included, and whether these are measured as primary energy or useful exergy.

To interpret the results as an aggregate production function, the variables must be co-integrated, the parameters must be non-negative and there must be evidence that the inputs 'cause'[10] the output. Santos et al. find that none of the specifications incorporating a TFP multiplier meet these criteria – suggesting that standard formulations are incorrect. Instead, the only specification that meets these criteria is when both capital and labour are quality-adjusted and useful exergy is included – again suggesting that standard formulations are incorrect. Moreover, this specification includes two co-integrating relationships, one of which is interpreted as a constraint on input combinations as a consequence of the essential contribution of useful exergy to economic production. Overall, this rigorous study provides strong support for the inclusion of useful exergy as a factor of production.

Rebound effects from improved exergy efficiency

Improvements in exergy efficiency make useful exergy cheaper, thereby boosting economic output and encouraging the substitution of useful exergy for capital and labour. This in turn reduces the exergy savings from those improvements – a form of rebound effect.

There is a large and growing literature on rebound effects, but most studies focus upon energy efficiency improvements by consumers rather than producers since these are easier to estimate. This is unfortunate, since rebound effects for producers may potentially be larger (Saunders, 2013; Sorrell, 2007). The development of economic models incorporating useful exergy opens up a new route for investigating such effects. Following Heun et al. (2017), Brockway et al. (2017) estimate aggregate three-input CES production functions for the US, UK and China over the period 1980–2010, replacing primary energy with useful exergy. Following Saunders (2008), Brockway et al. (2017) derive an expression for the rebound effect that relies upon the cost share theorem.[11] This leads to a mean estimate of the rebound effect 13 per cent for the US and UK, and 208 per cent for China. Or in other words, Brockway et al.'s (2017) results suggest improved exergy efficiency leads to significant exergy savings in the US and UK, but *increased* exergy consumption in China.

The confidence intervals on Brockway et al.'s (2017) estimates are large, and the method has other limitations such as the arbitrary choice of nesting structure and the continued reliance upon the cost share theorem. However, the use of an exogenous, thermodynamic measure of energy efficiency (ε_{PU}) represents an important step forward – and points the way to further work in this area.

Including useful exergy within energy–economy models

The next stage is to incorporate useful exergy into whole-systems models of the economy and to use these to develop projections of future economic growth and energy consumption. Such models can overcome some of the limitations of aggregate and sectoral production functions and can better capture the dynamic feedbacks that drive economic growth (Ayres and van den Bergh, 2005). Widely used computable general equilibrium (CGE) models are insufficient for this purpose, since they embody many of the problematic assumptions of orthodox theory and assume rather than estimate many of the parameters (Sorrell, 2014). A more promising approach is to employ macroeconometric models, consisting of a group of simultaneous equations that represent key macroeconomic relationships. These include identities (such as GDP being equal to the sum of public and private consumption, investment and exports) and behavioural relationships estimated from historical data (Fair, 1984).

The MAcroeconomic Resource COnsumption model (MARCO-UK) is the first attempt to integrate useful exergy within such a model and currently consists of 30 identities and 27 behavioural equations estimated from UK data over the period 1971–2013 (Sakai et al., 2018). Both useful exergy (B_U) and final to useful exergy efficiency (ε_{FU}) are endogenous variables, with the former being estimated from its lagged value, GDP and quality-adjusted labour and capital. Useful exergy is also an explanatory variable for other variables, including consumption, investment, exports, labour supply and final energy consumption. Capital and exergy are specified as complementary, with one being required to activate the other. Constructing the model in this way captures both the drivers of improvements in ε_{FU} and the contribution of those improvements to output growth – which occur through both the consumption and production side of the economy. For example, lower-priced useful exergy improves productivity and stimulates increased production through additional capital investment.

MARCO-UK allows the development of counterfactual scenarios in which the values of key variables are held at their base year values – thereby allowing the contribution of those variables to economic growth to be estimated. Initial results (Figure 8.6) suggest that improvements in final to useful exergy efficiency have contributed *one-quarter* of UK economic growth since 1971 – comparable in scale to that contributed by capital investment. In contrast, increases in labour inputs are estimated to have contributed only 10 per cent of the observed growth. Put another way, the results suggest that improved energy efficiency has played a far more important role in driving UK economic growth than is traditionally assumed.

These results are provisional and require further analysis and development. But the MARCO-UK framework is flexible and can be extended in a variety of ways, including to other countries.

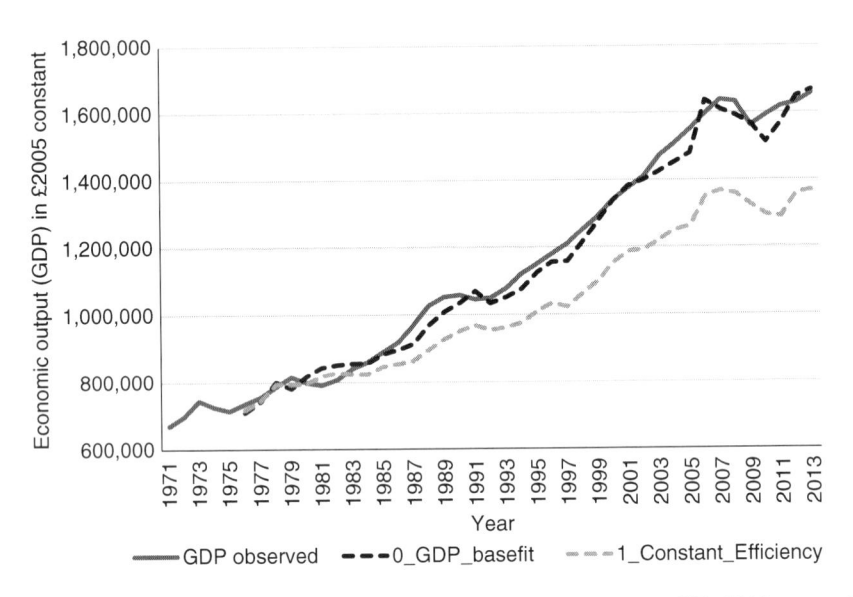

Figure 8.6 Counterfactual simulations of UK economic output 1971–2013 using the MARCO-UK model.

Source: the authors.

Conclusions, future directions and policy recommendations

Orthodox economics provides the lens through which most researchers and policymakers view the world, but that lens may obscure or distort some important features – such as the critical role of energy in economic production. The neglect of energy derives from the foundational assumptions of orthodox economics and could lead to misleading policy recommendations.

This neglect of energy has long been criticised, but alternative approaches lack a coherent theoretical framework and methodological approach. The developments described in this chapter could provide a more robust alternative, based around the concept of useful exergy. As illustrated above, a number of research topics are currently being pursued and the initial results suggest that improved exergy efficiency is a key driver of economic growth. However, the research is at an early stage and is handicapped by lack of data, and the unfamiliarity of the exergy concept inhibits wider acceptance.

Nevertheless, research is progressing in a number of directions, including: disaggregating the MARCO-UK model to the sector level; extending the co-integration approach to other countries and functions (Santos et al., 2018); extending the exergy framework to incorporate passive systems and energy services (Cullen et al., 2011); investigating the relationships between useful exergy, energy services and human needs (Brand-Correa and Steinberger, 2017); and incorporating useful exergy with input–output frameworks (Heun et al., 2018).

Taken together, these have the potential to significantly improve our understanding of the relationship between energy use and economic activity, and thereby the feasibility of absolute decoupling.

The policy implications of this work are mixed: if improved energy efficiency is essential to economic growth it should be given much greater policy support; but if rebound effects are large the environmental benefits of those improvements may be less than anticipated. But such conclusions are likely premature: what is more important – given the imperative of accelerated energy-GDP decoupling – is the willingness to question established assumptions, and to explore alternative ways of understanding the role of energy in the economy. Exergy economics is a step in that direction.

Acknowledgements

This research was funded by the United Kingdom's Engineering and Physical Sciences Research Council (EPSRC) through grants to the Centre on Innovation and Energy Demand (EPSRC award EP/K011790/1) and the Centre for Industrial Energy, Materials and Products (EPSRC award EP/N022645/1). In addition, Paul Brockway's time was funded as part of the research programme of the UK Energy Research Centre (UKERC), supported by the UK Research Councils under EPSRC award EP/L024756/1.

Notes

1 However, when TFP is estimated as a residual it also reflects measurement error and other factors such as omitted variables.
2 The exergy content of a quantity of materials is the amount of exergy required to produce the material from a reference environment by reversible processes.
3 The exergy and energy content of electricity are identical. But there are different ways of accounting for the primary energy/exergy content of nuclear and renewable electricity sources and no consensus on the preferred approach (Johansson et al., 2012).
4 Such as the requirement that the marginal productivity of an input should decline when the use of that input increases (Saunders, 2008).
5 A closer examination reveals an increase in useful exergy intensity after the Second World War in all countries except Portugal, followed by a modest decline after 1970. The first period coincides with the 'golden years' of post-war economic growth, while the second begins around the time of the first oil crisis. But taking the period as a whole, Serrenho et al. (2016) find no evidence of a time trend at the 5 per cent significance level.
6 The LINEX function was first introduced by Kümmel (1982) and imposes additional constraints on the allowed combinations of inputs.
7 A (KL)E CES function is specified as: $Y = \varphi e^{\lambda t} \left[\delta_1 \left[\delta K_t^{-\rho} + (1-\delta) L_t^{-\rho_1} \right]^{\rho/\rho_1} + (1-\delta_1) E_t \right]^{\frac{-1}{\rho}}$, where ρ and ρ_1 define the ease of substitution between inputs, δ and δ_1 define the contribution of each input to economic output, λ defines the rate of productivity growth and φ is a scaling factor. Estimation involves obtaining values for these six parameters.

8 Santos et al. (2018) assume a simpler 'Cobb Douglas' production function of the form: $Y = \varphi e^{\lambda t} K_t^{a_K} K_t^{a_E} E_t^{a_E}$. The α_i terms define the partial output elasticity of each input and λ defines the rate of productivity growth. The Cobb Douglas was chosen because it can be straightforwardly related to the co-integration specification, but it is restrictive because it assumes a unitary elasticity of substitution between each variable.
9 With time series data it is common for one or more of the variables to be non-stationary, creating the risk of 'spurious regressions'. But it is possible for two or more non-stationary variables to be co-integrated, meaning that certain linear combinations of these variables are stationary and that there is a stable long-run relationship between them.
10 This relies upon Granger causality tests. A time series (x_t) is said to 'Granger cause' another time series (y_t) if the prediction of y is improved by the inclusion of past values of x in addition to past values of y. Granger causality tests are designed to show whether one variable can meaningfully be described as dependent variable and the other as independent, or whether the relationship is bidirectional, or whether no relationship exists (Stern, 2011). This is test of 'statistical precedence' rather than causality as normally understood, since the fact that A precedes B need not necessarily mean that A causes B. For example, a meteorological forecast of rain can be shown to Granger cause rain!
11 Brockway et al. (2017) define energy rebound as: $R = 1 + \eta_{\varepsilon_{PU}}(B_P)$, where $\eta_{\varepsilon_{PU}}(B_P)$ is the elasticity of primary exergy consumption with respect to primary to useful exergy efficiency. They use the implicit function and cost share theorems to derive an expression for $\eta_{\varepsilon_{PU}}(B_P)$.

References

Aghion, P., Howitt, P., Howitt, P.W., Brant-Collett, M. and García-Peñalosa, C. (1998) *Endogenous Growth Theory*. MIT Press, MA, USA.

Allen, R.C. (2009) *The British Industrial Revolution in Global Perspective*. Cambridge University Press, Cambridge, UK.

Anderson, W. (1887) *On the Conversion of Heat into Work: A Practical Handbook on Heat-Engines*. Whittaker & Co, London, UK.

Ayres, R.U. and van den Bergh, J.C.J.M. (2005) A theory of economic growth with material/energy resources and dematerialization: Interaction of three growth mechanisms. *Ecological Economics* 55(1): 96–118.

Ayres, R.U. and Warr, B. (2005) Accounting for growth: The role of physical work. *Structural Change and Economic Dynamics* 16(2): 181–209.

Ayres, R.U. and Warr, B. (2010) *The Economic Growth Engine: How Energy and Work Drive Material Prosperity*. Edward Elgar Publishing, Cheltenham, UK.

Ayres, R.U., Ayres, L.W. and Warr, B. (2003) Exergy, power and work in the US economy, 1900–1998. *Energy* 28(3): 219–273.

Baierlein, R. (2001) The elusive chemical potential. *American Journal of Physics* 69: 423–434.

Berndt, E.R. (1978) Aggregate energy efficiency and productivity measurement. *Annual Review of Energy* 3: 225–273.

Berndt, E.R. and Wood, D.O. (1979) Engineering and econometric interpretations of energy-capital complementarity. *The American Economic Review* 69(3): 342–354.

Brand-Correa, L.I. and Steinberger, J.K. (2017) A framework for decoupling human need satisfaction from energy use. *Ecological Economics* 141: 43–52.

Brockway, P.E., Barrett, J.R., Foxon, T.J. and Steinberger, J.K. (2014) Divergence of trends in US and UK aggregate exergy efficiencies 1960–2010. *Environmental Science and Technology* 48(16): 9874–9881. DOI: 10.1021/es501217t. Creative commons license: http://creativecommons.org/licenses/by/4.0/.

Brockway, P.E., Saunders, H., Heun, M.K., Foxon, T.J., Steinberger, J.K., Barrett, J.R. and Sorrell, S. (2017) Energy rebound as a potential threat to a low-carbon future: Findings from a new exergy-based national-level rebound approach. *Energies* 10(1): 1–24.

Brockway, P.E., Steinberger, J.K., Barrett, J.R. and Foxon, T.J. (2015) Understanding China's past and future energy demand: An exergy efficiency and decomposition analysis. *Applied Energy* 155: 892–903. Available at: https://doi.org/10.1016/j.apenergy.2015.05.082 Creative commons license: http://creativecommons.org/licenses/by/4.0/.

Cleveland, C.J., Costanza, R., Hall, C. and Kaufmann, R. (1984) Energy and the US Economy: A biophysical perspective. *Science* 225(4665): 890–897.

Cleveland, C.J., Kaufmann, R.K. and Stern, D.I. (2000) Aggregation and the role of energy in the economy. *Ecological Economics* 32(2): 301–317.

Csereklyei, Z., Rubio, M. and Stern, D. (2014) Energy and Economic Growth: The Stylized Facts. CCEP Working Paper 1417, Crawford School of Public Policy, Australian National University.

Cullen, J.M. and Allwood, J.M. (2010) Theoretical efficiency limits for energy conversion devices. *Energy* 35(5): 2059–2069.

Cullen, J.M., Allwood, J.M. and Borgstein, E.H. (2011) Reducing energy demand: What are the practical limits? *Environmental Science & Technology* 45(5): 1711–1718.

Dickey, D.A., Jansen, D.W. and Thornton, D.L. (1994) *A Primer on Cointegration with an Application to Money and Income*. Springer, London and New York.

Fair, R.C. (1984) *Specification, Estimation, and Analysis of Macroeconometric Models*. Harvard University Press, MA, USA.

Foxon, T.J. (2017) *Energy and Economic Growth: Why We Need a New Pathway to Prosperity*. Routledge, London, UK.

Gentvilaite, R., Kander, A. and Warde, P. (2015) The role of energy quality in shaping long-term energy intensity in Europe. *Energies* 8(1): 133–153. DOI:10.3390/en8010133.

Giraud, G. and Kahraman, Z. (2014) How Dependent is Growth from Primary Energy. Output Energy Elasticity in 50 Countries (1970–2011). Available at: www.parisschoolofeconomics.eu/IMG/pdf/article-pse-medde-juin2014-giraud-kahraman.pdf.

Guevara, Z., Sousa, T. and Domingos, T. (2016) Insights on energy transitions in Mexico from the analysis of useful exergy 1971–2009. *Energies* 9(7): 488–517. DOI: https://doi.org/10.3390/en9070488.

Hammond, G. and Stapleton, A. (2001) Exergy analysis of the United Kingdom energy system. *Proceedings of the Institution of Mechanical Engineers, Part A: Journal of Power and Energy* 215(2): 141–162.

Hardt, L., Owen, A., Brockway, P., Heun, M.K., Barrett, J., Taylor, P.G. and Foxon, T.J. (2018) Untangling the drivers of energy reduction in the UK productive sectors: Efficiency or offshoring? *Applied Energy* 223: 124–133.

Henningsen, A. and Henningsen, G. (2012) On estimation of the CES production function – Revisited. *Economics Letters* 115(1), 67–69.

Heun, M.K., Ownn, A. and Brockway, P.E. (2018) A physical supply-use table framework for energy analysis on the energy conversion chain. *Applied Energy* 226: 1134–1162.

Heun, M.K., Santos, J., Brockway, P.E., Pruim, R., Domingos, T. and Sakai, M. (2017) From theory to econometrics to energy policy: Cautionary tales for policymaking using aggregate production functions. *Energies* 10(1): 203–247. DOI:10.3390/en10020203.

Jarvis, A.J., Leedal, D.T. and Hewitt, C.N. (2012) Climate-society feedbacks and the avoidance of dangerous climate change. *Nature Climate Change* 2: 668–671.

Johansson, T.B., Patwardhan, A.P., Nakićenović, N. and Gomez-Echeverri, L. (2012) *Global Energy Assessment: Toward a Sustainable Future.* Cambridge University Press, Cambridge, UK.

Kalimeris, P., Richardson, C. and Bithas, K. (2014) A meta-analysis investigation of the direction of the energy-GDP causal relationship: Implications for the growth-degrowth dialogue. *Journal of Cleaner Production* 67: 1–13. DOI: 10.1016/j.jclepro.2013.12.040.

Kander, A., Malanima, P. and Warde, P. (2014) *Power to the People: Energy in Europe over the Last Five Centuries.* Princeton University Press, New Jersey, USA.

Kilian, L. (2008) The economic effects of energy price shocks. *Journal of Economic Literature* 46(4): 871–909.

Kümmel, R. (1982) The impact of energy on industrial growth. *Energy* 7(2): 189–203.

Kümmel, R., Ayres, R.U. and Lindenberger, D. (2010) Thermodynamic laws, economic methods and the productive power of energy. *Journal of Non-Equilibrium Thermodynamics* 35(2): 145–179.

Lindenberger, D. and Kümmel, R. (2011) Energy and the state of nations. *Energy* 36(10): 6010–6018.

Loftus, P.J., Cohen, A.M., Long, J.C.S. and Jenkins, J.D. (2014) A critical review of global decarbonization scenarios: What do they tell us about feasibility. *Wiley Interdisciplinary Reviews Climate Change* 6(1): 93–112. DOI: 10.1002/wcc.324.

Magerl, A. (2017) The sociometabolic transition in India: An exergy and useful work analysis 1971–2012. Masters Thesis, Alpen-Adria-Universität Klagenfurt, Austria.

Miller, J., Foxon, T.J. and Sorrell, S. (2016) Exergy accounting: A quantitative comparison of methods and implications for energy-economy analysis. *Energies* 9(11): 947–969. DOI: 10.3390/en9110947. Creative commons license: http://creativecommons.org/licenses/by/4.0/.

OECD (2001) *Measuring Productivity–OECD Manual.* OECD, Paris, France.

Omri, A. (2014) An international literature survey on energy-economic growth nexus: Evidence from country-specific studies. *Renewable and Sustainable Energy Reviews* 38: 951–959.

Pomeranz, K. (2009) *The great divergence: China, Europe, and the making of the modern world economy.* Princeton University Press, NJ, USA.

Prywes, M. (1986) A nested CES approach to capital-energy substitution. *Energy Economics* 8: 22–28.

Rant, Z. (1956) Exergie, ein neues Wort fur 'Technische Arbeitsfaehigkeit' (Exergy, a new word for technical availability). *Forschung auf dem Gebiet des Ingenieurwesens A* 22: 36–37.

Reistad, G. (1975) Available energy conversion and utilization in the United States. *ASME Transactions Journal of Engineering Power* 97: 429–434.

Romer, P.M. (1994) The origins of endogenous growth. *Journal of Economic Perspective* 8: 3–22.

Romero, J.C. and Linares, P. (2014) Exergy as a global energy sustainability indicator. A review of the state of the art. *Renewable and Sustainable Energy Reviews* 33: 427–442.

Sakai, M., Brockway, P.E., Barrett, J.R. and Taylor, P.G. (2018) Energy efficiency is found to be a key engine of economic growth. In preparation.

Santos, J., Domingos, T., Sousa, T. and St Aubyn, M. (2018) Useful exergy is key in obtaining plausible aggregate production functions and recognizing the role of energy in economic growth: Portugal 1960–2009. *Ecological Economics* 148: 103–120.

Saunders, H.D. (2008) Fuel conserving (and using) production functions. *Energy Economics* 30(5): 2184–2235.

Saunders, H.D. (2013) Historical evidence for energy efficiency rebound in 30 US sectors and a toolkit for rebound analysts. *Technological Forecasting and Social Change* 80(7): 1317–1330.

Serrenho, A. (2013) Useful work as an energy end-use accounting method: Historical and economic transitions and European patterns. PhD Thesis, University of Lisbon, Lisbon.

Serrenho, A.C., Sousa, T., Warr, B., Ayres, R.U. and Domingos, T. (2014) Decomposition of useful work intensity: The EU (European Union)-15 countries from 1960 to 2009. *Energy* 76: 704–715.

Serrenho, A.C., Warr, B., Sousa, T., Ayres, R.U. and Domingos, T. (2016) Structure and dynamics of useful work along the agriculture-industry-services transition: Portugal from 1856 to 2009. *Structural Change and Economic Dynamics* 36: 1–21.

Solow, R.M. (1956) A contribution to the theory of economic growth. *The Quarterly Journal of Economics* 70(1): 65–94.

Sorrell, S. (2007) *The Rebound Effect: An Assessment of the Evidence for Economy-wide Energy Savings from Improved Energy Efficiency*. UK Energy Research Centre, London, UK.

Sorrell, S. (2014) Energy substitution, technical change and rebound effects. *Energies* 7(5), 2850–2873.

Sousa, T., Brockway, P.E., Cullen, J.M., Henriques, S.T., Miller, J., Serrenho, A.C. and Domingos, T. (2017) The need for robust, consistent methods in societal exergy accounting. *Ecological Economics* 141: 11–21.

Stern, D. (2011) From Correlation to Granger Causality. Crawford School Research Paper No. 13, Australian National University, Canberra, Australia.

Stern, D. and Kander, A. (2012) The role of energy in the Industrial Revolution and modern economic growth. *The Energy Journal* 33(3): 125–152.

Stern, D.I. (1997) Limits to substitution and irreversibility in production and consumption: A neoclassical interpretation of ecological economics. *Ecological Economics* 21: 197–215.

Stern, D.I. (2010) Energy quality. *Ecological Economics* 69(4): 1471–1478.

Szargut, J., Morris, D.R. and Steward, F.R. (1987) *Exergy Analysis of Thermal, Chemical, and Metallurgical Processes*. Hemisphere Publishing, Philadelphia, USA.

Valero, A. and Valero, A. (2011) A prediction of the exergy loss of the world's mineral reserves in the 21st century. *Energy* 36(4): 1848–1854.

Warr, B. and Ayres, R.U. (2012) Useful work and information as drivers of economic growth. *Ecological Economics* 73: 93–102.

Warr, B., Ayres, R., Eisenmenger, N., Krausmann, F. and Schandl, H. (2010) Energy use and economic development: A comparative analysis of useful work supply in Austria, Japan, the United Kingdom and the US during 100 years of economic growth. *Ecological Economics* 69(10): 1904–1917.

Warr, B., Schandl, H. and Ayres, R.U. (2008) Long term trends in resource exergy consumption and useful work supplies in the UK, 1900 to 2000. *Ecological Economics* 68(1–2): 126–140.

Williams, E., Warr, B. and Ayres, R.U. (2008) Efficiency dilution: Long-term exergy conversion trends in Japan. *Environmental Science and Technology* 42(13): 4964–4970.

Wrigley, E.A. (2013) Energy and the English industrial revolution. *Philosophical Transactions of the Royal Society Part A* 371: 20110568, 1–10. DOI: 10.1098/rsta.2011.0568.

9 Energy-saving innovations and economy-wide rebound effects

Gioele Figus, Karen Turner and Antonios Katris

Introduction

A common characteristic of human societies is the ongoing effort to achieve the same or better outcomes with less use of natural resources. Especially during the industrial revolution, steam engines were increasingly used to provide mechanical power and to increase the productivity of labour. Engines, however, required an energy source to operate and the fuel of choice at the time was coal. Even back then, engineers were seeking to improve the efficiency by which engines were using coal with a view to reduce the resource requirements and therefore the associated costs. However, as Jevons (1865) first identified, the improvements in the energy efficiency of steam engines made the use of those engines more attractive, thereby accelerating the use of coal. This was partly due to the fact that steam engines were used in the production of iron, so efficiency improvements reduced the cost of iron, thus increasing its consumption, and indirectly the use of coal. The net result was that continuous improvements in efficiency of steam engines throughout the industrial revolution were accompanied by continuous increases in the consumption of coal – the so-called 'Jevons' paradox'.

Jevons' paradox is an extreme example of a more general phenomenon, known as the 'rebound effect'. This is an umbrella term for a variety of economic responses to improved energy efficiency, whose net result is to reduce the energy savings achieved. For example, people may take the benefits of improved insulation in the form of warmer homes rather than realising the full potential reductions in energy consumption (a direct rebound effect). Alternatively, they may spend the cost savings on other goods and services that also require energy and emissions to be produced (an indirect rebound effect). One hundred and fifty years after Jevons, the rebound effect is still closely associated with energy efficiency improvements in both production and consumption (Khazzoom, 1980). Its existence makes some commentators sceptical towards the use of energy efficiency policies as a climate change mitigation tool.[1] The fact that a part of the technologically feasible energy savings is almost inevitably eroded creates the impression that energy efficiency improvements deliver less than what the allocated funds can theoretically achieve. However, this neglects the fact that reductions in energy consumption may not be the only goal.

In 2014, the International Energy Agency (IEA) published a study detailing how energy efficiency improvements can serve a multitude of purposes including, but not limited to, climate change mitigation (IEA, 2014).[2] According to IEA, energy efficiency improvements provide benefits in five main fields, namely: macroeconomic, public budget, health and wellbeing, industry and energy delivery. In each of these fields, IEA identified specific areas that could benefit from efficiency improvements, with the nature and magnitude of those benefits being closely related to how the efficiency improvements are implemented. This 'multiple benefits' approach not only emphasises the wide-ranging benefits of energy efficiency, but also crucially highlights the true source and nature of the rebound effect. The rebound effect is not an observable phenomenon that reduces the value of energy efficiency policies. Rather it is associated with the fact that improvements in energy use may generate a wide array of positive outcomes throughout the economy and society.

For example, improvements in the energy efficiency of domestic boilers will make heating cheaper and households may take advantage of this by enjoying higher levels of thermal comfort. This will increase their 'consumer surplus'[3] which contributes to aggregate social welfare – as will the impacts in other fields identified by the IEA. Energy consumption will normally be reduced, but not by as much as it would have been in the absence of the increased demand for heating. Since energy consumption contributes to climate change, it imposes costs on other people both now and in the future. But these 'external costs' must be set against the multiple benefits of the efficiency improvement, including the benefits to consumers of warmer homes. Provided the latter are larger than the former, energy efficiency improvements provide net benefits to society.

In our Engineering and Physical Sciences Research Council End Use Energy Demand (EPSRC EUED) research project 'Energy saving innovations and economy-wide rebound effects', we explore the potential, economy-wide socio-economic impacts of improved energy efficiency. In the case studies presented in this chapter we examine the impact of improved energy efficiency in households in both the UK and Scotland, as a devolved nation within the UK. Using economy-wide macroeconomic modelling we identify how a reduction in the physical energy use requirements of different household income groups impacts the UK and Scottish economies. Moreover, since energy efficiency policies could serve as a means to pursue multiple policy objectives, we explore options for funding energy efficiency programmes from the public budget and analyse the potential impact of these options.

Through these case studies we seek to present evidence that there are more important elements to energy efficiency policy than just the changes in energy use. Therefore, policy consideration needs to adopt a wider view of the impacts, rather than focusing solely on potential rebound effects. In each case, we estimate the rebound effects associated with the energy efficiency improvements. But as discussed above, these effects may be a barrier to implementing energy efficiency policies owing to the negative connotations of the 'lost' energy savings. We therefore propose an alternative metric for evaluating the effectiveness of energy efficiency policies in reducing energy use and/or carbon emissions. This is the Carbon (or Energy) Saving Multiplier, which indicates the total[4] carbon or energy savings for each unit of carbon or energy saved by the target of the energy efficiency policy.

This chapter is structured as follows. The next section summarises the results of our analysis of the impact of household energy efficiency improvements, focusing on UK and Scottish households. The following section proposes an alternative metric to evaluate energy efficiency policies and applies this metric to an illustrative household energy efficiency improvement example. The concluding section provides remarks and reflections on potential future steps.

Analyses of the economy-wide impact of improved household energy efficiency

When considering energy efficiency policies, significant attention has been allocated to the associated rebound effect. This has driven a growing literature focused upon estimating the direct and indirect rebound effects following energy efficiency improvements by households. By combining econometric analysis and Input–Output (IO) analytical techniques, a number of studies have estimated the effectiveness of energy efficiency policies in reducing energy use and/or carbon emissions at the economy-wide level (see for example Brännlund et al., 2007; Chitnis and Sorrell, 2015; Druckman et al., 2011; Freire-Gonzáles, 2011; Lenzen and Dey, 2002; Mizobuchi, 2008). This approach uses the IO models to estimate the energy use and emissions that are 'embodied' in different goods and services.

However, IO models rest upon a number of simplifying assumptions, including a fixed production structure, fixed market prices and fixed nominal wages. As a result, they are limited in their potential to capture the full macroeconomic impacts of improved energy efficiency. For example, reduced energy requirements will reduce the current cost of energy for consumers, who may seek a higher level of comfort by using a portion of the energy they originally saved. To assess these broader, economy-wide implications of improved energy efficiency, a Computable General Equilibrium (CGE) approach is preferred, since this relaxes some of assumptions that restrict other analytical methods. CGE models are based upon IO models, but (unlike the latter) are able to simulate adjustments to prices and other variables, together with substitution between different inputs. CGE models typically simulate a regional or national economy but can be extended to the multi-regional level.

CGE models are large-scale numerical economic models that capture the interaction among key economic actors within an economy, such as industries, final consumers, government, markets for factors of production and external transactors (imports/exports). The behaviour of an economic actor is described by mathematical functions based on rigorous economic theory. These models are parametrised on data from the real world and solved numerically with the help of computer software. The solution is found under the assumption that the represented markets within the economy are simultaneously in equilibrium. CGE models can simulate the impact of policies such as improvements in energy efficiency and capture, in principle, the impact and ramification of such policy in different components of the economy. Results from simulations may be sensitive to assumptions regarding the specification of the model and the availability of estimates for key exogenous parameters. Therefore, a sensitivity analysis of those assumptions is a key element of the majority of research works using CGE models.

A growing number of studies have used CGE models to identify the potential economy-wide impacts of energy efficiency improvements (see for example Allan et al., 2007; Anson and Turner, 2009; Broberg et al., 2015; Glomsrød and Taoyuan, 2005; Grepperud and Ramussen, 2004; Hanley et al., 2009; Koesler et al., 2016; Turner, 2009; Yu et al., 2015). The typical approach is to compare the economy-wide energy consumption in a baseline scenario to that in a scenario that includes an energy efficiency improvement in one or more sectors. A common characteristic of these studies is their primary focus on industrial energy efficiency improvements. Household energy efficiency has received much less attention, with only a handful of studies to date (Duarte et al., 2016; Dufournaud et al., 1994; Koesler, 2013; Lecca et al., 2014). This is one of the main drivers of our decision to focus on household energy efficiency.

Household energy efficiency improvements in the UK – can public support be justified?

In the first of two case studies presented in this chapter, we focus on the UK as a whole. We identify five household income groups based on their gross weekly equivalised income.[5] In the case of UK households, the majority of the energy purchases are for residential use, i.e. lighting, cooking and heating. However, as the weekly income rises, so does the portion of energy spending for mobility purposes. Furthermore, the lowest-income households spend a larger portion of their disposable income on energy than the more affluent ones (energy spending 7 per cent of total consumption for the lowest-income households and 4 per cent for the highest-income ones).

We explore the impact of increasing the efficiency of household energy use in the UK so that they can run their homes (heat, cook, light etc.)[6] while using 10 per cent less physical energy. We examine two cases, one where all households receive the energy efficiency improvement (Scenario a) and a second where only the lowest-income group (20 per cent of UK households with the smallest weekly income) becomes more energy efficient (Scenario b). We initially study the impact of energy efficiency in isolation, by neglecting the capital and other costs associated with enabling energy efficiency (e.g. the installation of new boilers, or insulation).[7] For the purpose of our work, the focus is solely on the energy use within the UK, not considering the energy embodied in imported goods.

Reducing the energy requirements of households frees up a portion of the disposable income of each household, which in turn can be spent elsewhere.

160 Gioele Figus et al.

Because of the higher purchasing power of the households, we observe an increase in the demand for UK and imported goods and services.[8] This leads to a demand-driven expansion of the UK economy. As suggested by the IEA Multiple Benefits framework (IEA, 2014), an energy efficiency improvement in households delivers a multitude of macroeconomic benefits for the UK. Table 9.1 summarises the estimated changes in some key macroeconomic indicators due to the improvement in household energy efficiency. These are presented for two conceptual periods, the short run (SR) where industry capital stocks are fixed, and the long run (LR) where industry capital stocks are fully adjusted to the new macroeconomic equilibrium.

As can be seen in Table 9.1, at least in the LR when the economy has reached a new equilibrium, we observe a gross domestic product (GDP) expansion regardless of the target of the efficiency improvement. This implies increased employment and investment, as the sectors where the households spend their realised savings adapt to meet the increased demand. There is a

Table 9.1 Percentage change in key macroeconomic variables, relative to the baseline scenario, following a costless 10 per cent increase in household residential energy efficiency

	Scenario a		Scenario b	
	SR	LR	SR	LR
GDP	0.03	0.16	0.00	0.02
Consumer Price Index (CPI)	0.32	0.21	0.03	0.01
Investment	1.14	0.79	0.15	0.11
Unemployment rate	–0.82	–2.08	0.04	–0.13
Employment	0.05	0.13	0.00	0.01
Nominal wages	0.42	0.45	0.02	0.03
Imports	0.70	0.58	0.07	0.05
Exports	–0.49	–0.37	–0.04	–0.02
Total energy use	–0.67	–0.89	–0.09	–0.11
Disposable income (excluding savings)	0.52	0.58	0.06	0.07
Household total energy consumption	–1.66	–1.87	–0.22	–0.24
Residential energy consumption	–2.35	–2.62	–0.30	–0.33
Household rebound in residential energy	76.53	73.82	79.03	76.71
Household rebound in total energy	78.89	76.33	80.65	78.50
Economy wide rebound[1]	69.86	59.68	71.94	63.91

Source: the authors.

Note
1 Rebound occurs when the potential energy savings from an increase in energy efficiency are bigger than the actual energy savings. In this study we calculate the rebound effect as

$\left(1 - \dfrac{AES}{PES}\right) \times 100$, where AES are actual energy savings and PES are potential energy savings.

Depending on what is included in actual energy savings, it is possible to obtain different definitions of rebound. For instance, the rebound in residential energy use only considers savings in the residential sector, while the economy wide rebound considers energy savings in the whole economy. See Figus et al. (2017b) Appendix D for details.

significant difference in the level of expansion when only the lowest-income households are targeted (Scenario b) compared to the case where all the households receive the efficiency improvement (Scenario a). The explanation for this difference is that low-income households only account for a rather small portion of UK household consumption. In addition, the lowest-income households benefit less from the increases in wages and capital income as they rely more on transfers from the government, which are fixed in real terms. However, increased energy efficiency in the lowest-income households and the subsequent increase in demand has some impact on the income of other household groups, and this gives some additional momentum to the economic stimulus.

However, improving energy efficiency does not imply that there are only positive outcomes. Examining the Consumer Price Index (CPI), it is clear that increased demand for goods and services leads to an increase of output prices. This reduces the competitiveness of UK production sectors and consequently the level of exports. Moreover, it can be seen that even though physical energy requirements have been reduced by 10 per cent, the actual reduction of residential energy used is much smaller (around 2.62 per cent in the case where all households are targeted). This implies a significant rebound (>70 per cent) that is driven by the fact that the price of energy[9] is relatively lower, thus creating an incentive to consume more energy which especially benefits those households that are under-heating their properties, or in general did not fully meet their energy needs. In addition, to produce the additional goods and services which households consume, additional energy use by industry is necessary. But while the energy used by non-energy firms increases, the energy used by energy industry itself falls. This is because the reduction in energy demand in the residential sector more than offsets the increase in demand in industries. The net result is a reduction in the total energy used by industry. For this reason, the economy-wide rebound is smaller than the household rebound.

Our findings clearly demonstrate that improving energy efficiency simultaneously delivers energy savings, albeit less than what was technically feasible, together with wider macroeconomic benefits. It is important to point out that the economic expansion depicted in Table 9.1 does not imply that all sectors are experiencing increased activity. The difference between the household rebound and the economy-wide rebound demonstrates that overall industrial energy demand falls as a result of the increased energy efficiency in the residential sector.

So far, we have analysed the impact of improved household energy efficiency in isolation, without accounting for the cost of implementation. Since the role of energy efficiency is gaining increasing policy attention (see for example the Energy Efficient Scotland programme[10]), it is likely that those implementation costs will be funded by the public budget. Especially for low-income households that are interested in adopting efficiency improvements but who lack the funds to invest in improving their energy efficiency, the intervention of public spending is necessary. Assuming that the government would be reluctant to increase its deficit, we explore two main funding options: reallocation of existing government spending and increased income tax.

162 *Gioele Figus et al.*

In the case of reallocation of existing spending, the government funds the energy efficiency improvements by spending less on other goods and services. This reallocation is only temporary, five years, assuming that by the sixth year the efficiency improvement programme has been completed and paid for. The temporary reduction in government spending leads to a SR contraction in GDP. However, we assume that UK producers have perfect foresight.[11] As a result, they anticipate that the reduction in government spending is temporary and adjust their investment strategy accordingly. This leads to a GDP contraction shorter than the five-year period. Once the full cost of the energy efficiency improvements has been covered, i.e. after the five-year period, the LR results are identical to the ones presented in the costless case. Thus, it is evident that a temporary disturbance in government spending, and therefore the economy as a whole, to fund energy efficiency improvements, ultimately leads to permanent positive outcomes across the economy.

The other funding option we explore is a temporary increase in income tax. This allows the government to continue spending at the same level as before, while generating additional revenue to fund energy efficiency improvements. Additionally, since households benefit from the energy efficiency improvement of their dwellings, an increased income tax is an indirect way to make households pay for those improvements. It is important to note that increased income tax has distributional effects as higher-income households pay more tax. Furthermore, in the case where only the lowest-income households receive the efficiency improvement, the implication is that all the other households are paying for actions that they receive no, or at least limited benefits from. Under this funding option, our findings indicate an initial contraction of economic activity. This is due to the fact that increased income tax reduces the take-home wage of workers who in turn demand higher wages, thereby raising the production costs of industrial sectors. When all households benefit from energy efficiency improvements, the LR results are close to those observed in the costless case. However, when only lowest-income households receive the energy efficiency improvements, the associated demand boost is insufficient to compensate for the increased income tax, with the result that LR GDP is marginally (−0.005 per cent) below the original level. One of the drivers of this observation is made clear when examining the changes in the disposable income of the different household groups. The poorest 20 per cent of households experience a SR increase in disposable income of 0.58 per cent. On the other hand, all the other household groups experience reductions in their disposable income, which persist even after the income tax is reverted to the original level.

Household energy efficiency improvements in Scotland – a regional economic policy tool?

Our second case study focuses on Scotland as a devolved nation within the UK. We use a CGE model[12] that simulates the structure of the Scottish economy to investigate the impacts across that economy of energy efficiency improvements in Scottish households. Since the movement of workers between Scotland and the rest of the UK regions is relatively free of frictions, we assume that workers can freely migrate in and out Scotland from/to the rest of the UK. We model the net interregional migration of workers in response to the difference between the national and the regional unemployment rate and real wage. We assume that workers will migrate to the region that has the lowest unemployment rate and the highest real wage. A second important difference is that instead of disaggregating households into income groups, we identify a single representative Scottish household category. We assume that a costless energy efficiency improvement takes place that allows these households to achieve the same level of comfort and/or services while using 5 per cent less physical energy.[13]

In the base case scenario, we assume that the Scottish government spending is fixed. Changes to tax revenues are transferred to the central government in Westminster. Essentially, this reflects the fiscal arrangement between Scotland and the UK before April 2016. We call this FIXGOV.

Simulation results are reported in the second and third column of Table 9.2. Simulation results show SR results that are qualitatively similar to the UK case above (Scenario a). The 5 per cent energy efficiency improvement leads to a small expansion of the Scottish GDP (0.04 per cent), driven by an increase of household consumption by 0.3 per cent. This drives a net increase in investment

Table 9.2 Percentage change in key macroeconomic variables following a 5 per cent increase in Scottish household energy efficiency under alternative fiscal regimes

Time period	FIXGOV		FIXBAL		TAX	
	SR	LR	SR	LR	SR	LR
GDP	0.04	0.17	0.05	0.26	0.05	0.39
Consumer Price Index (CPI)	0.08	0.00	0.10	0.00	0.11	−0.08
Unemployment rate	−0.24	0.00	−0.31	0.00	−0.34	0.00
Total employment	0.06	0.18	0.08	0.27	0.09	0.39
Nominal gross wage	0.11	0.00	0.14	0.00	0.12	−0.19
Real gross wage	0.03	0.00	0.04	0.00	0.04	0.00
Household consumption	0.30	0.42	0.35	0.48	0.40	0.66
Investment	0.12	0.17	0.17	0.23	0.22	0.38
Exports	−0.12	0.00	−0.14	0.00	−0.15	0.14
Non-energy industries output	0.07	0.19	0.09	0.27	0.09	0.39
Energy industries output	−0.41	−0.41	−0.41	−0.37	−0.40	−0.22
Energy use	−0.89	−0.57	−0.87	−0.51	−0.85	−0.36
Energy demand by producers	−0.22	−0.24	−0.22	−0.19	−0.21	−0.03
Energy demand by households	−2.70	−1.47	−2.65	−1.41	−2.60	−1.26
Government expenditure	−	−	0.06	0.24	−	−
Government budget	53.70	165.50	−	−	−	−
Income tax	−	−	−	−	−0.10	−0.45
Household rebound	46.03	70.53	46.94	71.82	47.97	74.82
Economy-wide rebound	27.65	53.62	29.01	58.14	30.69	70.61

Source: the authors.

164 *Gioele Figus* et al.

(0.12 per cent), total employment (0.06 per cent) and nominal wages (0.11 per cent). Like in the UK case, the increase in wages puts upward pressure on output prices, so that Scottish sectors lose in terms of international competitiveness.

However, the initial increase in the real wage (0.03 per cent) together with the fall in the unemployment rate (–0.24 per cent) triggers net in-migration. As workers migrate to Scotland, the real wage falls and the unemployment[14] rate increases until in the LR they are back to their baseline values. The latter is a key finding as the decrease of output prices (driven by movement of labour) means that over time the competitiveness of Scottish industries is gradually restored and any negative impact on export activity is eliminated by the time the Scottish economy reaches a new equilibrium. However, restoration of export competitiveness implies additional demand from abroad for the outputs of Scottish sectors, which in turn requires the use of additional energy compared to the SR. Therefore, the initial energy savings are gradually eroded as the Scottish sectors increase their production to meet the export demand.

Up until this point, we have assumed that any budget savings[15] realised by the Scottish government will be transferred to the central UK government. However, since April 2016 the devolved Scottish government has acquired the power to determine income taxes and use the revenue obtained. To explore what the potential impact of new fiscal powers could be in the case of increased household energy efficiency, we examine two ways in which the government could use the budget savings: they could be returned to the economy via increased government purchases (FIXBAL case) or via reductions in income taxes (TAX case). Table 9.2 summarises the key macroeconomic effects of these two uses of the budget savings, along with the standard case where savings are accumulated and returned to UK government (FIXGOV).

The results in Table 9.2 show that if the budget savings driven by the 5 per cent household energy efficiency improvement are returned to the economy via government purchases, this leads to increased government consumption of 0.06 per cent in the SR and 0.24 per cent in the LR. Essentially, the additional revenue obtained as a result of the energy efficiency-driven economic stimulus is recycled into the economy generating additional stimulus. As a result, we observe larger increases in GDP, employment, investment and household consumption compared to the case where the budget savings are accumulated. However, the additional economic stimulus also leads to further erosion of the economy-wide energy savings achieved via this energy efficiency improvement.

In the case where we assume that the budget savings are returned to the economy via income tax reductions, we find that these are sufficient for a 0.1 per cent tax cut in the SR and 0.45 per cent in the LR. However, as discussed in the UK case study, changes in the income tax have impacts on both the demand and the supply side of the economy. On the demand side, a lower income tax means increased household disposable income and therefore increased consumption (0.66 per cent in the LR). This is significantly larger than both the FIXGOV and the FIXBAL approaches. At the same time, the increased net-of-tax wage of households puts downward pressure on the demand for higher wages. This reduces the cost of labour and stimulating production and employment. Moreover, due to the competitiveness boost of reduced labour costs, the export activity of Scottish industries is also stimulated. Overall, a reduction in income tax delivers significantly greater economic stimulus than the other two cases, which in turn is associated with greater increases in employment, investment and other variables. However, the additional production also requires the use of additional resources including, but not limited to, energy.

In general, we find that greater fiscal autonomy allows for greater economic expansion from the efficiency improvements, when the government uses the additional revenue from taxes to increase current government spending or reduce the income tax rate. However, the extent to which those improvements reduce economy-wide energy consumption is inversely proportional to size of the economic expansion.

A saving multiplier as an alternative to rebound indicator

These case studies in the previous section demonstrate that the erosion of the energy savings achieved from improved energy efficiency largely depends on how the economy reacts to the increased disposable income of households, and/or to the reduced energy costs for industries. It has been also indicated that a number of studies (see Madlener and Turner, 2016; Sorrell, 2007; Turner, 2013 for reviews) have sought to identify the indirect and wider economic rebound of increased efficiency in both consumptive and productive energy use. IO analytical techniques were often used to conduct such studies (e.g. Chitnis et al., 2013, 2014; Druckman et al., 2011; Freire-Gonzáles, 2011; Lecca et al., 2014; Lin and Du, 2015; Pfaff and Sartorius, 2015; Thomas and Azevedo, 2013a, 2013b), but all of them have used the rebound effect to estimate the effectiveness of energy efficiency and gauge the impact across the different supply chains.

The problem with rebound as an indicator is that it solely focuses on what we fail to achieve from efficiency improvements, rather than what we actually achieve. It is not surprising, therefore, that the concept has generated resistance from policymakers. Moreover, there is no standardised approach to estimate rebound effects. Most studies calculate rebound as the ratio of the 'actual energy savings' over the 'potential energy savings'. While actual energy savings can be accurately calculated in an energy–economic modelling framework, problems arise when 'potential energy savings' need to be determined. The main issue revolves around the energy used by energy producers to produce output. Following an energy efficiency improvement, the demand for the output of energy producers falls and as a result we observe quantity adjustments on the energy used by those industries. Guerra and Sancho (2010) argue that these quantity adjustments need to be included in the 'potential energy savings', whereas Turner (2013) argues that they should be reflected in the actual and not the 'potential energy savings'. Different studies adopt different approaches in specifying the 'potential energy savings' that in turn contribute to the divergence in rebound estimates. This leads to conflicting messages to policymakers.

166 *Gioele Figus et al.*

The aforementioned issues could be partly addressed by using a 'multiplier approach', especially in cases where IO is used as the methodological framework. Multiplier analysis is commonly used by policymakers in a range of areas, such as estimating how many jobs are created across the economy for a set number of jobs created in a specific sector. Multiplier analysis relies upon IO models and is therefore limited by the assumption that prices and wages remain constant. However, the policy community is familiar with this approach and it relies upon relatively straightforward calculations. These are desirable qualities when studying the impacts of improved energy efficiency, since inconsistencies in calculation methods contribute to the existing confusion. As an alternative then to rebound, we propose the use of a Carbon (Energy) Saving Multiplier (CSM).

In our work, CSM is calculated using an interregional IO table from the World Input Output Database project (Timmer *et al.*, 2015). This version of the table we used for our work captures the economic interrelationships between 35 sectors in 41 countries and regions, together with the associated energy use and carbon emissions. We define as CSM the ratio of the direct and global supply chain carbon savings over the direct carbon savings. Direct are the carbon savings that occur at the point where the efficiency improvement takes place (e.g. Agriculture sector or households) and supply chain savings are the ones that occur in the domestic and international upstream supply chains, as a result of the reduction in energy demand. For household energy efficiency improvements, the CSM measures the domestic and international reduction in carbon emissions following a unit reduction in emissions at the household level.

To illustrate the use of the CSM, we use the example of a 10 per cent reduction in UK household demand for the outputs of the UK 'Electricity Gas and Water Supply' (EGWS) sector. This corresponds to $5,525.8 million less spending on the sector (Table 9.3) and reduces household CO_2 emissions by 6,172 kilotonnes (kt). Reduced demand for EGWS output means reduced EGWS production and therefore reduced demand from its upstream supply chain, both domestic and international. As a result, a total saving of 16,625 kt of CO_2 is achieved globally. This means that for each kt of CO_2 saved by UK households, 2.69 kt of CO_2 are saved globally – which is the CSM.

As demonstrated by this example, the CSM is defined in such a way that clearly shows what needs to be included in the numerator and the denominator, helping to avoid inconsistencies in its calculation. Moreover, it focuses on savings achieved (in this case carbon) rather than savings missed. The CSM remains constant even if we assume that the households, or any targeted sector, opt to use a part of the initially realised monetary savings for more heating, lighting, cooking or water (i.e. a direct rebound). As seen in Table 9.3, even if we assume a 10 per cent or a 50 per cent rebound (take back of original demand reduction) the direct and supply chain savings are eroded but the CSM remains the same.[16]

What actually changed the CSM are the re-spending decisions of households. To illustrate, we explored an alternative scenario where households spend all of the cost savings from the efficiency improvements on hotels and restaurants.

Table 9.3 Changes in CO_2 emissions associated with a decreased spending in UK households use of UK EGWS outputs following a 10 per cent energy efficiency improvement

	No direct rebound	10% direct rebound	50% direct rebound
Reduction in monetary spend on UK EGWS outputs ($million)	−5,525.8	−4,973.2	−2,762.9
Change in CO_2 emissions (kilotonnes)			
A Reduction in direct CO_2 emitted by UK households	−6,172	−5,554	−3,086
Reductions in CO_2 emissions in UK EGWS supply chains			
Total multiplier effect per $1m spend:	1.89	1.89	1.89
B Emissions in UK EGWS sector (1.67kt per $1m)	−9,202	−8,282	−4,601
C Emissions in other UK industries (0.08kt per $1m)	−422	−380	−211
Subtotal UK	−9,624	−8,662	−4,812
D Emissions in all overseas industries (0.15kt per $1m)	−829	−746	−414
Global total	−10,453	−9,408	−5,227
Total reduction in UK CO_2 emissions	−15,796	−14,216	−7,898
Total reduction in global CO_2 emissions	−16,625	−14,962	−8,312
Carbon saving multiplier (UK level)	2.56	2.56	2.56
Carbon saving multiplier (Global level)	2.69	2.69	2.69

Source: the authors.

168 *Gioele Figus* et al.

This leads to an increase of 514 kt in UK CO_2 emissions and 794 kt globally. These findings indicate a more significant impact in the domestic supply chains than the international ones, signalling that the hotel and restaurant sector relies more on domestic suppliers than overseas. This finding is also reflected in the erosion of the CSM. The domestic CSM is eroded by 4 per cent (from 2.56 to 2.48) while the global one is eroded by 3 per cent (from 2.69 to 2.57), demonstrating a larger impact within the UK compared to abroad.

In general, the methodological procedure used to calculate the CSM[17] not only allows us to estimate the effectiveness of energy efficiency improvements, but also enables the disaggregation of the impacts along different supply chains. This way we can gain a better understanding on which sectors and in which nations are more likely to be impacted by an efficiency improvement and any subsequent re-spending decisions.

Conclusions and policy recommendations

In this chapter, we have discussed how rebound as a standalone indicator of the effectiveness of energy efficiency policies can be misleading and ultimately discouraging for policymakers. A focus on rebounds highlights the failure to achieve the technologically feasible energy use reductions but neglects the wider range of economic and social benefits that energy efficiency improvements can deliver. As our work has shown, energy efficiency improvements can contribute to a range of policy objectives, beyond climate change alone.

As shown in our UK case study, household energy efficiency improvements can provide a stimulus to a country's economy, as IEA (2014) suggests is possible, leading to increases in employment, investment and wages, while achieving a substantial, yet smaller than anticipated, reduction in energy use. This observed rebound effect is not indicative of the failure of energy efficiency policy, rather it is a necessary companion to the broader improvements in social welfare that the efficiency improvements provide.

Through this work, we have also highlighted that the magnitude of the benefits largely depends on the number and the purchasing power of the beneficiaries of energy efficiency. When the policy targets all households in an economy we achieve the maximum socio-economic benefits – which are likely to be substantial enough to cover the public funds required to support the efficiency improvement. However, policy often targets less privileged households that are less likely to be able to afford the efficiency improvement. In this case, our findings indicate that the energy efficiency improvement provides a smaller stimulus to the economy, while improving the welfare of low-income households through, for example, warmer homes. In the case where only the lowest-income households benefit from the improvement, it is of paramount importance to carefully design the policy and how it will be funded. Our research has shown that funding via increased taxation can be disruptive for the economy, leading to a slight reduction in economic activity, despite the realised benefits for low-income households. On the other hand, a reallocation of existing funds ultimately leads to an economic stimulus.

At the regional level, energy efficiency can also have expansionary impact to the economy. However, the greater openness in the goods and labour markets leads to differences in some of the indicators. Most notably, at the national level we observed a LR reduction in export activity due to a rise in prices, which is not observed when studying a region. The in-flow of labour puts downward pressure on wages and therefore prices leading to LR export levels that are on par to the pre-efficiency ones. Apart from that, even at the regional level, energy efficiency improvements deliver benefits in terms of GDP, employment, investment and household consumption increases, while also fulfilling the climate change policy role by driving reductions in economy-wide energy use. It can be seen then that energy efficiency can be used as an instrument for regional development.

Part of the regional, and national, benefits gained by energy efficiency improvements are the increased revenue from taxes due to increased production. This additional fiscal space could prove to be a useful tool to achieve further economic stimulus for the regional economy, provided that the region has the authority to use the accumulated budget savings. As we saw by exploring two different options, recycling the realised budgets savings has the potential to provide not just a demand-driven stimulus to the economy, if the savings are returned as increased government purchases, but also a permanent boost to competitiveness if the savings lead to a reduction of income tax rates. In any case, the combination of energy efficiency improvement and recycling of the accumulated budget savings from this improvement are useful policy tools in achieving macroeconomic targets for the regional economy.

Our findings are in agreement with the IEA claims that energy efficiency can deliver, among other things, macroeconomic benefits for the wider economy. We have found this to be the case both for regional and national economies. It is important to keep in mind that this by no means implies that the entirety of the economy will be better off following an energy efficiency improvement. Instead, there will be winners and losers, with the energy sectors facing a drop in activity and, in some cases, particular household groups. Overall, energy efficiency improvement is beneficial for both the economy and the environment, and this should be a strong incentive for policymakers to support such policies, always following careful consideration of the funding and what its impact might be.

However, even though research such as ours demonstrates the macroeconomic benefits of energy efficiency, the use of rebound as an indicator of its effectiveness could still create barriers to the support of such policies. In an attempt to resolve this issue, we proposed and demonstrated the use of an alternative metric, the CSM. CSM makes use of multiplier analysis, a familiar analytical tool for policymakers, to focus attention on the carbon or energy savings achieved, rather than failed to be achieved, while providing a fuller set of information to policymakers. This includes the spatial breakdown of the savings and/or any impacts from re-spending, along with the full set of information provided by the rebound indicator. Of course, there are limitations on the way in which the CSM is calculated and used, but the combination, in future research projects, of CSM with the CGE models used for our two case studies, could resolve the existing issues.

In summary, our project leads us to four main conclusions.

1   The rebound effect could be a misleading indicator in that it only accounts for the 'negative' portion of the outcome of an efficiency improvement in energy use. Evaluations of energy efficiency programmes should adopt a holistic approach and carefully assess the full range of benefits and costs of such programmes without focusing solely on a single indicator.
2   Improving household energy efficiency delivers both reduced energy use and increased economic activity. However, there is typically an inverse relation between the energy savings achieved and the size of the economic expansion. A bigger economy requires more inputs such labour and capital but also energy and other intermediate inputs.
3   Government-funded energy efficiency programmes can help low-income households who are not able to heat their homes properly. In addition, they can be used as a means of economic stimulus. However, the way in which the necessary funds are raised must be weighed carefully against the benefit of a more efficient use of energy.
4   Alternative measures such as the CSM can highlight the positive impact of energy efficiency improvements by focusing on the achieved savings in energy use and carbon emissions. This can be used as alternative to the rebound indicator and, in conjunction with other macroeconomic indicators, provide policymakers with a more comprehensive picture of the likely impact of energy efficiency measures.

Acknowledgements

The authors acknowledge support from the UK Engineering and Physical Sciences Research Council (EPSRC grant ref. EP/M00760X/1), Economic and Social Research Council (ESRC PhD Studentship ref. 1562665) and Natural Environment Research Council (ESRC-NERC PhD Studentship, ref. [ESRC] 1207166). We also acknowledge the support of ClimateXChange, the Scottish Government-funded Centre for Expertise in Climate Change. Please note that the opinions in this chapter are the sole responsibility of the authors, and not necessarily those of the ClimateXChange or the Scottish government.

Notes

1   Reducing the need for energy would reduce the emissions associated with the generation and use of this energy.
2   While this specific terminology originates with the IEA (2014), arguments and evidence that energy efficiency will enhance economic welfare in a range of ways, including as a result of macroeconomic expansion, have been considered in other studies, notably (in terms of reflecting on the recent dominant focus on rebound effects) in the recent contribution by Gillingham et al. (2016).

3 Consumer surplus is the difference between what people are willing to pay for a good and what they actually pay.
4 As will be discussed in the relevant section, it is possible to spatially disaggregate the total carbon or energy savings, so that we observe the impact in a national or international level.
5 To disaggregate the households in the CGE model, data from the Office for National Statistics (ONS) Living Costs and Food Survey have been used. The methodology used by ONS is described in the technical reports that can be found here: www.ons.gov.uk/peoplepopulationandcommunity/personalandhouseholdfinances/incomeandwealth/methodologies/livingcostsandfoodsurvey#technical-report. Detailed information on the CGE model used for this case study can be found at Figus et al. (2017).
6 The energy required to run a house constitutes what we refer to as residential energy. The key difference between a household's residential energy use and total energy use is that in total energy use we include the energy required for private transportation purposes.
7 Note that, while the enabling phase constitutes a temporary cost, the achieved efficiency can be considered as permanent, at least throughout the lifetime of the accommodation.
8 We assume that UK and imported goods and services are imperfect substitutes. As a result, UK consumers are more likely to turn to UK outputs rather than imported ones to spend their energy savings.
9 Note that here we refer to both the reduction in the effective price of residential energy services driven by the increase in energy efficiency, and the reduction in the market price driven by the fall in demand for energy.
10 Details on the programme and its route map are available online at: www.gov.scot/Resource/0053/00534980.pdf.
11 This reflects a situation where the government announces in advance its intention to divert some of the current spending to fund energy efficiency improvements for only 5 years. This allows firms to have a clear vision of how future government spending are going to be allocated and to plan investment accordingly.
12 For a detailed exposition of the CGE model used please refer to Figus et al. (2018).
13 Differently from the UK case above, here we consider household energy efficiency improvements in all household energy use, including private transport. Hence, we call this household energy efficiency rather than residential.
14 We assume that wages respond immediately to changes in the economy. On the other hand, we assume that there is a single modelling period (year) lag in the migration response as workers observe the economic circumstances in the previous period and decide on whether to move or not in the current one. Essentially, migration occurs from year two onwards, until the labour market returns to equilibrium.
15 In our central scenario (FIXGOV) we assume fixed government spending, as also seen in Table 9.2. Therefore, any budget savings come from additional revenue from income taxes, indirect taxes on consumption goods etc.
16 This finding is accurate in an IO framework modelling where we assume no changes in prices and wages. In a more sophisticated modelling approach such as CGE, which considers a wider set of changes within the economy, this finding might not be the same.
17 For a detailed discussion of the methodology used please see Turner and Katris (2017).

References

Allan, G., Hanley, N., McGregor, P., Swales, K. and Turner, K. (2007) The impact of increased efficiency in the industrial use of energy: A computable general equilibrium analysis for the United Kingdom. *Energy Economics* 29(4): 779–798.

Anson, S. and Turner, K. (2009) Rebound and disinvestment effects in refined oil consumption and supply resulting from an increase in energy efficiency in the Scottish commercial transport sector. *Energy Policy* 37(9): 3608–3620.

Brännlund, R., Ghalwash, T. and Nordström, J. (2007) Increased energy efficiency and the rebound effect: Effects on consumption and emissions. *Energy Economics* 29(1): 1–17.

Broberg, T., Berg, C. and Samakovlis, E. (2015) The economy-wide rebound effect from improved energy efficiency in Swedish industries: A general equilibrium analysis. *Energy Policy* 83: 26–37.

Chitnis, M. and Sorrell, S. (2015) Living up to expectations: Estimating direct and indirect rebound effects for UK households. *Energy Economics* 52(S1): 100–S116.

Chitnis, M., Sorrell, S., Druckman, A., Firth, S.K. and Jackson, T. (2013) Turning lights into flights: Estimating direct and indirect rebound effects for UK households. *Energy Policy* 55: 234–250.

Chitnis, M., Sorrell, S., Druckman, A., Firth, S.K. and Jackson, T. (2014) Who rebounds most? Estimating direct and indirect rebound effects for different UK socioeconomic groups. *Ecological Economics* 106: 12–32.

Druckman, A., Chitnis, M., Sorrell, S. and Jackson, T. (2011) Missing carbon reductions? Exploring rebound and backfire effects in UK households. *Energy Policy* 39(6): 3572–3581.

Duarte, R., Feng, K., Sanchez-Choliz, J., Sarasa, C. and Sun, L. (2016) Modelling the carbon consequences of pro-environmental consumer behaviour. *Applied Energy* 184: 1207–1216.

Dufournaud, C.M., Quinn, J.T. and Harrington, J.J. (1994) An Applied General Equilibrium (AGE) analysis of a policy designed to reduce the household consumption of wood in the Sudan. *Resource and Energy Economics* 16(1): 67–90.

Figus, G., Lecca, P., Turner, K. and McGregor, P.G. (2017a) Energy Efficiency as an Instrument of Regional Development Policy: Trading off the Benefits of an Economic Stimulus and Energy Rebound Effects. Discussion Paper 17.02, Department of Economics, University of Strathclyde, Scotland. Available at: www.strath.ac.uk/media/1newwebsite/departmentsubject/economics/research/researchdiscussionpapers/17.02.pdf.

Figus, G., Lecca, P., McGregor, P. and Turner, K. (2018) Energy efficiency as an instrument of regional development policy? The impact of regional fiscal autonomy. *Regional Studies*: 1–11. DOI: 10.1080/00343404.2018.1490012.

Figus, G., Turner, K., McGregor, P.G. and Katris, A. (2017b) Making the case for supporting broad energy efficiency programmes: Impacts on household incomes and other economic benefits. *Energy Policy* 111: 157–165.

Freire-González, J. (2011) Methods to empirically estimate direct and indirect rebound effect of energy-saving technological changes in households. *Ecological Modelling* 223(1): 32–40.

Gillingham, K., Rapson, D. and Wagner, G. (2016) The rebound effect and energy efficiency policy. *Review of Environmental Economics and Policy* 10(1): 68–88.

Glomsrød, S. and Taoyuan, W. (2005) Coal cleaning: A viable strategy for reduced carbon emissions and improved environment in China? *Energy Policy* 33(4): 525–542.

Grepperud, S. and Rasmussen, I. (2004) A general equilibrium assessment of rebound effects. *Energy Economics* 26(2): 261–282.

Guerra, A.I. and Sancho, F. (2010) Rethinking economy-wide rebound measures: An unbiased proposal. *Energy Policy* 38(11): 6684–6694.

Hanley, N., McGregor, P.G., Swales, J.K. and Turner, K. (2009) Do increases in energy efficiency improve environmental quality and sustainability? *Ecological Economics* 68(3): 692–709.

IEA (2014) *Capturing the Multiple Benefits of Energy Efficiency: A Guide to Quantifying the Value Added*. IEA, Paris, France.

Jevons, W.S. (1865) *The Coal Questions-Can Britain Survive?* Macmillan and Co., London, UK.

Khazzoom, J. (1980) Economic implications of mandated efficiency in standards for household appliances. *The Energy Journal* 1(4): 21–40.

Koesler, S. (2013) Catching the Rebound: Economy-wide Implications of an Efficiency Shock in the Provision of Transport Services by Households. Centre for European Economic Research. Discussion Paper No. 13–082.

Koesler, S., Swales, K. and Turner, K. (2016) International spillover and rebound effects from increased energy efficiency in Germany. *Energy Economics* 54: 444–453.

Lecca, P., McGregor, P.G., Swales, J.K. and Turner, K. (2014) The added value of a general equilibrium analysis of increased efficiency in household energy use. *Ecological Economics* 100: 51–62.

Lenzen, M. and Dey, C.J. (2002) Economic, energy and greenhouse emissions impacts of some consumer choice, technology and government outlay options. *Energy Economics* 24(4): 377–403.

Lin, B. and Du, K. (2015) Measuring energy rebound effect in the Chinese economy: An economic accounting approach. *Energy Economics* 50: 96–104.

Madlener, R. and Turner, K. (2016) After 35 Years of Economic Energy Rebound Research: Where Do We Stand? In: Santarius, T., Walnum, H.J. and Aall, C. (Eds) *Rethinking Climate and Energy Policies*. Springer International Publishing, Cham, Switzerland.

Mizobuchi, K. (2008) An empirical study on the rebound effect considering capital costs. *Energy Economics* 30(5): 2486–2516.

Pfaff, M. and Sartorius, C. (2015) Economy-wide rebound effects for non-energetic raw materials. *Ecological Economics* 118: 132–139.

Sorrell, S. (2007) The Rebound Effect: An Assessment of the Evidence for Economy-wide Energy Savings from Improved Energy Efficiency. Technical Report 4, UK Energy Research Centre, London, UK.

Thomas, B.A. and Azevedo, I.L. (2013a) Estimating direct and indirect rebound effects for U.S. households with input-output analysis part 1: Theoretical framework. *Ecological Economics* 86: 199–210.

Thomas, B.A. and Azevedo, I.L. (2013b) Estimating direct and indirect rebound effects for U.S. households with input-output analysis part 2: Simulation. *Ecological Economics* 86: 188–198.

Timmer, M.P., Dietzenbacher, E., Los, B., Stehrer, R. and de Vries, G.J. (2015) An illustrated user guide to the world input–output database: The case of global automotive production. *Review of International Economics* 23(3): 575–605.

Turner, K. (2009) Negative rebound and disinvestment effects in response to an improvement in energy efficiency in the UK economy. *Energy Economics* 31(5): 648–666.

Turner, K. (2013) 'Rebound' effects from increased energy efficiency: A time to pause and reflect. *The Energy Journal* 34(4): 25–42.

Turner, K. and Katris, A. (2017) A 'Carbon Saving Multiplier' as an alternative to rebound in considering reduced energy supply chain requirements from energy efficiency? *Energy Policy* 103: 249–257.

Yu, X., Moreno-Cruz, J. and Crittenden, J. C (2015) Regional energy rebound effect: The impact of economy-wide and sector level energy efficiency improvement in Georgia, USA. *Energy Policy* 87: 250–259.

Part IV

Policy mixes and implications

10 Political acceleration of sociotechnical transitions

Lessons from four historical case studies

Cameron Roberts and Frank W. Geels

Introduction

Chapters in this book have already established that transitions in energy demand are inherently complex affairs, which resist attempts at straightforward modelling, forecasting or policy influence. This is to be expected, given that the sociotechnical approach that is core to this book emphasises complex interactions between different factors in the transition process. But it also creates a huge problem for policymakers looking for ways to make transitions happen quickly. By fully embracing the complexity and heterogeneity of transitions in energy demand, it becomes difficult to fully understand the implications of any one transition policy *ex ante*, much less to actively design policies to achieve particular outcomes. This chapter aims to address this problem using past transitions, which can give us a valuable window into how the sociotechnical complexities of transitions and their deliberate acceleration play out in the real world.

The historical research on transitions that has been conducted so far has given transition scholars plenty of discouraging news. Smil's (2010) study of past energy transitions, for example, argues that they are necessarily long and arduous processes – a finding that will be discouraging to anybody who understands the urgency of reducing carbon emissions. Some scholars, unsurprisingly, have therefore tried to challenge this view, arguing that there is scope for transitions to sustainability to occur more quickly (Sovacool, 2016; Sovacool and Geels, 2016). Typically, however, these arguments depend critically on *intentionality*, citing the fact that the past energy transitions cited by Smil had no powerful actors deliberately trying to speed them up, but that transitions to low-carbon energy systems, by virtue of their urgency, will be deliberately accelerated by governments and other powerful actors.

This suggests another kind of historical investigation: What, in practice, occurs when policymakers deliberately try to accelerate sociotechnical transitions? What political and policy strategies are most effective at doing this, what effects do they have, and under what circumstances are they most likely to be adopted? To answer these questions, this chapter uses four historical case studies of transitions that were deliberately, and successfully, accelerated by policymakers. We define a deliberately accelerated transition not necessarily as one

178 *Cameron Roberts and Frank W. Geels*

that was *initiated* by policy action, but one in which the rate of change was deliberately increased as a result of government policies.

Our four case studies were selected to consider deliberately accelerated transitions occurring in a wide variety of contexts. First, we considered two different kinds of motivation for the deliberate acceleration of transitions: Problem-driven transitions, motivated by a dramatic crisis; and opportunity-driven transitions, driven by attempts to capitalise on a technological or economic opportunity. Second, we considered drivers accounting for the development of a new system: Commercial development, in which most of this work comes from private companies (though, in line with the focus on deliberate acceleration, much of this is encouraged by policy incentives); and institutional development, in which government institutions plan and implement the transition directly. These criteria give us a 2x2 matrix of transitions (Figure 10.1). We have identified a transition to roughly occupy each of the four quadrants in this matrix:

1   The British transition from a railway-based transport system to a road-based transport system (1918–1972). This transition was mediated mainly by a private sector market for cars and travel, and exploited the opportunity provided by a new technology. It was accelerated by the public construction of highways in the 1960s.

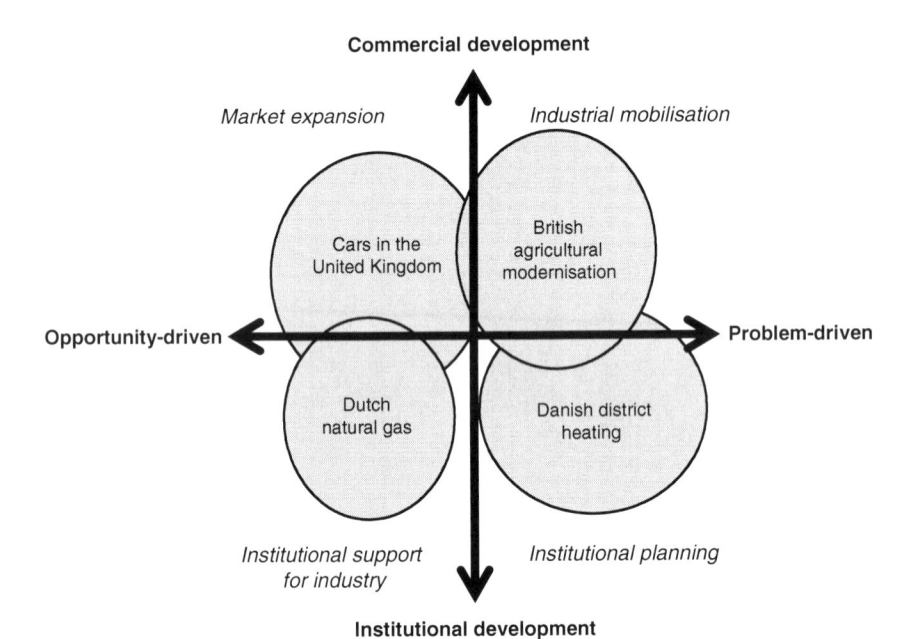

Figure 10.1 The positioning of the four case studies.
Source: the authors.

2   The British transition from Victorian mixed agriculture to modern, specialised grain agriculture (1920–1970). This occurred in response to an urgent wartime food shortage and was implemented by individual farmers and agricultural supply firms in the context of supportive policies.
3   The Dutch transition to a natural gas-dominated heating system (1945–1973). This occurred in response to a major opportunity in the form of the Slochteren gas field and was implemented mainly through state-coordinated infrastructure projects.
4   The Danish transition to an urban heating system dominated by district heating (1973–1990). This occurred in response to the 1973 and 1979 oil crises and was implemented primarily by state and municipal coordination efforts.

In addition to the two dimensions outlined in Figure 10.1, this list of cases also allows us to study four countries, two sectors and several historical contexts. Commonalities occurring in these diverse transition contexts are likely to be important for deliberately accelerated transitions more generally. To study these case studies, we used secondary historical sources, chosen to reflect diverse aspects of the transition, including technologies and infrastructures; policies; changes in user preferences and markets; cultural discourses; and business developments.

The following four sections present brief historical narratives of these case studies, broken into two phases: The formative phase, when the relevant technologies, institutions and networks of actors were taking shape, and the acceleration phase, occurring after policymakers intervened in the transition and the rate of change sped up accordingly. In each phase, we describe both the development of the niche technology, and the state of the incumbent regime with which it competed. These case studies inform an analysis section which looks for common patterns that might be hallmarks of successful policy acceleration of accelerated transitions towards reduced energy demand. It also considers the differences between the case studies, to look for ways in which their diverse contexts led to different strategies and outcomes. A final section draws conclusions from this analysis and proposes specific policy implications.

Case study 1: the transition from British Rail to road transport

Formative phase (1918–1945)

Prior to the First World War, the railways were the dominant form of transport in the United Kingdom. Cars were mainly an upper-class luxury. The war, however, had a major impact on both systems. The railways, which had been nationalised for war use, came out of the war faced with an unanswered policy question about how best to return them to the private sector (Aldcroft, 1975; Dyos and Aldcroft, 1974). Due to fears about monopolism, the answer that policymakers arrived at was to return the railways to private hands only on the condition that they abide by a strict programme of fare regulations, which among other measures subjected all fare changes to government review (Dyos and Aldcroft, 1974; Scott, 2002). Partly due to war damage, the railways struggled to provide adequate service. The railways' strategy to overcome these difficulties largely consisted of a political campaign, arguing for a relaxation of the regulations facing them. Despite some small victories in this campaign, the railways' ability to compete with road transport was badly hampered during this period.

Road transport, meanwhile, had benefited from the war. Wartime improvements in vehicles and production technologies and infrastructures allowed price–performance improvements that put car ownership within reach of some of the upper-middle classes. Members of the working classes who had learned to drive during the war were also able to start lorry and bus businesses, often using military surplus vehicles (Aldcroft, 1975). Road transport thus rapidly moved from an upper-class luxury to a utilitarian transport system. New users developed new uses for cars, such as road touring (Jeremiah, 2007). Investors also took an interest in road transport businesses, including bus and lorry firms, car manufacturers, and road builders (Church, 1994). These businesses were able to compete successfully with the railways, due largely to the aforementioned regulatory imbalances (Roberts, 2015; Scott, 2002). Because the railways were unable to easily adjust their fares, bus or lorry operators could simply show up at railway stations before a train was leaving, and undercut it (Bagwell, 1988). Strategies like this, along with general public frustration with the railways, allowed these businesses to grow rapidly at the expense of the railway companies.

The road industry, in turn, began to develop political power in the form of a well-organised motor lobby, as organisations such as the Royal Automobile Club, the Automobile Association, the Society of Motor Manufacturers and Traders, oil companies and road construction interests (Hamer, 1987). While these groups initially disagreed about their policy preferences and advocacy strategies, by the end of the 1930s they had coalesced behind the British Roads Federation, which advocated aggressively for greater government investment in motor roads. The British Roads Federation was supported in these efforts by persistent growing pains in the road transport system, such as rural blight and the increasing rate of road accidents, both of which seemed to demand an infrastructural solution (Dyos and Aldcroft, 1974; Jeremiah, 2007). They also benefited from increasing public enthusiasm, which portrayed it as an exciting, efficient, and modern way to escape the tyranny of the railways. The result of all these pressures was that by the end of the 1930s, the British government was investing in a major programme of trunk road construction, with the slogan 'the quickest way to safety' (Roberts, 2015).

Acceleration phase (1945–1973)

At the end of the Second World War, the railways were once again faced with a backlog of wartime repairs. This time, however, they were never returned to private hands: The post-war Labour government nationalised the trains under the British Transport Commission in 1948 (Aldcroft, 1975). In 1955 they launched a modernisation programme, designed to address wartime damage and bring new technologies, such as diesel traction, into use on the railways (Bonavia, 1981). This plan, however, was late to arrive, poorly designed, slow to be implemented, underfunded, and, ultimately, inadequate (Aldcroft, 1975). It also faced considerable public and political criticism; a problem which only worsened as it faced repeated cost overruns. Politicians and the media began criticising the railways as a Victorian system, out of step with modern times and therefore undeserving of state support (Roberts, 2015). In 1961, as policymakers became more and more reluctant to invest additional money into the railways, Transport Minister Ernest Marples appointed Richard Beeching to lead British Rail, giving him political cover as he made major cuts to the system, pruning it back only to its most profitable lines.

The car industry, meanwhile, boomed after the Second World War (Church, 1994), as middle-class incomes surged and cheaper cars, such as the Morris Minor, entered the market. Cars became fashionable, seen as a symbol of progress, prosperity and freedom. While the growth in car ownership was slowed during the immediate post-war years by an export-focused industrial policy, car ownership nevertheless increased rapidly during the 1950s and 1960s (Church, 1994). Commercial road transport also flourished, as did related businesses, such as road builders. Politicians and the media began talking excitedly about a coming 'motor age' (Roberts, 2015).

These developments, however, increased pressure on existing infrastructures, as the nation's limited network of A roads was threatened with severe congestion. Motorways quickly emerged as the favourite solution to this: The government passed a Special Roads Act to permit their construction in 1949 (Merriman, 2011). The motor lobby, meanwhile, had come out of the war more united and well resourced than ever, and had developed an effective strategy of targeting both local and national policymakers to advocate for the construction of specific motorways (Hamer, 1987). They scored an early victory with the completion of the Preston Bypass in 1958. That same year, the government began the construction of the M1, which was completed a year later. In 1962, Transport Minister Marples promised 1,000 miles of motorway in the next ten years; a target which was narrowly met in 1972. By then, the roads had become the country's dominant transport infrastructure, with cars and buses together accounting for the majority of all travel.

Case study 2: the UK transition from traditional mixed farming to specialised wheat agriculture

Formative phase (1918–1938)

Traditional British agriculture was a Victorian system, in which farms used a mixture of crops and animals to create an interdependent system. Cleaning crops helped with pest removal and were then fed to animals whose manure provided fertiliser. Most motive power, meanwhile, came from animal or human labour. This system, however, was in crisis during the agricultural depression of the interwar period, during which foreign imports undercut British grain production, starving farms of income and investment. Workers moved to the cities, farmers abandoned their properties, and many fields were simply left to go fallow. Britain's crop area declined from 147,000 acres in 1870 to 54,700 acres in 1930.

This slowed the adoption of agricultural innovations. Tractors, combines, chemical fertilisers and pesticides were all first developed in the nineteenth century, and had diffused widely in grain-exporting countries such as the United States and Canada, by the 1930s. British farmers in the midst of the agricultural depression, however, were unable to invest in them (Martin, 2000). Machinery in particular also suffered technical teething problems, such as the challenges associated with using tractors on relatively small British farms. British farmers during this period were also broadly sceptical of technology (Holmes, 1985). Land improvements such as field drainage, a Victorian innovation that is particularly important in making heavy soils suitable for grain agriculture, were scaled back during this period.

This situation created a reaction that set the stage for the later development of British agriculture. Aesthetic and cultural concerns about the impacts of the agricultural depression on the rural landscape became prominent in public discourse (Martin, 2000; Rooth, 1985). Farmers concerned about their situation also began agitating against the government (Wilt, 2001). This led the Conservative government, concerned about the potential loss of important rural seats, to begin looking for ways to placate them. The immediate result of this was the passage of the 1932 Wheat Act, which set a fixed price for wheat, financed by a levy on flour (Grigg, 1989). Efforts to address the root of the problem, however, were blocked by international trade politics. British industrial strategy during this period emphasised the export of industrial goods and free trade, which gave grain-exporting countries, such as Canada, the USA and Australia, a strong influence over British trade policy (Rooth, 1985). Thus, at the 1932 Ottawa Conference, the same government that had passed the Wheat Act gave commonwealth nations privileged access to British grain markets, making Britain a dumping ground for Canadian and Australian grain (Wilt, 2001).

Farmers continued to organise in light of this, with the farm lobby adopting a long-term strategy of cultivating deep connections with important government ministers (Wilt, 2001). The farm lobby, normally split between conflicting goals of the National Farmers' Union (NFU), the Central Landowners Association and the agricultural labour unions, started to unite behind the NFU in favour of policies that would ensure British farmers' income and farm production through mechanisms such as guaranteed prices, subsidies, and modernisation (Wilt, 2001). Government ministers built relationships with the new lobby, both to look for solutions to the agricultural depression, and to shore up their rural support.

Acceleration phase (1945–1970)

In 1938, the Czech crisis made clear the urgency of securing the nation's food supply in the event of a war. The war posed two threats to the food supply: First, German submarines threatened to block off food imports, and second, ships were needed to transport war material (Cox et al., 1987; Grigg, 1989). To cope with these problems, policymakers consulted with the farm lobby, and particularly with the NFU (Wilt, 2001). This relationship was further strengthened in 1939 by the appointment of Sir Reginald Dorman-Smith, former President of the Farmers' Union, as Minister of Agriculture (Wilt, 2001). He developed a new policy paradigm that was largely in line with what the NFU and its allies had been advocating for years already, aiming to maximise production at reasonable prices for consumers, while protecting stable incomes for farmers (Bowers, 1985). To accomplish this, Dorman-Smith established a subsidised guaranteed price for wheat, as well as a far-reaching programme of agricultural modernisation. Because the additional horses needed to power the expanded production would take several years to breed, the only solution was to purchase Fordson tractors from the United States. Cheap loans were provided for machinery and land improvement, and the government established local War Agricultural Executive Committees, all of which included NFU representation, to assist with the modernisation and with putting more land into cultivation (Bowers, 1985; Grigg, 1989; Martin, 2000).

These measures facilitated the successful expansion and modernisation of agriculture during the Second World War. They also entrenched the NFU in an increasingly strong position as the 'right hand' of the Ministry (Wilt, 2001). Wartime investment in farms, meanwhile, had knock-on effects. The increased use of machinery, in particular, demanded more space on farms for vehicles to manoeuvre, forcing farmers to simplify farm layouts and thus abandon mixed agriculture in favour of chemical fertilisers and pesticides (Whetham, 1974).

These changes continued to take hold after the war, as farming evolved into an agri-business. In the policy sphere, institutional arrangements and problem definitions, particularly the fear of another war, meant that many policies to support farmers continued in the post-war period in order to maintain domestic food security (Cox et al., 1987). The NFU, meanwhile, became an important part of the agricultural policy process, fiercely defending agricultural subsidies (Cox et al., 1987). Institutional arrangements around farming became an 'iron triangle' linking the Ministry of Agriculture, NFU, agricultural supply industries and research organisations, which locked in a system of specialised agriculture during the post-war era.

Case study 3: the Dutch transition from coal to natural gas

Formative phase (1948–1959)

At the end of the Second World War, the heating sector in the Netherlands was dominated by coal. This Dutch coal industry, however, faced competition on two fronts. Domestically, the price of coal was being undermined by oil heating, which was cleaner, more convenient, and easier to control precisely than coal furnaces (Correljé et al., 2003; Correljé and Verbong, 2004). Cheaper American coal was also undercutting Dutch coal, which had trouble competing due to the difficult geology of Dutch coal seams, as well as a strong labour market that forced Dutch coal mines to pay high wages (Correljé and Verbong, 2004). The result was that while coal remained the dominant form of heating in the Netherlands, the commercial and political power of the coal industry was weakening. The government, meanwhile, had no energy policy to speak of, and public discourse about energy issues was rather limited.

While this was happening, two industries provided a niche for the development of gas heating. The first of these was city gas produced locally in cities using gasified coal, or piped in from coke furnaces, and used primarily for cooking (Correljé and Verbong, 2004). The second was a growing oil industry in the north of the country, which produced natural gas as a by-product. Oil companies saw this gas as a distraction from their primary product and looked for ways to get rid of it as reliably as possible, rather than looking for a very high profit. They were therefore happy to sell it at cheap, prearranged prices to public utilities (Correljé et al., 2003). This allowed the northern village of Coevorden to be the site of the first Dutch experiments with natural gas for heating, cooking, and lighting (Raven and Verbong, 2007).

The government, interested in creating a national gas supply system, brokered an agreement in 1954 between the State Gas Company (SGB) and the Dutch Petroleum Company (abbreviated in Dutch as NAM), a business alliance of the oil companies Exxon and Shell. The agreement said that SGB would have a monopoly to distribute NAM's gas on a cost-plus basis (Correljé and Verbong, 2004). While local city gas interests blocked the further expansion of natural gas infrastructure (Raven and Verbong, 2007), this agreement establish some important principles for the future development of natural gas, most importantly that it would be produced on the basis of monopoly production, with fixed prices (Correljé et al., 2003).

Acceleration phase (1959–1973)

In 1959, the Dutch oil industry discovered the Slochteren gas field in the north of the country. The sheer volume of gas in this field threatened to overwhelm existing institutions and infrastructures (Correljé and Verbong, 2004). The government and the oil industry therefore began negotiations to determine how best to exploit the Slochteren field. While these negotiations were facilitated by the principles established in the 1954 agreement and the existence of a broadly pro-business political culture, there were tensions between different opinions and goals. Shell proposed a segmented market approach. Exxon, however, had had problems with such an approach in the United States, when cheap natural gas had undermined their oil business. They therefore proposed a full-scale national transition to natural gas for heating, cooking, and many industrial uses.

The state, which was mainly interested in maximising tax revenue, was eventually persuaded of the value of Exxon's approach, and hammered out the specifics of the relevant legislation using a 'pre-baked' negotiation process, in which different segments of Dutch society were represented in closed-door meetings (Correljé and Verbong, 2004).

Full agreement was reached in 1962: A nationwide network of gas infrastructure would be built, connecting all Dutch cities to the Slochteren field. NAM would have a monopoly on gas in the Netherlands, and the gas would be sold at a market value price, set at the same level as heat from coal or oil. Consumers, as a result, would neither lose nor gain financially from the deal. This principle would guarantee considerable profits for NAM, due to the relative cheapness of natural gas extraction and distribution, while approximately 70 per cent of the gas's value would be recouped by the state due to various taxes and fees (Correljé et al., 2003). City gas interests, meanwhile, were simply side-lined from the negotiations.

Once this agreement was reached, the explicit alliance of the state with prominent businesses allowed a rapid expansion of the natural gas system. Pipeline construction began in 1963, and despite the delicate negotiations to overcome the obstacles put up by various local interests, the pipelines had reached every mainland municipality by 1968 (Correljé et al., 2003). Meanwhile, a nationwide publicity campaign, portraying natural gas as clean, efficient, and modern, combined with a rebate programme for appliance conversions, successfully persuaded consumers to switch to natural gas heating (Correljé and Verbong, 2004). Often, city gas workers were hired to make the necessary retrofits to consumer appliances (Correljé et al., 2003).

What remained of resistance from coal interests was disarmed using a similar strategy of compensation. Coal workers were offered retraining schemes, while coal companies were given lucrative positions in the natural gas regime (Correljé and Verbong, 2004). Coal interests attempted a 'cosy coal' advertising campaign, but this was largely ineffective. By the 1970s, natural gas accounted for the vast majority of Dutch heating; the system had achieved political lock-in due to significant tax revenues that helped fund the Dutch welfare state; and the last coal mines were being closed (Correljé et al., 2003).

Case study 4: the Danish transition from oil to district heating

Formative phase (1945–1973)

Denmark became heavily dependent on oil for most forms of energy after the end of the Second World War, as cheap foreign supplies quickly displaced indigenous energy sources (Chittum and Østergaard, 2014; Johansen, 1986). By 1973, imported fuels, mostly oil, accounted for more than 90 per cent of Danish energy supply. As with the Netherlands, there was no meaningful government energy policy during this period, nor was there any major public discourse about energy issues. Most policies influenced the energy system only inadvertently.

Post-war housing policies and local government reforms, for example, gave municipalities increasing control over local infrastructures.

Not all Danish heating, however, used individual household boilers. District heating had been developing as a niche technology in Denmark since 1903, mostly using waste heat from power plants or industry, and run by either municipalities or local cooperatives (Eikeland and Inderberg, 2016). The consolidation of the electricity system after the war allowed the expansion of this system, as urban generating systems made increasing use of combined heat and power. District heating in Denmark also benefited from some cultural and institutional facts about Danish society, such as an affinity for cooperative enterprises, often seen as a core part of Danish identity (Campbell et al., 2006). This found policy expression in municipal *kommunekredit* arrangements, in which municipal governments would underwrite loans to cooperative enterprises, including those involved with expanding the district heating system. It also ensured that many early district heating systems would be run on a non-profit basis, contributing to their social, cultural, and political legitimation. By the time the energy crisis hit in 1973, Danish district heating accounted for 30 per cent of Danish homes, and had a strong set of established practices, technologies and networks.

Acceleration phase (1973–1990)

The 1973 oil crisis hit Denmark especially hard, due to its heavy dependence on foreign oil. Danish cities had to resort to extreme measures, such as turning off every other street light and banning Sunday driving, to conserve energy (Johansen, 1986). High indirect taxes and public austerity imposed by the government in response to the crisis created high unemployment and a negative growth rate (Johansen, 1986). This created a sense of urgency among Danish policymakers and the public, as they debated ways to avoid a future energy crisis. The Danish government quickly overturned its old laissez-faire approach to energy policy, and began planning interventions in the energy system, the first of which was the creation of the Danish Energy Agency in 1975 (Hawkey, 2016).

While the lack of a domestic fossil fuel energy source made it easy to reach agreement about the need to transition away from oil, there was controversy over what exactly should be used as an alternative. Early policy papers preferred a nuclear–electric system, in which a series of new nuclear power plants would power heat pumps to heat Danish homes and buildings (Nyborg and Røpke, 2015). This plan, however, ran into a series of obstacles. Public concern about the environmental effects of nuclear power led to widespread opposition (Nyborg and Røpke, 2015). Electrical utilities, who might have been the key industrial supporters about the plan, were instead lukewarm about it: The plan called for just one power company to be given the license to operate nuclear power plants, and neither of Denmark's two biggest utilities wanted to risk being the one left out of the deal (Hawkey, 2017). Because the nuclear–electric heating plan relied so heavily on a few very large and controversial pieces of infrastructure, these issues became massive handicaps.

An alternative approach, supported by several prominent academic and activist groups as well as many on the political left (Johansen, 1986), was to heat buildings using district heating and combined heat and power. While it did not initially benefit from state support, this plan had many political, technological, and institutional advantages over nuclear–electric heating. District heating already enjoyed favourable public opinion and it could also be easily expanded incrementally, meaning that it required less political consensus to go ahead. As public concern about nuclear power increased, it became a popular option. While the political divide over the solution to the energy problem meant that policies explicitly supporting district heating were slow to be implemented during the 1970s, it made far more incremental progress than the nuclear plan. The 1976 Electricity Supply Act, for example, mandated that all new electricity production, apart from renewables, must be combined heat and power. This created a ready supply of heat that could be used for future district heating projects (Chittum and Østergaard, 2014; Hawkey, 2016).

In 1979, a second energy crisis, combined with public resistance to nuclear power, finally settled the debate, and the Heat Supply Act of that year used several aggressive policy measures to accelerate the uptake of district heating. It required municipalities to produce five-year heating plans and granted broad legal powers with which to enforce them. It established different heating zones, some of which were designated only for district heating expansion (Chittum and Østergaard, 2014; Hawkey, 2016). It also included strong consumer protections: All district heating systems were to be run on a non-profit basis; consumers were to be given representation on their boards; and in the event of any district heating system being sold, its customers would be given the right of first refusal to purchase it and form a cooperative (Chittum and Østergaard, 2014). Policies passed throughout the 1980s further accelerated the system's growth, sometimes deliberately and sometimes unintentionally. Energy taxes, for example, passed mainly for fiscal reasons, made oil even less competitive with district heating, and other legislation from that period simply required buildings to connect to district heating networks, or forbade electric power in new buildings (Klok *et al.*, 2006). Less coercive contributions to the acceleration of the transition to district heating came from measures like the Danish Energy's technical catalogues, which spread technical knowledge, or *kommunekredit* arrangements, that ensured the availability of capital.

The combined impacts of these policies were a rapid acceleration in the development of district heating. While district heating networks are capital-intensive, government policies ensuring stable markets and low financial risks ensured that the district heating systems enjoyed cheap financing (Danish Energy Agency, 2012). This, combined with strong consumer confidence and ever-increasing technical competence among Danish district heating businesses and engineers, allowed the system to grow rapidly and continuously. In the present day, 60 per cent of Danish houses are connected to district heating networks; twice as many as before the energy crisis (Statistics Denmark, 2016).

Analysis

On their most basic level, these case studies show that it is possible for governments to deliberately increase the pace of sociotechnical change. They also show that successful strategies for deliberately accelerating transitions do not typically rely on market mechanisms alone, as the policies used in every single one of our case studies involved some level of direct government intervention. This could take the form of direct investment in infrastructure, as was the case in the British transport and Dutch natural gas. It could also involve intervention in markets, as was the case with the British government's fixed price for wheat, the Dutch government's fixed price for gas, or the Danish government's non-profit requirement for district heating companies. The Danish and Dutch cases also show the value of direct, prescriptive planning of new systems, by either national or municipal governments. This finding has important implications for the role of market-based mechanisms (such as carbon taxes) versus more direct intervention strategies for the acceleration of transitions in energy demand.

This level of intervention, however, brings with it another problem. Because each of the case studies considered here depended on highly aggressive policy interventions, they also necessarily depended on a political context that facilitated these interventions, whether it took the form of a major public crisis, as in the British agricultural and Danish district heat cases, public enthusiasm for a new technology, as in the British transport case, or the government's ability to conduct negotiations behind the scenes, as in the Dutch heating case. This finding, that deliberate acceleration is not just a *policy* problem, but is also a *political* problem, is not new. Past research on sociotechnical transitions has often highlighted the political difficulties inherent in accelerating them (Geels, 2014; Meadowcroft, 2011). The comparative historical approach advanced in this chapter, however, can help address this issue by suggesting ways that these political challenges can be addressed.

Both the policy and political questions are addressed by three key similarities. Because these occur reliably in four diverse case studies, there is a strong case that they are more general attributes of successfully accelerated transitions. The first of these is the lack of strong resistance to deliberate acceleration policies from regime actors. In some cases, this was partly due to exogenous conditions. Deliberate acceleration in the Dutch and British cases, for example, relied on the structural weakness of the coal and rail industries respectively, while in the British agriculture case, deliberate acceleration owed much of its success to a major landscape disruption of the food importation regime. In the Danish heating case, incumbent resistance was avoided by the absence of a domestic oil industry in the first place. This does not suggest, however, that policymakers should simply sit and wait for the right conditions to weaken regime opposition. The right policies can help disarm regime resistance, either during the formative phase, as was the case with British rail regulations which created a structural weakness in the incumbent rail industry, or in the acceleration phase, as was the case with Dutch government policies to compensate coal incumbents during the transition to natural gas.

The second similarity is the lack of strong public opposition to the deliberate acceleration. This sometimes benefited from explicitly positive discourses supporting the new system, as was the case with British transport and Dutch natural gas. In the case of Danish district heating, it arose from a deeper cultural affinity for the new system, due to its suitability for cooperative enterprise. As with the previous point, however, policies could also be designed to encourage public support, or at least to disarm public opposition. In the case of Danish district heating this was done by including strong consumer protections in the legislation aimed at accelerating the transition. In the Dutch natural gas case it was done by protecting the public from any *adverse* effects of the transition by limiting prices and then simply shutting them out of the process by conducting negotiations behind closed doors.

The final similarity in all four cases is the existence of a viable niche innovation before the transition was deliberately accelerated. In Danish district heating and British agriculture, a mature niche innovation developed independently of government interventions. In the latter case, in fact, most of the technological capabilities for industrial wheat agriculture were perfected in foreign countries. In the other two cases, however, the mature niche innovation and its supporting networks emerged partly as a result of policies implemented during the formative phase. In the case of Dutch natural gas, this occurred with the help natural gas networks in Coevorden in the 1950s, and in the case of British transport, this occurred due to road construction programmes of the 1920s and 1930s.

The common theme to all these similarities is that they all remove potential sources of friction that could impede deliberate acceleration policies. This friction can come from incumbents, from the public or from technological teething problems. Another point that comes out from this analysis is that in many cases of successful acceleration, these sources of friction do not exist at the outset due to contextual conditions. There is also, however, evidence that where they do exist, they can be deliberately removed through strategic policies, often in a way that supplements favourable contextual conditions. This suggests that, while context matters, policymakers' hands are not tied when it comes to accelerating transitions in energy demand, and that they can, with the right strategy, create their own luck by disarming possible barriers.

It is a salient point in and of itself that the main similarity between all four cases is the lack of certain key impediments to accelerating the transition, rather than any specific positive drivers. This suggests that while positive drivers can vary depending on the specificities of each political and transition context, the absence of a few key sources of friction, whether this occurs by happenstance or policy design, is important. It is, nevertheless, worthwhile to consider what common drivers of the transition occur between the four case studies. One has already been mentioned above: The absence of technical obstacles to block the transition depends on the positive presence of a viable niche innovation. This can be seen clearly in all four cases, none of which involved attempts to develop and upscale a brand-new technology or system from scratch.

Beyond this aforementioned point, there are two important positive similarities between the cases. The first is the involvement in each case of organised actors outside of government, not just as political lobbyists (though this was an important function, particularly in the UK transport case), but as on-the-ground enactors of change that shared the government's vision. This was particularly obvious in the Dutch case, with the importance of the oil and gas industry in negotiating and implementing the transition plan, and in the British transport case, where the car and road industries played such a key role. It played a slightly different role in the other two cases. In the British agricultural case, farm organisations, particularly the NFU, played a key role during the war, when they were even given spots on the War Agricultural Executive Committees. In the Danish district heating case, this role was played by rural energy cooperatives and municipal utilities, many of whom were already invested in district heating, and who helped develop the relevant infrastructure. These alliances with practitioners outside of government played a key role in each case, due both to the political leverage and the on-the-ground implementation skills that they provided.

A final major driver that appears in all four cases is relative consensus within political institutions. Not one of these cases included any significant opposition to the deliberate acceleration by opposition parties or large activist groups, and indeed many of them are marked by a high degree of unity between actors in the policy process. In the British agricultural case, this took the form of the NFU uniting disparate farm groups, which represented different and often conflicting interests within the agricultural system (labour, landowners, farmers themselves), to fight the wider collective threat posed by imports and the agricultural depression. This allowed a very clear assertion of agricultural interests to the government, which was itself relatively united, with the governing Conservative Party having a strong incentive to listen to the farmers, and the opposition Labour party doing little if anything to oppose the changes.

Similar policy consensus was formed in the Dutch case by the 'pre-baking' of the transition plan, in a closed-door meeting between representatives of different political groups. This enabled a heat bill with enough compromises that the opposition did not oppose it. In the British transport case, enthusiasm for motorway construction and road transport more generally occurred in both major parties, with the post-war Labour government writing legislation to provide for motorways that the Conservatives then built. The Danish case is perhaps the strongest demonstration of this point, because it was precisely the lack of a strong political consensus that held up any energy transition for six years between the energy crisis in 1973 and the passage of the Heat Supply Act in 1979. Once the government gave in to those who opposed the nuclear–electric plan and embraced district heating, things were able to change quickly. The importance of these kinds of united political fronts was most likely due to the fact that the deliberate acceleration of transitions is a big, ambitious policy that is easily derailed by political opposition. Therefore, support from opposition parties and other political factions is critical in ensuring that efforts to accelerate change can be sustained.

There are also, however, important differences in how each of these transitions occurred. These relate primarily to the role of national and historical context in shaping what kinds of interventions were viable in each case. These case studies identify three important contextual factors that influence this. The first of these has to do with the political economy of the country in which the transition takes place. In the United Kingdom, a liberal market economy, the most viable acceleration strategies depended on price interventions, subsidies or infrastructural projects, whose second-order effects accelerated the transition through market mechanisms (although the fixed price for wheat during the Second World War was an exception to this). Denmark and the Netherlands, both of which are coordinated market economies, were also able to supplement market strategies with aggressive and sometimes coercive interventions such as price fixing, regulations, and detailed planning, often to impressive effect.

Another important difference is the magnitude of the crises that spurred government action. Political momentum for policy changes emerged very quickly in the two cases involving sudden, urgent problems: British agriculture and Danish district heating. In each of these cases, the political status quo shifted almost overnight from support of incumbents to support of radical change. The Dutch district heating and British transport cases, meanwhile, relied on opportunities rather than problems for political momentum, meaning that this momentum was built up more gradually.

A final important difference is the existence or non-existence of a policy consensus in the formative phase. This had important implications for the temporality of the acceleration phase. Sometimes, expert consensus emerges around desirable policy interventions in advance of the necessary political conditions to implement these interventions (Kingdon, 1984). This was the case in British agriculture, where government ministers became increasingly sympathetic to the NFU's suggestions before the outbreak of war but could not implement them due to international politics. It also occurred in the Dutch case, where the government maintained an interest in a nationwide natural gas infrastructure throughout the 1950s. In both of these cases, the prior consensus allowed policy action to follow very quickly after the crisis. In the Danish and British transport cases, however, the crisis provided political consensus in advance of any consensus on policy, resulting in a period of urgent debate that delayed policies for deliberate acceleration. In the British transport case, this took the form of piecemeal attempts to improve rail transport and integrate it with the roads during the 1950s. In the Danish case, it took the form of a six-year long period of debate between the proponents of nuclear–electric heating and district heating.

Conclusions and policy recommendations

Despite taking place in very different contexts, our four case studies have three important things in common. In each one, there is a lack of regime resistance to the acceleration of the transition, a lack of public resistance to the transition and a mature niche technology that can be exploited. The fact that these occur in such a diverse set of cases suggests they are features of deliberate acceleration more generally. Our research also shows that both the creation of favourable conditions, and the exploitation of pre-existing favourable conditions, are reliable options for deliberate acceleration policies.

Our cases also show that factors such as governance structure, different kinds of crises and the role of policy consensus can lead to different temporal patterns of acceleration, and to different kinds of policies being effective at accelerating transitions. It is therefore important for policymakers to pay attention to their context in order to design the right kinds of interventions for deliberate acceleration.

These findings suggest a few key policy implications for the deliberate acceleration of transitions in energy demand:

- Successful policy interventions are likely to take advantage of technologies that are already well developed, with strong supporting networks. These can be developed using innovation policies in advance of deliberate acceleration.
- If there is a strong regime that might block deliberate acceleration of the transition, policy interventions may find a way to weaken its influence, either by empowering its opponents or by introducing regulations that will weaken its power.
- Successful acceleration policies are likely to promote new technologies as public goods, avoiding monopolies or coercive policies that will aggrieve users.
- The odds of success at deliberate acceleration will be greatly increased if the government promoting these policies can forge a strong united front, both with industry and practitioner communities, and with other factions within political institutions.
- Finally, though these four case studies are instructive, policymakers should be cautious trying to copy lessons from foreign examples, and should take into account the political, economic, cultural and technological particularities of the systems they wish to influence.

References

Aldcroft, D.H. (1975) *British Transport since 1914: An Economic History*. David and Charles Publishers, Newton Abbot, Devon, UK.

Bagwell, D.P. (1988) *The Transport Revolution 1770–1985*. Routledge, London, UK.

Bonavia, M.R. (1981) *British Rail, the First 25 Years*. David and Charles Publishers, Newton Abbot, Devon, UK.

Bowers, J.K. (1985) British agricultural policy since the Second World War. *The Agricultural History Review* 33(1): 66–76.

Campbell, J.L., Hall, J.A. and Pedersen, O. (2006) *National Identity and the Varieties of Capitalism: The Danish Experience*. McGill-Queen's Press, Montreal, Quebec and Kingston, Ontario.

Chittum, A. and Østergaard, P.A. (2014) How Danish communal heat planning empowers municipalities and benefits individual consumers. *Energy Policy* 74: 465–474.

Church, R.A. (1994) *Rise and Decline of the British Motor Industry*. Revised Edition. Cambridge University Press, Cambridge, UK.

Correljé, A., van der Linde, C. and Westerwoudt, T. (2003) *Natural Gas in the Netherlands: From Cooperation to Competition?* Oranje-Nassau Groep, Amsterdam.

Correljé, A. and Verbong, G. (2004) The Transition from Coal to Gas: Radical Change of the Dutch Gas System. In: Elzen, B., Geels, F., Green, K. (Eds) *System Innovation and the Transition to Sustainability: Theory, Evidence and Policy*. Edward Elgar, Cheltenham, 114–138.

Cox, G., Lowe, P. and Winter, M. (1987) Farmers and the state: A crisis for corporatism. *The Political Quarterly* 58(1): 73–81.

Danish Energy Agency (2012) District Heating – Danish and Chinese Experiences. Danish Ministry of Energy, Utilities and Climate, Copenhagen. Available at: https://ens.dk/en/our-responsibilities/global-cooperation/experiences-district-heating.

Dyos, H.J. and Aldcroft, D.H. (1974) *British Transport: An Economic Survey from the Seventeenth Century to the Twentieth*. Penguin Books, London, UK.

Eikeland, P.O. and Inderberg, T.H.J. (2016) Energy system transformation and long-term interest constellations in Denmark: can agency beat structure? *Energy Research and Social Science* 11: 164–173.

Geels, F.W. (2014) Regime resistance against low-carbon transitions: Introducing politics and power into the multi-level perspective. *Theory Culture Society* 31(5): 21–40.

Grigg, D. (1989) *English Agriculture: A Historical Perspective*. Wiley-Blackwell, Oxford, UK and New York, USA.

Hamer, M. (1987) *Wheels within Wheels: A Study of the Road Lobby*. Routledge, London, UK.

Hawkey, D. (2016) European Heat Policies and Practices. In: Hawkey, D., Lovell, H. and Webb, J. (Eds) *Sustainable Urban Energy Policy: Heat and the City*. Routledge, London, 47–67.

Hawkey, D. (2017) Personal Correspondence. Interviewed by: Roberts, C. (6 February 2017).

Holmes, C.J. (1985) Science and Practice in Arable Farming, 1910–1950. In: Oddy, D.J. and Miller, D.S. (Eds) *Diet and Health in Modern Britain*. Croom Helm, London, UK, 5–31.

Jeremiah, D. (2007) *Representations of British Motoring*. Manchester University Press, Manchester, UK.

Johansen, H.C. (1986) *The Danish Economy in the Twentieth Century*. Routledge, London, UK.

Kingdon, J.W. (1984) *Agendas, Alternatives, and Public Policies*. Longman, New York, USA.

Klok, J., Larsen, A., Dahl, A. and Hansen, K. (2006) Ecological tax reform in Denmark: History and social acceptability. *Energy Policy* 34(8):: 905–916.

Martin, J. (2000) *The Development of Modern Agriculture: British Farming Since 1931*. Palgrave Macmillan, London, UK.

Meadowcroft, J. (2011) Engaging with the politics of sustainability transitions. *Environmental Innovation and Societal Transitions* 1(1): 70–75.

Merriman, P. (2011) *Driving Spaces: A Cultural-Historical Geography of England's M1 Motorway*. John Wiley and Sons, Oxford, UK.

Nyborg, S. and Røpke, I. (2015) Heat pumps in Denmark – From ugly duckling to white swan. *Energy Research and Social Science* 9: 166–177.

Raven, R. and Verbong, G. (2007) Multi-regime interactions in the Dutch energy sector: The case of combined heat and power technologies in the Netherlands 1970–2000. *Technology Analysis and Strategic Management* 19(4): 491–507.

Roberts, C. (2015) *The Evolution of Discursive Story-lines during Socio-technical Transitions: An Analytical Model Applied to British and American Road and Rail Transport during the Twentieth Century.* University of Manchester, Manchester, UK.

Rooth, T. (1985) Trade agreements and the evolution of British agricultural policy in the 1930s. *The Agricultural History Review* 33(2): 173–190.

Scott, P. (2002) British railways and the challenge from road haulage: 1919–39. *Twentieth Century British History* 13(2): 101–120.

Smil, V. (2010) *Energy Transitions: History, Requirements, Prospects.* ABC-CLIO, Santa Barbara, CA, USA.

Sovacool, B.K. (2016) How long will it take? Conceptualizing the temporal dynamics of energy transitions. *Energy Research and Social Science* 13: 202–215.

Sovacool, B.K. and Geels, F.W. (2016) Further reflections on the temporality of energy transitions: A response to critics. *Energy Research and Social Science* 22: 232–237.

Statistics Denmark (2016) Occupied dwellings by region, heating and time. Statbank Denmark. Available at: www.statbank.dk/statbank5a/default.asp?w=1920.

Whetham, E.H. (1974) The Agriculture Act, 1920 and its repeal – the 'Great Betrayal'. *The Agricultural History Review* 22(1): 36–49.

Wilt, A.F. (2001) *Food for War: Agriculture and Rearmament in Britain before the Second World War.* Oxford University Press, Oxford, UK.

11 The challenge of effective energy efficiency policy in the United Kingdom

Janette Webb

Introduction

'It's all been about cost and not use' said a former member of UK Parliament, now a member of the House of Lords. He went on to suggest that political sensitivities over energy prices are damaging the prospects for effective policies to save energy and reduce bills. This was, he observed, because such policies are regarded as increasing the price of energy. This is true in the immediate sense, because current government practice is to add the costs of policy to energy tariffs, rather than meeting them from tax revenues. There is however ample evidence that even when the costs of energy efficiency policy *are* added to tariffs, this has resulted in lower bills (Rosenow et al., 2017). Moreover, beyond savings on bills, reducing the need for energy in buildings, particularly for heating, has integral societal benefits for welfare, climate protection and jobs (Deasley and Thornhill, 2017; Rosenow et al., 2017; Washan et al., 2014). Using less energy is also a means to improve system security and resilience: it reduces reliance on imports and should lower the total costs of developing a low-carbon system (Watson et al., 2017).

The UK government however seems slow to respond to the evidence of multiple benefits, despite the best efforts of researchers, advisory bodies and lobbyists. This chapter explores reasons for this paradox. It considers the impact of a political focus on the short-term cost of energy efficiency policies and asks whether this is effective even within its own terms of cost reduction. It also considers whether its short-termism downgrades public goods of climate protection and societal welfare. The account is informed by a sociological perspective and is based on ethnographic study of policymaking practices in the UK. It departs from the more familiar economic evaluation of policy to consider the experiences of a sample of policymakers, government advisers and industry practitioners. This approach is adopted because there is already consistent evidence about the societal value of upgrading our building stock and reducing the need to use energy for heating. What is lacking is the more interpretative exploration of why such apparently clear evidence does not lead to more concerted policy action. Instead, there is continuing contest and struggle over policy costs relative to societal benefit, resulting in policy instability and uncertainty. Investment in upgrading buildings and appliances has been uneven and has declined significantly in the last few years. Current policy is also regarded as too limited to meet climate protection targets, even after any additional policies implied by the UK Clean Growth Strategy 2017 are taken into account (CCC, 2018).

The chapter begins with a summary of recent patterns of energy use in buildings and the impacts of policy changes, before moving to discussion of contemporary energy efficiency policymaking. The qualitative data are derived from interviews, policy documents, observation and participation in policy forums. Specifically, the account is based on: 12 interviews with 15 expert representatives of the main parties to energy efficiency policymaking in the UK (6 UK government officials, 1 representative of a government advisory agency, 7 representatives of energy efficiency trade, business and campaigning organisations, and one senior researcher and former energy advisory body manager). I also attended, and participated in, trade conferences and government and advisory body meetings concerned with energy efficiency and low-carbon heat policy. Lastly, I was chair of the Advisory Group contributing to the UK Committee on Climate Change (CCC) 2016 advice to UK government: *Next Steps for Heat and Energy Efficiency*. The different data sources are used as a means of triangulation between perspectives and help to minimise over-reliance on a single source. The material is part of a larger dataset from comparative research on heat and energy efficiency policy and practice in Denmark, Germany and UK.

Energy use in UK buildings and energy efficiency policy

Energy use has been decreasing during the early part of the twenty-first century. In households for example there has been a 29 per cent reduction in average energy use since 2004, and the average annual energy bill has reduced by £490 between 2004 and 2015 despite increasing energy prices (Deasley and Thornhill, 2017; Rosenow, 2012; Rosenow *et al.*, 2017). Much of the change is attributed to the effects of policies, in particular the successive variants of the Energy Supplier Obligation (ESO) although this was first introduced in 1994 (Mallaburn and Eyre, 2014; Rosenow *et al.*, 2017). Regulatory requirements to use condensing boilers for gas central heating, and incremental increases in European Union (EU) efficiency standards of electrical appliances have also played a critical role (CEBR, 2011; Elwell *et al.*, 2015). In this period then, policies to improve energy performance of buildings and appliances have been associated with both reduced energy use and cost savings. Furthermore, analysis demonstrates that *socially* cost-effective energy efficiency measures could reduce household energy use by a further 25 per cent by 2030, with benefits including lower bills, better health, and fewer households in fuel poverty, as well as climate protection (Rosenow *et al.*, 2017).

The evidence has however failed to convince the UK government to maintain public investment in energy-saving policies. Instead, since 2012 an increasingly politicised debate on energy prices has resulted in withdrawal of all public funding for energy efficiency and significant reductions in the ESO budget (CCC, 2017). Public funding for the Warm Front scheme for low-income households in England has ended, and ESO energy-saving targets have been lowered. The 2016 zero carbon homes policy in England was also withdrawn in 2015, despite cross-sector support for its implementation. Annual UK investment in energy efficiency has declined by 53 per cent and there has been an 80 per cent to 90 per cent reduction in number of household measures installed since 2012, with the sharpest declines in England (ACE, 2016; CCC, 2016).

Scotland is the exception to this pattern, as is discussed further in the recommendations and conclusions section of the chapter. Other UK policies have also been removed or scaled back, and energy savings and emission reductions in public and commercial buildings have stalled since 2008 (CCC, 2017). The CRC Energy Efficiency Scheme for large energy users in public and private sectors was designed originally as an energy-saving incentive but became a form of energy tax and has been on hold since a review of business energy taxes was announced in 2015. The UK Clean Growth Strategy 2017 includes aspirations for improved energy efficiency in buildings, but the necessary policies are missing (Carbon Brief, 2017; CCC, 2018). The UK CCC 2018 review of the Strategy also concludes that even if the promised heat and energy efficiency policies materialise, these will be insufficient to meet the legislated 4th (2023–2027) and 5th (2028–2032) carbon budgets for buildings.

A sociological perspective on UK energy efficiency policy

> There are quite a lot of reports out there now ... that show a really clear story that there is a very positive net benefit from energy efficiency at the national level, and it's getting us nowhere.
>
> (CEO Trade Association A)

Given the gap between evidence of societal benefits of reducing energy use and policy to secure such benefits, the sociological conjecture explored here is that effective policy is hampered by contemporary political adherence to classical economic theory of efficient markets as the core solution to societal problems. The commitment to extending the sphere of markets as a means to govern the allocation of resources has characterised neoliberal modes of UK government since the 1980s (Bowman et al., 2014; Crouch, 2011; Hodges and Lapsley, 2016). Sociologists of scientific knowledge argue that such economic theory is not simply derived from empirical findings about existing markets but is instrumental in attempts to format such an economy and bring it into being (Beunza et al., 2006; Callon, 1998). It is a theory that is being tested *in vivo*, and sociologists are studying its material impacts on societies (Mackenzie et al., 2007).

In line with the sociology of science and economic sociology, dynamics of energy supply and use are interpreted not as a result of universal laws of markets, but as governed by societal institutions (including in the UK those of consumer capitalism), theories and technical expertise, codes of conduct and understandings of normal practice (Caliskan and Callon, 2009; Shove, 2017). Energy prices are interpreted as socially organised facts, and not as abstract summations of market information (Beunza et al., 2006). Energy efficiency policies are components of the sociotechnical arrangements that constitute markets. Tensions over policy costs relative to energy prices are interpreted as politically inflected and entangled with competing economic theories about value (Fligstein, 1996).

The privatisation and liberalisation of the UK energy system during the 1980s and 1990s is a central example of the application of classical economic theory to the redesign of the energy system. Energy is now constituted as a market commodity, and price is used as the critical indicator of value. Energy use is defined as a matter of the financial economics of supply and demand. The model consumer is expected to understand the concept of economic self-interest and to respond by, for example, switching energy suppliers or upgrading property, where this results in a calculable financial return. Policy and regulation are defined *ex ante* as sources of market costs and inefficiencies, rather than as a means to secure public goods. Government's role becomes that of 'supporting the market to discover solutions', including shaping the market 'where necessary to overcome barriers' (DECC, 2013, p. 7).

Applied to energy saving, the efficient markets hypothesis assumes that rising energy prices will increase economic opportunities to invest in building or appliance upgrades which deliver the same, or better, material standards from using less energy. All property owners, whether domestic or commercial, are expected to act on the same logic: when the modelled financial savings from insulation or appliance upgrades, discounted over the notional lifetime of the measures, are greater than the capital costs of installing them, then investment should follow. On the supplier side, where a service can be provided more cheaply by investing in energy efficiency rather than selling energy, then in theory the supplier should choose the efficiency option. The value of energy saving is hence equated primarily to a monetary sum recoverable over a relatively short term which demonstrates its cost-effectiveness compared to other spending options. As one interview respondent said, 'in the UK we do seem to take quite a strict definition of what we see our benefits to be. So, it's very much based around what … can be financialised…' (Policy Manager, UK Trade Association B).

According to such economic theory, investing in energy saving has value when the costs are paid back from energy bill reductions in a relatively short period. The corresponding political perspective is that policy imposes costs that damage market efficiencies. Any policy measure therefore has to pass a cost-efficiency test: is policy cost outweighed by evidence of its beneficial market impact in overcoming something defined as a barrier? The existence of barriers to an efficient market is usually inferred from low demand for retrofit from property owners, despite evidence of monetary benefit. Property owners are defined as economically rational consumers whose failure to act on evidence of economic advantage is caused by 'barriers' such as lack of information about opportunities, perceptions of inconvenience and disruption, lack of trust in suppliers, or perverse incentives (CCC, 2016, 2018). Whether the costs of a policy designed to ameliorate these factors are justified by evidence of value in expanding the energy efficiency market is however subject to political contest. There is for example disagreement over what counts as relevant factors to be included in a calculation of costs and benefits, and over the value and weighting to be attached to them.

The argument about the material consequences of this approach to energy efficiency policy is developed in the following three sections, first in relation to the ESO, then in relation to the Green Deal, and lastly in relation to the UK response to the EU Energy Performance of Buildings Directive. In the subsequent sections of the chapter, these examples inform discussion of the counter-efficiencies of using policy to make efficient markets, and reflection on the structural problems of institutionalising energy-saving policy in UK government. Finally, I consider some of the ways forward and conclude with a summary of the argument.

The ESO: designing policy to address market failure?

> The Carbon Emissions Reduction Target, which was a supplier obligation, treated something like eight million homes, came in under cost, was very effective in terms of treating carbon emissions.
>
> (UK BEIS Official, economist)

The ESO is a major example of the application of efficient markets theory to domestic energy efficiency policy. It was originally designed to overcome a perceived flaw in the design of liberalised energy markets, following privatisation of gas and electricity. The disaggregated market structure had provided motivation for energy suppliers to increase their profitability by increasing sales, while keeping wholesale market prices down and minimising the costs of customer services. The SO, introduced in 1994, aimed to change the market calculus by enforcing an energy-saving target on large suppliers. It was intended to simulate the economic rationality of an energy services model, based on the USA concept of Least Cost Planning. This argues that an energy supplier should choose to save a unit of energy, rather than supplying it, when this is the cheaper option (Rosenow, 2012). The SO was envisaged as a means to stimulate such a market in energy services: suppliers could work with private contractors or with managing agents such as local authorities to meet the target at notional least cost. As explained by a former energy advice agency manager:

> That argument was bought by, for example, Stephen Littlechild as [energy markets] regulator ... It came out of ... a neoclassical economic framework: 'This is a market failure that we need to address; this is the way to do it'.
>
> (Senior Researcher)

In the 1990s, the gas market regulator was an important voice of dissent about the principle of adding energy efficiency costs to tariffs on the grounds that this would distort the energy price mechanism and damage a competitive market in energy efficiency (Rosenow, 2012). The SO was however adopted and became increasingly ambitious as it became clear that suppliers met the obligations easily, despite repeated claims about their difficulties. Successive versions of the SO (from Energy Efficiency Standards of Performance (1994–2002) to the Energy Efficiency Commitment (2002–2008) and Certified Emissions Reduction Target (CERT) and Community Energy Savings Programme (CESP) 2008–2012) increased the energy-saving target and allowable programme costs. The funds supported growth of a significant energy efficiency industry, with around £1 billion per year being invested by 2012, particularly in low-cost measures such as loft and cavity wall insulation and heating system upgrades.

In the direct sense of number of measures installed at relatively low cost, and industry development, the SO has been evaluated as cost-effective, but its mass market approach has weaknesses. Energy savings are deemed, using government-defined metrics, rather than measured, and the lack of direct evidence about *actual* energy savings has lead to questions about the long-term systemic value of the scheme. This has been linked to suggestions that there is an underlying lack of public confidence in the potential for major energy saving from building upgrades, which further erodes the foundations for comprehensive policy. The 'one size fits all' model can also lead to inappropriate solutions and poor-quality work: 'particularly with vulnerable people, you need to worry about how it's being done and the advice … and much more contextual issues than just whack in some cavity wall insulation' (Senior Researcher). Local authorities working with ESO suppliers have criticised the growing complexity, and lack of continuity, of the scheme rules, which add hidden costs to project management (Webb et al., 2017). Neither has the ESO stimulated the envisaged energy services market, which is a labour-intensive industry requiring skilled analysis of appropriate building retrofit, coordination of advice, finance and supply chain, and performance guarantees. Instead the post-privatisation energy retail business model, driven by short-term price competition, has continued.

Latterly the political focus on energy prices has led to significant scaling back of the ESO: in comparison with the former £1 billion per annum budget, the revised Energy Company Obligation (ECO) has a budget ceiling of £640 million per annum at 2017 prices (BEIS, 2018). The energy-saving target is correspondingly lower, and the scheme is focused on low-income households. A UK BEIS official explained that policy action is directed to maximising the value from less money in relation to a 'triangle of things that we've got to do' (UK BEIS Official, domestic policy). The triangle entails balancing the combined demands of a Conservative Party manifesto commitment, an obligation to lessen fuel poverty, and the political requirement to reduce policy costs: 'the reduced £640m a year, to hit the "one million homes improved through insulation" target in the manifesto over the life of this parliament, and we've got to try and take forward progress against our fuel poverty targets' (UK BEIS Official, domestic policy). The reduced ESO budget, resulting from political sensitivities over perceived escalating bills, has however resulted in loss of energy efficiency industry capacity. The National

Audit Office (NAO) 2016 criticised the political focus on short-term price, at the cost of public value. In line with practitioner perspectives expressed in this research, NAO conclude that future costs of rebuilding supply chain competence and skills are likely to be higher than necessary, with essential industry innovations, such as those for solid wall insulation, stalled.

The changing rules and objectives of the ESO since its introduction in 1994 demonstrate the difficulties of using an individual policy mechanism to build a market which values energy saving, without changing the underlying design of the market. Policy could instead pursue a macroeconomic strategy, combining taxation, regulation and market incentives to govern systemic renovation of building stock to meet the 2050 zero carbon buildings target (European Commission, 2016).

The Green Deal: designing policy to address market barriers?

> [The] Green Deal didn't work. Even ministers are prepared to say that.
> (UK BEIS Official, domestic policy)

The Green Deal (GD) was a UK policy device intended to tackle 'market barriers' through provision of loans to pay for investment in energy-saving measures. Loans were made available to owner–occupiers and private landlords defined as 'able to pay' to retrofit their property. The debt was expected to be repaid through energy bills using the 'golden rule' that savings on energy costs would exceed loan repayments. Initial modelling had suggested that 14 million households would take up the offer between 2013 and 2020 (DECC, 2011). In practice around 14,000 were signed up between 2013 and 2016. The UK government decided to cease funding the Green Deal Finance Company in 2015, because the low take up of loans made the framework too costly to justify further support (Rosenow and Eyre, 2016). The GD proved to be an extremely expensive and inefficient experiment in energy market economics: the NAO 2016 report concluded that the

> £240 million expenditure on the GD has not generated additional energy savings, because its design and implementation of the scheme did not persuade people that energy efficiency measures are worth paying for. The Green Deal has therefore not been value for money.
> (NAO, 2016, p. 12)

The GD mechanism sought to use economic theory to create a model consumer responsive to financial self-interest. In some ways it achieved exactly that, but with the opposite outcome to that intended. Households were expected to procure an unfamiliar, and potentially difficult-to-manage, energy services contract, without clear quality assurance and with uncertain costs. In addition, the cost of the GD Assessment, and the interest rate attached to GD finance, meant that the target population evaluated the offer as poor value for money, and as adding new costs arising from the mechanism to attach debt to their property.

This led to perceived risk of damage, rather than improvements, to the market value of their house (Marchand et al., 2015). On the same calculus, the finance sector assessed the scheme rules and concluded that the risk of default on GD debt following a house sale was high, further increasing the cost of loans. The policy focus on economic theory, rather than on the value of the retrofit for quality of life, meant that the GD failed to engage either households or private landlords in a meaningful way. Poor quality control over the process for managing GD assessments and contracts also allowed bogus companies and scams to operate, further undermining trust among a public already highly distrustful of financial institutions (Rosenow and Eyre, 2016). Overall, the scheme demonstrated that using public resources to devise market metrics to persuade owners to retrofit their property is a very inefficient way to save energy.

The UK government response to the EU Energy Performance of Buildings Directive: measuring the optimal level of investment in energy efficiency

The EU Energy Performance of Buildings Directive (EU EPBD) requires Member States to decide the optimal level of investment in energy performance of buildings, and to devise policy to ensure that this standard is met. The Directive was expected to be a foundation for new policy measures, but in practice the results appear limited, leading to concerns that the use of a concept of cost optimal investment is reducing the application of the EU EPBD to 'a tick box exercise' (Senior Manager, Energy Services Company). Findings from the larger research project have shown that the UK formula for estimating cost optimal performance is relatively conservative in attributing value to energy saving, in comparison with other European countries (Hawkey, 2016). Appraisal methodology follows Treasury supplementary Green Book guidance (DECC, 2015), which applies a long-run variable cost (LRVC) formula to assign a value to welfare gains from increases in energy performance standards of buildings. The LRVC of energy is calculated on the basis of existing gas and electricity networks. Most of the maintenance costs are excluded from estimated future energy costs of buildings, because these costs are regarded as unaffected (in aggregate) by changes in demand. The resulting calculation attributes lower benefit to investment in energy efficiency than in other European countries, justifying lower building standards than if the full costs of networks were included in the LRVC formula. Over their lifetime, occupants of such buildings are likely to use more energy 'making the retrofitting issue even worse' (Policy Manager, UK Trade Association B).

The use of a cost optimality formula also affects the pace and standards of building retrofit. The concept of an optimal *present* cost is related to assumptions that future prices will fall, and/or that economic growth will lead to wealthier future populations who are better able to pay. In the short term the argument is that the available finance could have a higher return from investment in other opportunities. Policymakers are conscious of the potential conservative implications of the formula, which may be used as a reason for delay, as explained by this official:

> This is the big crux of it … what is the definition of cost optimal? Is it what's cost optimal for the person that's living in the home today? Or is it thinking about what are the measures that we need to do over time?
> (UK BEIS Official, heat and business)

Current UK policy adopts a cautious view of the value of retrofit. A senior official explained that government is not setting an optimal path for all building stock upgrade, in line with a 2050 zero carbon budget. There is, he said, no plan 'that goes from here to, say, 2050 and says "We've got to do these homes now and these homes then" other than in terms of fuel poverty' (UK BEIS Senior Official, energy). The cautious stance was further exemplified by the UK Conservative government decision to cancel the 2016 zero carbon housing standard, which was perceived by UK CCC, trade associations and campaigning organisations as exerting effective pressure on the housing construction industry. Its cancellation has reduced industry investment in innovation and skills (CCC, 2016). Despite the UK commitment to the EU EPBD target for all new buildings to be 'Nearly Zero-Energy' from 2020, the UK Clean Growth Strategy 2017 lacks any policy to deliver this. As a result, higher carbon emissions from buildings are locked in, and Department for Business, Energy and Industrial Strategy (BEIS), as the government department responsible for meeting carbon budgets, has to find equivalent carbon savings elsewhere, potentially at a higher cost. Limited or cautious investment and delay are normalised, on the expectation that action will be more affordable in future.

The short-term cost minimisation approach is perceived by energy efficiency advocates and by at least some government officials as a false economy: 'costs don't fall by waiting, costs fall by actually doing stuff' (Senior Researcher); 'if you don't do anything we're not going to see the costs come down' (UK BEIS Official, heat and business). At present, the dominant position taken by government is however that early action results in a higher total cost, because measures are installed before their costs have decreased. This is ignoring the risk that inaction now is contributing to increasing climate disruption in future, with ultimately high costs. Minimising this risk requires action based on the precautionary principle and a societal welfare approach, which sets annual retrofit targets to ensure a low-energy building stock by 2050. Such a strategy is likely to be essential in a system reliant on renewable energy and would set a very different benchmark for effective policy.

The counter-efficiencies of managing market efficiencies: cost metrics and policy performance

> We have an anti-regulatory agenda now in general. And in the previous administration, our Secretary of State for Communities and Local

Government was almost pathologically opposed to anything, well anything, but certainly anything like that.
(CEO, Trade Association A)

The examples of the ESO, the GD and the UK application of cost optimal formulae to the EU EPBD demonstrate the displacement of policy to ensure a low-energy building stock onto the problem of devising cost metrics to justify policy to ameliorate market barriers. The approach eschews use of mandates and directives to govern change, which was depicted by a senior government official as 'not … the way these things have been done in the UK historically'. Politics of deregulation are in turn making it harder 'to use some of the levers that we might want to do otherwise … for those to be politically acceptable we need to show that they're not adding any cost to consumers' (UK BEIS Official, heat and business). Each element of investment is assessed separately, and officials 'score small victories quite highly' (UK BEIS Official, domestic policy). One of the side effects is that of obscuring the potential socio-economic value of a whole area, or even whole building, solution:

> Everyone has been worrying about some correlation factor in SAP … instead of stepping out and thinking about the house …
> (CEO, Buildings and Construction Organisation)

> We look at the effectiveness of energy efficiency interventions … in isolation … what's cost effective in this very narrow window? … So, all you're left with is some kind of big components … like solid wall insulation in tower blocks, air source heat pumps off the gas network. And then you design separate programmes … you don't start looking at the integration of those things across different types of schemes and initiatives.
> (Senior Policy Manager, Third Sector Organisation)

Turning energy efficiency policy benefits into a monetary sum is also an imprecise art: 'It's fiendishly difficult to work out what the energy and carbon savings might be [from a revised ESO]' (UK BEIS Official, economist). Officials aim to model not just the direct costs and benefits of any policy, but also its 'additionality' over and above the status quo. This includes analysing its expected interaction with other policies, which may be either counter-productive or synergistic, and which are subject to considerable uncertainty and complex contingencies. What counts as a factor in the cost–benefit equation is also ultimately a matter of value judgement. Societal benefits of public subsidies for example were considered under-valued:

> We never think about spill-over, either within the household in terms of other action once they've taken that first one, or to other households who see their peers doing things and think 'Oh, we'll do that.' You know, we have a very, very narrow definition of additionality.
> (CEO, Trade Association A)

The monetising of costs and benefits in turn leads to a further demand for policy to anticipate and avoid the risks of gaming the rules: 'do we see suppliers increasing the cost artificially, because it's subsidised?' (UK BEIS Official, energy policy evaluation). The result is the search for ever more precise policy rules:

> as failures are happening, the inclination of the public sector is to manage risk more and more carefully. If you manage risk more carefully ... you're forced down to ever lower levels of detail, and ever more levels of bureaucracy to manage risk, which frustrates the private sector even more.
> (CEO, Buildings and Construction Organisation)

There is then a risk of a vicious circle where increasingly refined technical formulae reinforce business dependence on government to make the market work, while government officials are increasingly frustrated with the lack of initiative by businesses.

The same counter-efficiencies are present in the commercial sector, where discrete market instruments 'to nudge corporations into thinking about what their ongoing energy costs are and making the case that they could be better off if they made improvements ... has been largely unsuccessful' (Policy Manager, Trade Association B). The Climate Change Levy and the CRC Energy Efficiency Scheme (due to be abolished in 2019) were perceived by trade organisations as using overly complex rules, reinforcing a focus on short-term payback, and resulting in least possible action. With energy costs a small proportion of total commercial costs in most cases, property owners have invested very little in upgrading the performance of their buildings (Mallaburn, 2016). An emphasis on corporate social responsibility, health and productivity are regarded by trade organisations as more effective, but these are not integral to UK energy efficiency policy. The current approach is experienced as one that presents a spurious accuracy with counter-productive effects. Instead, a senior independent adviser to government suggested the need to 'stop predicting to the pound outcomes on what energy saving will do. Recognise that however good the science, when you put humans in charge, the chance of getting outcomes predicted are low'.

The problem of institutionalising energy efficiency policy in UK government

> It always feels like there's lots of good work going on, dotted around all over the place. But trying to join it up into one place...
> (UK BEIS Official, domestic energy)

Departmental structures within UK government are themselves part of the challenge of devising progressive policy for a low-energy building stock. Energy saving is relevant to multiple specialisms, none of which focus on the problem of low-energy, low-carbon buildings. Within BEIS there is a structural division between energy efficiency and energy markets, which has persisted following the 2016 merger of the UK Department of Energy and Climate Change (DECC) and UK Business, Innovation and Skills. DECC was itself set up only in 2008, under a UK Labour government, because energy privatisation had been expected to remove the need for a specialist energy function. Energy efficiency had formerly been treated as an environmental issue and was part of the UK Government Department of Environment, Food and Rural Affairs (Defra). A short-lived DECC Energy Efficiency Deployment Office was introduced in 2012 by the UK Conservative–Liberal Democrat coalition government to accelerate energy saving; it was shut down following the 2015 election of the Conservative government, and its work dispersed. Housing, Communities and Local Government (England), Health and Treasury also have an interest in the potential for energy saving to meet health, welfare and economic goals, but none have a direct remit.

Establishing common metrics of value across departments for energy-saving policy is a potential means to establish a more integrated approach, but it is regarded as contentious because it could lead to requirements on departments to cede control over components of their budget. Attempts to integrate health metrics, such as Quality Adjusted Life Years, into cost–benefit formulae for energy saving in housing, for example, have not succeeded: 'I don't think we'll ever get to a place where the Department of Health says, "Well, you can have a billion pounds of our health funds, because you can then go away and reduce the queues at doctor's surgeries"' (UK BEIS Official, domestic energy). The problem is exacerbated by political commitment to reducing public spending, such that money invested in one area is lost from another. The resulting fragmented structural location of energy efficiency policy contributes to perceived constant change in scheme rules, which tend to weaken both the skills and capabilities available for policy implementation, and the willingness of trade, industry and third sector organisations, as well as local authorities, to invest in capacity.

The difficulty of establishing secure and effective policy to reduce energy use is arguably a historical legacy of the precedence given to supply side technology policy (Steward, 2012). The result is a circumscribed view of policy options, marginalising the socio-economic logic for a whole-systems analysis to compare costs of investment in efficiency with investment in supply for a low-carbon system (Hawkey et al., 2016). Historical neglect of energy use is regarded by trade and third sector organisations as resulting in repeated overestimation of future supply needs, and failure to value energy saving on an equal basis with supply in energy and capacity market policies. In the following quote from a third sector enterprise, the manager references the economic concept of the 'marginal abatement cost curve' (MACC). The curve plots the alternative options for reducing emissions from energy against a metric for cost per tonne of carbon abated. The marginal cost calculation is used to indicate the cost of reducing each additional unit of emissions:

> When you look at the MAC curve, and you look at demand versus supply options ... energy efficiency always looks like the no-brainer, but what we

like to do in this country is then look at the cost effectiveness profile associated with all the things that are in that cost-effective bracket, and then we start saying, 'Well this is really expensive.' No, it's not, because if you look at the wider MAC curve, this isn't expensive. This is more cost effective to do than all the stuff that you're already doing up here, Hinkley Point etc.
(Policy Manager, third sector organisation)

The MACC concept is used by government in energy and climate policy but is subject to contention over the impact of different cost assumptions on the cost ranking of different technologies and solutions (Grubb, 2014). The financial value attributed to supply vs. demand measures is contested, with industry suggestions that the carbon abatement price set by policy is more advantageous for supply side technologies: 'carbon saving from supply side technologies such as wind or nuclear [is set] ... at far higher value than that for energy saving in buildings' (Senior Manager, energy services business and former market analyst). Consistency in carbon-saving metrics across supply and demand remains elusive, perpetuating the lack of consensus over the substantive value of a systemic focus on reducing energy use.

Conclusions and policy recommendations

Despite stalled progress, there is much that can be done to improve the building stock across the UK. These actions will however require a valuation framework structured by acceptance of societal responsibility for welfare and sustainable prosperity, rather than a sole focus on the theory of efficient markets and short-term price metrics. Achieving such a change in perspective needs greater consensus on the societal value of improving every building and systematically reducing energy use. Such strategies have precedents, particularly in countries with limited indigenous fossil fuel reserves such as Denmark or Japan. It is however usually oil crises or accidents, such as the nuclear accident at Fukushima in Japan, which have prompted systematic action. There is a question about whether we can learn from past innovations in practice, without the immediate crisis conditions. Part of the answer to this question lies in recognising the sociality of markets where rules of conduct, beliefs and values are formed and practised. This sociality is recognised by businesses and consultancies who are concerned to galvanise action: 'what it takes is people on the ground organising to do something' (CEO, Non-Domestic Energy Efficiency Consultancy); 'we've got to create ... an outcome-oriented movement of people who share a common cause, and they want all this to work' (CEO, Buildings and Construction Organisation). Recommendations for policy are discussed below.

An accessible starting point for consensus building, and the consequent development of new social norms, is the use of regulatory standards. There is however political opposition to new regulatory standards, based on belief in the economic efficiencies of *de*regulation. The importance of regulation to cost-effective policy is however exemplified by UK regulation to enforce use of gas condensing boilers. The regulation was announced in 2003 for a start in 2005; 80,000 gas fitters were trained in 2 years, and an initial period of financial support was followed by exclusion of less efficient boilers from the market once the technology was established (Mallaburn and Eyre, 2014). The value of such regulation is formally recognised by at least some government officials:

> the changes that were made in 2005 … were really successful in terms of bringing industry on board who were really originally resistant and sceptical and were saying it was all going to be far too difficult and the costs are all too high. And then within literally a couple of years, the cost of the technology had come right down, installers were happy with doing it, and the penetration of the technology … was really quick.
> (UK BEIS Official, heat and business)

A refurbishment standard for every building, and progressively increasing standards for appliances, could achieve the same effect. Regulatory standards make sociotechnical innovation in businesses and supply chains a matter of necessity. There is evidence for example that the announcement of a (very modest) 2018 minimum energy efficiency standard for the private rental sector prompted immediate action. Enforcement of standards is critical to their effectiveness and requires trained building inspectors to assess quality of work, including solid wall insulation, air tightness and heat and ventilation system efficiencies. Prominent publication of tables to compare the energy standards of public and commercial buildings can also prompt investment, because of reputational concerns (Mallaburn, 2016). The CRC Energy Efficiency Scheme for example originally included annual publication of league tables, which, according to one trade association, worked effectively because businesses disliked being exposed as 'doing really poorly in comparison to your peers'. Perceived reputational damage, he argued 'really does drive … a lot of activity amongst businesses' (Policy Manager, UK Trade Association B). Such tables would also enable public procurement of leasehold buildings to require the highest energy performance standards.

In the domestic sector, the Bonfield Review, 'Each Home Counts' (Bonfield, 2016), is a potential building block to improve standards and performance guarantees, backed by regulation, for all retrofit and renewable energy installations. It proposes making any subsidies available only to businesses that comply with increasing standards, making businesses active in resolving cost and quality problems. The cost of loans to property owners is also likely to decrease as a result of such systematic standards. Developing a professionalised supply chain to retrofit all buildings is likely to require development of energy services suited to different sectors, whether private home owners, social or private sector landlords, commercial or public buildings. The smart meter programme is arguably the current policy mechanism intended to stimulate an energy services market in the domestic and small business sectors (see Chapter 6). The smart meter data is expected to provide market opportunities for new businesses to sell 'comfort' as an energy service. Such companies would have an incentive to make a profit from selling less energy by retrofitting the property, where this is the lower cost option. The Energy Systems Catapult has received public funding to pilot a Home Energy Services Gateway, connecting households with businesses to test technologies, services and business models. The NAO is however investigating the smart meter programme, which is behind schedule, and subject to criticism of costs and inflexible operating systems. Overall the low public trust in existing energy utilities makes the development of a self-regulating market for energy services unlikely. Public investment to set high standards in all buildings, to provide independent and impartial advice on options, and to enforce the quality of work by service providers, is needed.

Beyond regulation and standards, trade associations and campaigning organisations advocate the treatment of energy efficiency as a 'national infrastructure'. This should, they suggest, ensure an integrated cost–benefit analysis of investing in decreasing energy use vs. increasing supply. In England, the National Infrastructure Commission (NIC) is one potential route to a stable cross-departmental government strategy for energy efficiency retrofit. The NIC is an executive agency of UK Treasury and has the means to secure affordable finance: 'Just by having that long-term strategy you're attracting the large growing green bonds market' (Policy Manager, Trade Association B). The Commission is required to assess UK long-term infrastructure needs, and to make recommendations to government, once in each Parliament. Its first assessment is due to be published in 2018, but its original chair, and Labour peer, Andrew Adonis, has resigned over fundamental disagreement with government on its approach to infrastructure, particularly in the context of the UK exit from the EU. The NIC recommendations around energy use and their impact are therefore uncertain. In addition, UK Treasury will not necessarily provide finance, as a senior official explained:

> Just because something is infrastructure doesn't suddenly mean there's vast amounts of money available for it. I think sometimes the two things are seen as automatically going hand in hand.
> (UK BEIS Senior Official, energy)

Financial and business models, as well as engagement of commercial and residential property owners, are critical unresolved challenges for any comprehensive retrofit strategy (CCC, 2016; Washan et al., 2014). There is however scope to learn from the increasing jurisdictional and policy diversity within the UK. Unlike England, public investment in energy saving has been maintained in Scotland, with a small increase since 2012 from £190 to £192 million. The focus has been area-based home energy-saving schemes targeted on low-income households. In 2015 however, the Scottish government made energy efficiency a national infrastructure priority; further plans published in 2018 outlined a programme of work to 2032.

The aim is to provide an offer of support to all building owners to improve the energy efficiency of their property. An investment of around £0.5 billion is promised during the current Parliament, which is likely to run until 2021. Policy development has proceeded since 2015 through three pilot phases, with cross-sector work coordinated by local authorities to test project management for area-based building retrofit. The programme is a core part of the first Scottish Energy Strategy (2017), which, in line with restricted devolved powers over energy, focuses on reducing energy use. A critical device for planning, prioritising and carrying out retrofit is a proposed local government duty to develop a comprehensive local heat and energy efficiency strategy (LHEES) and implementation plan. Strategies will be customised to the circumstances of each area and will be based on options appraisal using socio-economic metrics encompassing health, climate protection and local economic benefits.

The Scottish programme has already prompted a new cross-departmental government focus on energy efficiency in health, housing, social policy, climate change, economy and enterprise. Methods for developing and implementing strategies, securing finance, managing contracts and procurement, engaging the public and commercial property owners are as yet under-specified, but there is public commitment to proceeding, calling on cross-sector problem solving to set a long-term framework for investment (Scottish Government, 2018). It may be argued that this comprehensive strategy is more feasible in smaller devolved governments than in the UK government. If that is true however, then it is an argument for more devolution of powers across the nations and regions of the UK.

Whatever the governance structures, it is critical that long-term ambitious energy-saving programmes are implemented across the UK, because of the urgency of action in the face of climate change and existing legislative commitment. The UK Clean Growth Strategy does set out intentions for future action on buildings in England, but policies need to be specified. There is for example a target of retrofitting houses to Energy Performance Certificate (EPC) Band C by 2035, but this is qualified on grounds of practicality and cost, which will blunt its effectiveness. There is also no plan to reinstate the zero carbon homes standard, despite the EU EPBD requirement for near-zero carbon new buildings by 2020. An effective starting point for a comprehensive strategy in UK government is the use of a White Paper. This was suggested by the academics appointed to advise the UK CCC on its heat and energy efficiency advice to UK government (Webb, 2016). A White Paper on heat and energy efficiency, combining building and technology regulation, taxation and incentives for every building owner, could be used to secure significant social, economic and environmental benefits. It would work to trigger Treasury review and reform of cost–benefit formulae and tax and accounting frameworks to support new low-carbon heat and energy efficiency policy. This would in turn be a means to integrate low-carbon buildings and heating into industrial strategy. Demonstrators of options for low-carbon heat and low-energy buildings are already being initiated through the UK Industrial Strategy challenge funding for local energy system demonstrators (Innovate UK, 2018).

A policy focus on heat and energy efficiency is the necessary catalyst for innovation in investment and energy services; it provides a means to initiate high standards in the energy efficiency, building and heat supply trades. As proposed in Scotland, local authorities could be mandated to develop an area-based energy efficiency strategy to deliver the retrofit of buildings and to decarbonise heat supply. This could combine funding sources (Energy Company Obligation (ECO) budgets, grants, gas and electricity distribution network operator funds) to capture systemic benefits from energy efficiency such as avoided or deferred costs of network reinforcement. It would also serve as a means to establish the governance process for achieving a low-carbon heat supply for a low-energy building stock, using measured rather than modelled energy consumption.

In conclusion, I have argued that present political sensitivities over energy prices, combined with a belief in deregulation of markets, are undermining effective policy for saving energy. Devising energy efficiency policy through progressive refinement of market metrics is proving to be an uneven and slow, if not counter-efficient, route to energy saving in buildings. Wider societal benefits from policy to upgrade the whole building stock for a low-energy system become a secondary consideration, acknowledged in principle (as in the Clean Growth Strategy, 2017), but not prioritised in practice. The current stalled investment in insulating homes, lack of progress in non-residential buildings and lack of potential for long-term planning of area-based programmes in England is further undermining confidence in the actual energy savings which could be achieved.

Given the considerable uncertainty in scenarios for the future UK energy supply, and the fact that scenarios are not predictors of actual future systems or of energy costs, there are societal reasons for maximising energy saving through a comprehensive programme for low-energy buildings. From a sociological perspective, patterns of energy use, technologies, market institutions and strategies are societally produced; 'energy efficiency' is itself embedded in assumptions about material standards of living and political-economic commitment to economic growth, with implications for resource-intensive consumption (Shove, 2017). This sociotechnical system can be remade to serve changed needs for a low-carbon system, but at present the wider perspective on societal prosperity is marginalised by political commitments to market mechanisms, deregulation and a small state.

The chapter has explored the sociology of policy processes for energy efficiency; it raises questions about the current politics of energy price, which are a reminder of the inevitable interconnection between energy and political values and beliefs. It highlights the necessity for a changed societal perspective on value, which is not reliant primarily on neoliberal belief in market mechanisms as universal solution to societal problems. The latter is proving to be an inefficient means to a low-energy building stock and is likely to impose higher costs on society, as we fail to comprehend the necessity to act on climate science rather than classical economic theory.

Acknowledgements

The Reframing Energy Demand research project was funded by Engineering and Physical Sciences Research Council (EPSRC) under the *Working with the End Use Energy Demand Centres* Programme Grant Ref EP/M008215/1. I am greatly indebted to interviewees from UK government, energy efficiency industry, research and advisory bodies, all of whom generously gave me their time and shared their considerable experience and insights. Thank you to Edinburgh University Reframing Energy Demand research colleagues Ronan Bolton (Co-Investigator), David Hawkey (Research Fellow) and Mark Winskel (Co-Investigator), and to David McCrone, Edinburgh University for comments on an earlier draft.

References

Association for the Conservation of Energy (ACE) (2016) Home Energy Efficiency 2010–2020. Available at: www.energybillrevolution.org/wp-content/uploads/2016/05/ACE-briefing-note-2016-03-Home-energy-efficiency-delivery-2010-to-2020.pdf.

BEIS (2018) Energy Company Obligation ECO3: 2018–2022 Consultation. HM Government, London, UK. Available at: www.gov.uk/government/consultations/energy-company-obligation-eco3-2018-to-2022.

Beunza, D., Hardie, I. and MacKenzie, D. (2006) A price is a social thing: Towards a material sociology of arbitrage. *Organization Studies* 27(5): 721–745.

Bonfield, P. (2016) Each Home Counts. HM Government, London, UK. Available at: www.gov.uk/government/publications/each-home-counts-review-of-consumer-advice-protection-standards-and-enforcement-for-energy-efficiency-and-renewable-energy.

Bowman, A., Erturk, I., Froud, J., Johal, S., Law, A., Leaver, A., Moran, M. and Williams, K. (2014) *The End of the Experiment: From Competition to the Foundational Economy*. Manchester University Press, Manchester, UK.

Caliskan, K. and Callon, M. (2009) Economization, part 1: Shifting attention from the economy towards processes of economization. *Economy and Society* 38(3): 369–398.

Callon, M. (ed.) (1998) *The Laws of the Markets*. Blackwell, London, UK.

Carbon Brief (2017) How the Clean Growth Strategy Hopes to Deliver UK Climate Goals. Available at: www.carbonbrief.org/in-depth-how-the-clean-growth-strategy-hopes-to-deliver-uk-climate-goals.

CCC (2016) Next Steps for UK Heat Policy. HM Government, London, UK. Available at: www.theccc.org.uk/publication/next-steps-for-uk-heat-policy/.

CCC (2017) Meeting Carbon Budgets: Closing the Policy Gap. Report to Parliament. HM Government, London, UK. Available at: www.theccc.org.uk/wp-content/uploads/2017/06/2017-Report-to-Parliament-Meeting-Carbon-Budgets-Closing-the-policy-gap.pdf.

CCC (2018) An Independent Assessment of the UK's Clean Growth Strategy: From Ambition to Action. HM Government, London, UK. Available at: www.theccc.org.uk/publication/independent-assessment-uks-clean-growth-strategy-ambition-action/.

CEBR (2011) An Assessment of the Drivers of Domestic Natural Gas Consumption. British Gas Home Energy Report, London, UK.

Crouch, C. (2011) *The Strange Non-Death of Neo-Liberalism*. Wiley, London, UK.

Deasley, S. and Thornhill, C. (2017) Affordable Warmth, Clean Growth: Action Plan for a Comprehensive Buildings Energy Infrastructure Programme. Frontier Economics and Energy Efficiency Infrastructure Group. Available at: www.frontier-economics.com/documents/2017/09/affordable-warmth-clean-growth.pdf.

DECC (2011) Final Stage Impact Assessment for the Green Deal and Energy Company Obligation. Available at: https://assets.publishing.service.gov.uk/government/uploads/system/uploads/attachment_data/file/42984/5533-final-stage-impact-assessment-for-the-green-deal-a.pdf.

DECC (2013) The Future of Heating: Meeting the Challenge. HM Government, London, UK. Available at: www.gov.uk/government/publications/the-future-of-heating-meeting-the-challenge.

DECC (2015) Green Book Supplementary Guidance: Valuation of Energy Use and Greenhouse Gas Emissions for Appraisal. HM Government, London, UK. Available at: www.gov.uk/government/publications/valuation-of-energy-use-and-greenhouse-gas-emissions-for-appraisal.

Elwell, C., Biddulph, P., Lowe, R. and Oreszczyn, T. (2015) Determining the impact of regulatory policy on UK gas use using Bayesian analysis on publicly available data. *Energy Policy* 86: 770–783.

European Commission (2016) The Macro-Economic and Other Benefits of Energy Efficiency. Available at: https://ec.europa.eu/energy/sites/ener/files/documents/final_report_v4_final.pdf.

Fligstein, N. (1996) Markets as politics: A political cultural approach to market institutions. *American Sociological Review* 61(4): 656–673.

Grubb, M. (2014) *Planetary Economics: Energy, Climate Change and the Three Domains of Sustainable Development*. Routledge, London, UK.

Hawkey, D. (2016) Reframing Energy Performance Requirements in Building Standards. Paper presented at DEMAND Centre Conference, Lancaster, 13–15 April 2016.

Hawkey, D., Webb, J., Lovell, H., McCrone, D., Tingey, M. and Winskel, M. (2016) *Sustainable Urban Energy Policy: Heat and the City*. Routledge Earthscan, London, UK.

Hodges, R. and Lapsley, I. (2016) A private sector failure, a public sector crisis – Reflections on the Great Recession. *Financial Accountability and Management*, 32(3): 265–280.

Innovate UK (2018) Prospering from the Energy Revolution. Available at: www.gov.uk/government/news/prospering-from-the-energy-revolution-full-programme-details.

Mackenzie, D., Muniesa, F. and Siu, L. (2007) *Do Economists Make Markets? On the Performativity of Economics*. Princeton University Press, New Jersey, USA.

Mallaburn, P. (2016) A New Approach to Non-Domestic Energy Efficiency Policy. UK CCC, London, UK. Available at: www.theccc.org.uk/publication/a-new-approach-to-non-domestic-energy-efficiency/.

Mallaburn, P. and Eyre, N. (2014) Lessons from energy efficiency policy and programmes in the UK from 1973 to 2013. *Energy Efficiency* 7(1): 23–41. DOI 10.1007/s12053-013-9197-7.

Marchand, R., Koh, L. and Morris, J. (2015) Delivering energy efficiency and carbon reduction schemes in England: Lessons from Green Deal Pioneer Places. *Energy Policy* 84: 96–106.

Rosenow, J. (2012) Energy savings obligations in the UK: A history of change. *Energy Policy* 49: 373–382.

Rosenow, J. and Eyre, N. (2016) A post mortem of the Green Deal: Austerity, energy efficiency, and failure in British energy policy. *Energy Research and Social Science* 21: 141–144.

Rosenow, J., Eyre, N., Sorrell, S. and Guertler, P. (2017) Unlocking Britain's First Fuel: The Potential for Energy Savings in UK Housing. UKERC/CIED Policy Briefing.

Available at: www.ukerc.ac.uk/publications/unlocking-britains-first-fuel-energy-savings-in-uk-housing.html.

Scottish Government (2018) Energy Efficient Scotland Routemap. Available at: www.gov.scot/Resource/0053/00534980.pdf.

Shove, E. (2017) What is wrong with energy efficiency? *Building Research and Information* 46(7): 779–789.

Steward, F. (2012) Transformative innovation policy to meet the challenge of climate change: Sociotechnical networks aligned with consumption and end-use as new transition arenas for a low-carbon society or green economy. *Technology Analysis & Strategic Management* 24(4): 331–343. Available at: http://dx.doi.org/10.1080/09537325.2012.663959.

UK National Audit Office (2016) Green Deal and Energy Company Obligation. HM Government, London, UK. Available at: www.nao.org.uk/wp-content/uploads/2016/04/Green-Deal-and-Energy-Company-Obligation.pdf.

Washan, P., Stenning, J. and Goodman, M. (2014) Building the Future: Economic and Fiscal Impacts of Making Homes Highly Energy Efficient. Cambridge Econometrics and Verco. Available at: www.energybillrevolution.org/wp-content/uploads/2014/10/Building-the-Future-The-Economic-and-Fiscal-impacts-of-making-homes-energy-efficient.pdf.

Watson, J., Ekins, P., Gross, R., Anable. J., Barrett, J., Bell, K., Bradshaw, M., Brand, C., Darby, S., Demski, C., Pidgeon, N. and Webb, J (2017) *Review of Energy Policy*. London: UKERC. Available at: www.ukerc.ac.uk/publications/review-of-energy-policy-2017.html.

Webb, J. (2016) Heat and Energy Efficiency: Making Effective Policy. Advisory Group Report for the UK Committee on Climate Change. HM Government, London, UK. Available at: www.theccc.org.uk/publication/heat-and-energy-efficiency-advisory-group-report-making-effective-policy/.

Webb, J., Tingey, M. and Hawkey, D. (2017) What We Know About Local Authority Engagement in UK Energy Systems. UKERC, London, UK. Available at: www.ukerc.ac.uk/publications/what-we-know-about-local-authority-engagement-in-uk-energy-systems.html.

12 Policy mixes for sustainable energy transitions
The case of energy efficiency

*Florian Kern, Paula Kivimaa,
Karoline Rogge and Jan Rosenow*

Introduction

Any new policy goals pertaining to sustainable energy transitions and associated policy instruments to help foster such change will not exist in a vacuum. Rather, they will become embedded in pre-existing policy contexts with legacies of goals and instruments already in place (Kern and Howlett, 2009). It is this messy reality which ultimately influences policy outcomes instead of theoretical considerations around 'first best' policy options and 'optimal' policy design. It is therefore increasingly important to explicitly study policy mixes, how they can be designed and how they can be implemented in order to promote deliberate sustainable energy transitions (Rogge et al., 2017). The policy mix literature is an attempt to make sense of this empirical complexity.

This chapter therefore focuses on policy mixes for sustainable energy transitions, an emerging area of research at the interface of policy sciences and sustainability transition studies. However, definitions of what constitutes a policy mix vary widely in the literature. For example, while economists focus on the interactions of multiple instruments (Lehmann, 2012), the policy design literature goes beyond that by also including policy goals (Kern and Howlett, 2009). In addition, innovation studies have called for a reconceptualisation of policy mixes to better capture their complexity in a 'real-world' context, including the underlying policy processes through which policy mixes develop (Flanagan et al., 2011, Rogge and Reichardt, 2016). Within The Centre on Innovation and Energy Demand (CIED), we have built on these various streams of literature to further conceptual and empirical insights on real-world policy mixes for sustainable energy transitions (Rogge et al., 2017).[1]

Research on policy mixes started initially with an interest in multiple policy instruments targeting a given policy field (such as energy policy). In contrast to earlier proposals to address each policy goal via one policy instrument, this early literature on policy mixes – typically grounded in economics – acknowledges the existence of situations in which 'several – instead of one – policy instruments are used to address a particular environmental problem' (Braathen, 2007, p. 186). Further, '[p]olluting sources may be affected directly or indirectly by several policies addressing the same pollution problem. This is referred to as a policy mix' (Lehmann, 2012, p. 1). The aim of this literature is mainly to understand how different instruments interact to avoid negative effects.

Accordingly, much of the early research on policy mixes for energy transitions has focused on the analysis of interactions of policy instruments designed to affect the operation of energy systems (e.g. Sorrell and Sijm, 2003; Spyridaki and Flamos, 2014). This line of thinking in terms of mixes of policy instruments has also been picked up by organisations like the International Energy Agency (IEA) which published a report on 'Interactions of Policies for Renewable Energy and Climate' (IEA, 2011a). In another publication it argued that

> [t]he need for a policy mix has been recognised by many governments, but experience to date has been that the interactions among multiple policies are often not well understood nor well-coordinated, which can lead to policy redundancy or policies undermining one another, reducing the effectiveness and efficiency of the overall package.
> (IEA, 2011b, p. 60)

However, definitions of policy mixes since have extended beyond instrument interactions. Consequently, there is a range of interesting strands of research on policy mixes for energy transitions (see the section below). These may combine attention to the instrument mix with corresponding policy strategies with their long-term targets, and/or with the associated policy processes; the analysis of overarching policy mix characteristics such as consistency, coherence or credibility; and policy design considerations. Such a broad perspective on policy mixes for energy transitions draws influences from multiple areas, including governance arrangements for policy mixes (e.g. Howlett and Rayner, 2006), instrument mixes in energy policy (e.g. del Río, 2010; Sorrell and Sijm, 2003) and innovation policy mixes (e.g. Flanagan et al., 2011). It also connects more explicitly to the sustainability transitions literature (Markard et al., 2012). The policy relevance of such a consideration of broader policy mixes is also evidenced by the interest of the IEA that published a report on 'Real-world Policy Packages for Sustainable Energy Transitions' (IEA, 2017).

While policy mix thinking is relevant generally in all policy fields, it is specifically important in the context of energy transitions. Public policy is expected to heavily contribute to sustainable change in energy systems, not only by internalising externalities but also by addressing a range of other structural and transformational system failures (Weber and Rohracher, 2012). However, energy systems are a complex web of sub-systems, including a diversity of fuel supply, conversion and use systems, which are often addressed by a range of different policies – making the overall policy mixes large, complicated and most likely incoherent.

Much of the emerging research on policy mixes for energy transitions has, however, focused on energy supply, while there is much less research on energy efficiency policy mixes. This is so despite the fact that energy efficiency is considered as critically important to achieving an energy transition in line with the pledges made in the Paris Agreement (IEA, 2015) (see Chapter 1 and

Chapter 2). In addition, the respective policy literature has often pointed out that due to the variety and complexity of end-users of energy, there are no 'silver bullet' policies that can stimulate action across this variety of actors. Instead of single instruments, it has been argued that there is a need to design comprehensive energy efficiency policy mixes which address the various challenges different actors are facing in advancing energy efficiency (Nilsson, 2012). This is reflected in policy strategies aimed at influencing energy efficiency but is not studied much in the existing energy efficiency policy literature.

This constitutes the main empirical gap in the literature that our research was trying to address. The aim of this chapter is to summarise some of the empirical research on energy efficiency policy mixes conducted as part of CIED in order to draw out overall academic insights and avenues for further research. We also provide policy reflections on policy mixes for sustainable energy transitions in which energy efficiency plays a key role. The next section introduces recent conceptual and empirical advances in the interdisciplinary literature on policy mixes for energy transitions. The following section discusses selected research on energy efficiency policy mixes in the UK, Finland and at the EU level conducted by CIED. The final section summarises what overall lessons we have learned from this work, develops policy recommendations and suggests a number of avenues for future research.

Advancing research on policy mixes for energy transitions

Given the rapid increase in interest in the topic of policy mixes for energy transition two CIED authors (jointly with Michael Howlett) guest-edited a recent special issue in the journal of *Energy Research and Social Science* (November 2017, Vol. 33)[2] which goes beyond looking at instrument mixes. As summarised in Rogge *et al.* (2017) the contributions in this special issue are clustered around five themes: policy mix rationales, interactions and coordination of policy instruments, designing effective policy mixes, policy mixes for creative destruction and the role of actors and institutions in shaping energy transition policy mixes. Below we will discuss a number of selected contributions to the special issue in order to illustrate these different strands of work.

In terms of *policy mix rationales*, Jacobsson *et al.* (2017) argue that European Union (EU) interventions in the context of decarbonisation mainly rest on neoclassical economics assumptions. They propose that this approach neglects important insights about the non-linear nature of technical change and industrial dynamics that are very relevant in the context of energy transitions. They propose an innovation system approach as a rationale for intervention and draw lessons for how effective instrument mixes can be designed which pay greater attention to dynamic efficiency and the structural build-up of innovation systems.

Contributing to the theme of *instrument interactions*, del Río and Cerdá (2017) analyse the impact of instruments to promote renewable electricity on CO_2 prices established through a cap and trade scheme or carbon tax. Their research shows that negative interactions can be mitigated through coordination, and that the adaptability depends on the choice of instruments and design features of each tool. They also find that the negative impact on CO_2 prices is more likely under quantity-based than under-price-based instruments.

In terms of *designing effective policy mixes*, Falcone et al. (2017) provide an analysis of policy mixes in the Italian biofuel sector. They explore different crises scenarios in order to identify and recommend the most effective policy combinations to foster a sustainable energy transition using a fuzzy inference. Their findings show that the most effective policy mixes vary across the scenarios and according to different pursued objectives.

Under the theme of *policy mixes for creative destruction*, Rogge and Johnstone (2017) analyse the effect of deliberate phase-out policies of established technological regimes on the development and diffusion of low-carbon technologies. Based on the case of the German transition towards renewable electricity, they show through a survey of innovation activities of German manufacturers of renewable power generation technologies that Germany's nuclear phase-out policy had a positive influence on manufacturers' innovation expenditures for renewables.

In terms of the *role of actors and institutions in shaping energy transition policy mixes*, Bahn-Walkowiak and Wilts (2017) undertake a closer analysis of the institutional background of policy mixes. Their contribution raises questions about the potential impact of different institutional settings on the consistency and coherence of policy mixes in the field of resource efficiency. They map the distribution of institutional responsibilities in 32 EU countries and find that resource efficiency policies are still mainly disconnected from energy issues. The paper stresses the need to include institutional and multi-level governance considerations into the design and development of policy mixes.

These five themes are showcasing the variety of strands of research on policy mixes for energy transitions but empirically much of this research has focused on energy supply policy mixes. However, little work within these themes focuses specifically on energy efficiency.

Applying policy mix thinking to the case of energy efficiency: what have we learned?

Having reviewed recent research on policy mixes for sustainable energy transitions and identifying some key themes, this section will summarise selected empirical analyses of energy efficiency policy mixes which have been conducted as part of CIED and which contribute to discussions in a number of these strands.

How do complex policy mixes develop over time and how consistent and coherent can they be? The cases of UK and Finnish energy efficiency evolution

Much of the existing energy policy mix literature only captures the policy mix at one point in time. We argue that in the context of long-term energy transitions, further analysis is needed to investigate how real-world policy mixes develop over time and how their characteristics change. In line with existing literature in the field of policy design, we claim that this is important as it influences the potential performance of such complex mixes.

In Kern et al. (2017) we therefore adopted the definition of policy mixes as 'complex arrangements of multiple goals and means which, in many cases, have developed incrementally over many years' (Kern and Howlett, 2009, p. 395). Conceptually, we drew on the work of Howlett and colleagues who have foregrounded two relevant characteristics of policy mixes: consistency and coherence. Howlett and Rayner (2013) define consistency as 'the ability of multiple policy tools to reinforce rather than undermine each other in the pursuit of policy goals' (p. 174). They define coherence as the 'ability of multiple policy goals to co-exist with each other and with instrument norms in a logical fashion' (ibid.).

However, such characteristics of mixes are never static since goals and instruments may be added to and subtracted from the mix over time. Existing research has distinguished four processes through which policy mixes typically change: *layering*, *drift*, *conversion* and *replacement* (Howlett and Rayner, 2007, 2013; Kern and Howlett, 2009; Table 12.1).

Layering refers to adding new policy goals and instruments to the mix without discarding previous ones (Howlett and Rayner, 2013). Howlett and Rayner (2007) argue that this often results in incoherence among goals and inconsistency of instruments. *Drift* refers to changing policy goals without 'changing the instruments used to implement them. These instruments then can become inconsistent with the new goals and most likely ineffective in achieving them' (Kern and Howlett, 2009, p. 395). *Conversion* involves the reverse situation in which instrument mixes evolve while the old goals are retained: 'If the old goals lack coherence, then changes in policy instruments may either reduce levels of implementation conflicts or enhance them, but are unlikely to succeed in matching means and ends of policy' (ibid.). *Replacement* refers to a process in which a conscious effort is made by policymakers to fundamentally restructure

Table 12.1 Relationship between policy development processes and the expected coherence and consistency of a policy mix

Instruments goals	Consistent	Inconsistent
Coherent	*Replacement*: conscious effort to restructure goals and instruments by sweeping aside the old mix and designing a new one from scratch	*Conversion*: instruments evolve while the old goals are retained
Incoherent	*Drift*: changing policy goals without changing the instruments used to implement them	*Layering*: adding new policy goals and instruments to the mix without discarding previous ones

Source: based on Kern and Howlett (2009, p. 396).

both goals and instruments in a coherent and consistent manner by sweeping aside the old mix and designing a new one from scratch (Howlett and Rayner, 2007). However, most policy mixes develop through either *layering, conversion* or *drift*, often resulting in inconsistent and incoherent policy mixes (Howlett and Rayner, 2013).

We applied this framework to the development of building energy efficiency policy in Finland and the UK between 2000–2014. The analysis was based on a systematic review of existing databases,[3] policy documents and IEA country reviews to identify current building-related policy goals and instruments at the national level, as well as identifying goals and instruments, which had been added, amended and removed during this timeframe. This information allowed us to trace policy developments over time. We utilised 19 stakeholder interviews to check the list of policy instruments and elicit information about the development of the policy mixes (including insights related to their coherence and consistency).

In the case of Finland, the development of the policy mix tended to follow a *replacement* process in the form of coherent long-term policy goals and (increasing) consistency of the instrument mix used to implement them. These processes have led to a policy mix with some promise of effectiveness. In contrast, the UK analysis revealed a pattern more akin to *drift* as the introduction of social and carbon reduction goals into traditional energy efficiency ambitions led to a set of partly incoherent goals. The goals are combined with a relatively consistent and prior to 2015 largely well-targeted instrument mix but which also displayed some gaps. The case also showed a rapid accumulation of new instruments (*layering*).

Overall, the analysis showed that both countries have developed extensive policy mixes to address building energy efficiency, including a variety of goals and instruments and making use of many different instrument types. In both countries, more new goals and instruments have been added over time than have been, which poses increasing challenges in terms of policy coordination as well as evaluating such increasingly complex mixes. Our analysis also showed that while in the UK there has been a lot of 'churn' in policy instruments, Finland has had a somewhat more stable policy mix, where the added policies have not as radically altered the mix. This is important in the context of policy mixes for energy transitions as a rapidly fluctuating policy environment can slow down innovation processes, as companies generally prefer a more stable climate for investment. This means that the UK policy mix may deter low-energy innovations and their diffusion.

What kinds of comprehensive and well-targeted instrument mixes are needed for stimulating energy efficiency improvements?

The comprehensiveness of policies has long been argued to be a relevant success factor of environmental and energy policies (Sovacool, 2009; Walls and Palmer, 2001). However, in these studies, comprehensiveness has remained a loosely defined concept. Drawing on conceptualisations of comprehensiveness in the field of marketing and environmental management systems (Atuahene-Gima and Murray, 2004; Miller, 2008), Rogge and Reichardt (2016) have concretised policy mix comprehensiveness as a characteristic which 'captures how extensive and exhaustive its elements are' (p. 1627). While they also include the degree to which policymaking and implementation are based on extensive decision-making, in this paper (Rosenow et al., 2017), we focused on the comprehensiveness of the instrument mix in the area of energy efficiency in selected EU Member States.

In line with Rogge and Reichardt (2016) we argue that instrument mix comprehensiveness can be assessed according to the degree to which it considers relevant failures and barriers (Lehmann, 2012; Sorrell, 2004; Weber and Rohracher, 2012). More specifically, it can be captured by assessing whether the instrument mix includes technology push, demand pull and systemic instruments (Cantner et al., 2016). We developed an analytical framework that can be used for the empirical assessment of energy efficiency instrument mixes and their degree of comprehensiveness (Rosenow et al., 2017). The main building blocks of the framework we applied in this chapter are (a) technological specificity, (b) types of policy instruments and (c) sector specificity.

Technological specificity can be assessed using two dimensions: the cost of supported technology and the complexity of supported technology. Instrument types are critical for comprehensiveness as there is great variety of policy instrument types. Depending on the type, policy instruments also support specific technologies or are technologically neutral. A sector specific analysis may reveal important gaps in the instrument mix. This is important since the ambitious energy efficiency targets required for sustainable energy transitions mean that all sectors have a significant contribution to make (Braungardt et al., 2014). While comprehensiveness may also be assessed through the lens of additional dimensions (for example the degree to which all relevant actors are addressed or the degree of geographical coverage etc.), we argue that our analytical framework covering three key aspects offers an approach that can be applied relatively easily to existing instrument mixes within the energy efficiency policy domain.

We applied this concept of instrument mix comprehensiveness to the field of energy efficiency. The empirical analysis focuses on national energy efficiency policies that have been notified by EU Member States to the European Commission as part of their transposition of Article 7 of the EU Energy Efficiency Directive (EED). The EED establishes a framework of measures to ensure the achievement of the EU's 20 per cent energy savings target by 2020 (EU, 2012). Data on instrument mixes in selected EU Member States was obtained from national experts from Austria, Belgium, Bulgaria, Denmark, Estonia, France, Germany, Greece, Italy, Netherlands, Poland, Spain, Sweden and the United Kingdom.[4]

The data shows that in the selected EU Member States none of the instrument types utilised by these countries specifically target highly complex and capital-intensive technologies, but instead focus on technologies characterised by relatively moderate costs and complexity. We argue that a comprehensive energy efficiency instrument mix needs to cover the full range of technologies regarding complexity and costs. The limited focus on more complex and costly technologies indicates that further policy development is required in order to achieve deeper energy efficiency improvements across all sectors.

This finding may partly be a function of the focus on existing commercialised technologies (rather than innovative technologies or technology combinations) that characterises Article 7 policies. However, it also indicates a possible gap in the instrument mix supporting deeper energy efficiency improvements, whereby the next set of mass market efficiency measures are not being sufficiently supported or incentivised. This gap needs to be addressed if ambitious EU targets are to be met. However, adding such instruments may be costly and, therefore, politically contested.

Future research should identify more precisely (through ex post analyses) the degree of comprehensiveness of the instrument mix. In particular, one focus of such studies should be the types of technologies targeted within the energy efficiency space as this becomes increasingly important given the diversity of national approaches to delivering EU energy-savings targets.

Accelerating sustainable energy transitions by fostering creative destruction through policy mixes?

When major transformations of energy systems are needed, particularly at a rapid pace, it is not sufficient that a policy mix aims at incremental improvement and innovation support. In such cases, the policy mix also needs to entail more disruptive instruments to overturn unsustainable energy regimes based on fossil fuels and high levels of energy consumption. What elements such policy mixes should comprise, was the focus of a study published by Kivimaa and Kern (2016) that this sub-section is based on. Empirically, it analysed whether contemporary policy mixes for low-energy innovation in the UK and Finland have the characteristics proposed.

The core idea proposed in the study is that well-designed policy mixes for sustainable energy transitions would include elements of 'creative destruction', involving both policies aiming for the 'creation' of new and for 'destabilising' the old. However, creating such policy mixes is by no means easy as it is dependent on the prevailing political climate (Howlett and Rayner, 2007). Yet, we argue that energy transitions benefit from analyses of the degree to which this takes place, and how existing policy mixes could be improved in this regard.

The framework developed in Kivimaa and Kern (2016) proposes to conceptualise policy mixes from the perspective of creative destruction. It draws on multiple innovation and transition concepts, including disruptive innovation (Abernathy and Clark, 1985; Christensen, 1997), technological innovation systems (Bergek et al., 2008; Jacobsson and Bergek, 2011; Suurs and Hekkert, 2009), Strategic Niche Management (Hoogma et al., 2002; Smith and Raven, 2012) and transition management (Kemp and Rotmans, 2004; Rotmans et al.,

2001). Drawing from technological innovation systems and Strategic Niche Management literatures, it is clear that to support the emergence of new innovative niches, policy mixes need to address the following functions:

1 knowledge creation, development and diffusion (e.g. R&D funding schemes, innovation platforms, educational policies);
2 new market formation (e.g. regulation, economic policy instruments, public procurement);
3 price–performance improvements (e.g. deployment and demonstration subsidies enabling learning-by-doing);
4 entrepreneurial experimentation (e.g. policies stimulating entrepreneurship and diversification of existing firms);
5 resource mobilisation (e.g. R&D funding subsidies, low-interest loans, labour-market policies);
6 support from powerful groups/legitimisation (e.g. foresight exercises, labelling); and
7 influence on the direction of search (e.g. strategic goals, targeted R&D funding, regulations, tax incentives, voluntary agreements).

In addition, Kivimaa and Kern (2016) argue, with support from the literature on transitions management and disruptive innovation, that in cases of unsustainable energy regimes, policy mixes also need to address the following regime destabilising functions:

1 control policies that internalise environmental costs (e.g. pollution taxes, carbon trading);
2 significant changes in regime rules (e.g. structural reforms in legislation or significant new overarching legislation);
3 reduced support for dominant regime technologies (e.g. withdrawing support for unsustainable technologies by cutting/removing R&D funding and other subsidies, or technology bans); and
4 changes in social networks and replacement of key actors (e.g. increasing the number of niche actors in advisory councils and forming new organisations and networks with key roles in system change).

Empirically, we examined the policy mixes in Finland and the UK addressing energy efficiency and energy demand reduction. The analysis covered three regimes – mobility, electricity and heating of buildings – cutting across multiple policy domains including innovation, energy, fiscal and transport policies. The method utilised was a policy instrument mapping exercise systematically going through four international data sources: the IEA's reviews of energy policies and policies and measures databases on energy efficiency, the European Environmental Agency's database on climate change mitigation policies and measures, the European Commission's Erawatch research and innovation policy database and the IEA Sustainable Buildings Centre's Building Energy Efficiency Policies

224 *Florian Kern et al.*

database. In addition, the data was supplemented with searches made on governmental websites to get descriptions of the objectives, justifications and main content of the policy instruments, and to identify new organisations and networks.

The study identified 73 policy instruments in the UK and 65 in Finland. In both countries there was an imbalance of policy instruments between niche support and regime destabilisation, although several 'control policies' were found in both countries. This imbalance was not only reflected in the number of instruments but also in policy content. Specifically, significant changes in regime rules and in policy networks and actors were rare. In addition, in the UK, we did not find reducing policy support for high-energy technologies beyond EU requirements. This is not surprising given the political difficulties of such changes but highlights gaps in the existing policy mixes.

Some of the destabilising functions proposed here connect to a recent and emerging debate on exnovation policies that aim to end use of given technologies by deliberately removing the infrastructure it relies on (David, 2017). In addition, our study links to the debate on the need for explicit phase-out policies to support sustainable energy transitions (Rehner and McCauley, 2016; Rogge and Johnstone, 2017). Further research could examine the development of 'creative destructive' policy mixes over time, how the instruments function in practice and their impact on the strategies of different policy target groups.

How to expand policy mix studies to the analysis of policy processes?

As mentioned above, recent research on policy mixes for energy and sustainability transitions has developed a broader perspective on policy mixes which includes policy strategies, policy mix characteristics (such as consistency and credibility), and increased attention to actors and institutions (Rogge *et al.*, 2017). One of the additional aspects highlighted as particularly worthy of further consideration within broader policy mix research are policy processes and the role they play for advancing sociotechnical transitions towards sustainability (Flanagan *et al.*, 2011; Rogge and Reichardt, 2016). Research shedding greater light on the process dimension of policy mixes calls for extending the interdisciplinary nature of policy mix research by more explicitly incorporating theories of the policy process (Sabatier and Weible, 2014).

As argued in Kern and Rogge (2018), such a greater attention to policy processes promises three advantages. First, policy processes can have direct impacts on innovation rather than just an indirect impact by shaping policy strategies and instrument mixes (Reichardt *et al.*, 2017). Second, studying the co-evolution of policy and sociotechnical change calls for more explicit attention to policy processes, which in turn may enable a better understanding of the dynamic nature and causal links between the two (Hoppmann *et al.*, 2014; Reichardt *et al.*, 2016). Third, a more sophisticated conceptualisation of policy processes may allow for a more proactive consideration of the underlying politics when drafting policy advice regarding policy design and procedural aspects and may thus have a greater chance of being adopted (Edmondson et al., 2018; Rogge and Reichardt, 2016).

Only few studies in the field of sustainability transitions have so far substantively drawn on theories of the policy process (Kern and Rogge, 2018). Exemptions include Markard et al. (2016) drawing on Sabatier's advocacy coalition framework (ACF) (Sabatier, 1988), Geels and Penna (2015) drawing on Baumgartner's punctuated equilibrium theory (Baumgartner and Jones, 1993) as well as Normann (2015) drawing on Kingdon's multiple streams approach (Kingdon, 1995). However, most of these contributions rather loosely build on theories of the policy process and typically refrain from justifying their choice vis a vis alternatives. In addition, they often rely on the 'classic' version of these analytical frameworks, neglecting more recent debates and further conceptual developments in the policy sciences literatures.

Therefore, in Kern and Rogge (2018) we provide a critical review of five well-established theories of the policy process. These include Sabatier's ACF, Kingdon's multiple streams approach, Baumgartner's punctuated equilibrium theory, Hajer's discourse coalitions framework (Hajer, 1995), and Pierson's policy feedback approach (Pierson, 1993). For each of these theories we provide an overview of the origin, key concepts, empirical applications, recent theoretical advances and most important criticisms (see Table 12.2). Perhaps most importantly we also offer reflections on their suitability for answering research questions of interest to scholars in the field of sociotechnical transitions towards sustainability. Overall, we find a great potential for cross-fertilisation of ideas across transition and policy studies, but we also identify two important shortcomings.

The first shortcoming is that these theories are often applied to study the emergence of single policy instruments or purposively designed policy programmes, rather than to explain the evolution of messy, real-world policy mixes. However, as such policy mixes are particularly important in the context of energy transitions, we argued that the reviewed theories of the policy process may have to be adapted to the logic of thinking in terms of policy mixes. For example, greater attention should be paid to policy changes which guide the direction of change, e.g. towards low-carbon solutions (Kern and Rogge, 2016).

The second shortcoming is that analyses often stop short at the output of policy processes and do not study policy outcomes and impacts, which are of particular importance in studying sustainable energy transitions. Indeed, many of the reviewed theories only help to explain how and why policies were adopted, with little attention to how these policy outputs impact the sociotechnical system. In Kern and Rogge (2018) we differentiated between direct and indirect links between policy processes and sociotechnical change which both should be taken into consideration in future policy mix studies. While the indirect link manifests itself through policy outputs (e.g. changes to the instrument mix) leading to impacts on the sociotechnical system, the direct link suggest that the nature of policy processes, such as a participatory policymaking style, can also directly influence sociotechnical change, e.g. through influencing perceptions and beliefs of innovators (Reichardt et al., 2017).

Table 12.2 Overview of policy process theories and their application in transition studies

	Advocacy coalition framework	Multiple stream approach	Punctuated equilibrium theory	Discourse coalition framework	Policy feedback theory
Scope and level of analysis	Advocacy coalition interaction, learning and policy change Coalitions and subsystems	Policy choice under ambiguity System, but implicit, and focus is on actors coupling streams	Political system towards stability and periodic major change System	Discourse coalition interaction, discursive struggles Coalitions and subsystem	How policies shape politics and subsequent policymaking System, but implicit
Model of the individual	Bounded rational; emphasis that individuals are motivated by beliefs and prone to devil shift	Challenges assumptions of comprehensive rationality; focus on ambiguity	Boundedly rational, particularly related to attention	Not explicitly discussed	Not explicitly discussed; suggests individual choice is shaped by policies and institutions
Actors making choices	Policy actors who form coalitions, act strategically, learn and so forth	Policy entrepreneurs and policymakers	Broadly, interest groups and other organisations, as well as individuals within groups and different venues	Policy actors who form coalitions and engage in a set of practices	Implicitly, actors who are affected by policy may in turn become policy actors
Relationship among key concepts	Factors that influence coalition formation, policy learning, and policy change	Broadly, three streams that come together during 'windows of opportunity' to cause major policy change	Factors that lead to major policy change and those that constrain change or produce incrementalism	Discourses are reproduced through practices and influence the policy response to policy problems as well as whether certain situations are seen as public policy problems	The effects of public policy on the meaning of citizenship, form of governance, power of groups, and political agendas – all of which affect future policy

Most promising aspects from transition studies perspective	Reconceptualise policy regime, incl. competing advocacy coalition(s) and integrate how beliefs can change over time and with what effects on dominant advocacy coalition. Incorporate role of public opinion on policymaking	Focus on policy entrepreneurs sheds light on role of individual agency vis-à-vis systems. Idea that solutions/policies look for problems is useful in studying agency of niche actors pushing their respective solutions, but also requires detailed analysis of developments in politics stream	Clear parallels in conceptualising equilibrium and path dependency which can be disrupted by crises in both punctuated equilibrium theory (PET) and multi-level perspective (MLP): scope for mutual learning about mechanisms of change. Approach can be also applied to diffusion of public policy innovations which is useful in the context of technological innovation system (TIS) studies trying to account for transnational factors and institutional political contexts	Focus on interpretative processes and the roles of ideas in shaping policy, especially in situations of Knightian uncertainty, is very promising. Foregrounding discursive struggles between competing discourse coalitions is a useful conceptual tool to focus on the politics of transition processes (e.g. on how actors interpret sustainability and the goals of potential transitions differently)	Promising approach to conceptualise the co-evolution between policy mixes and sociotechnical systems (via policy effects and feedback processes). Extend notion of path dependency of policy regime by paying attention to various policy effects as well as feedbacks. Utilise approach to conduct analyses of effects of policies on mass publics and their political mobilisation in transitions
Application in transitions literature	Markard et al. 2016; Geels and Penna 2015	Normann 2015; Elzen et al. 2011	Geels and Penna 2015	Smith and Kern 2009; Rosenbloom et al. 2016	Edmondson et al. 2018

Source: Kern and Rogge (2018), with permission.

228 *Florian Kern et al.*

Within CIED we have started to tackle both shortcomings by developing an interdisciplinary conceptual framework for investigating the co-evolution of policy mix and sociotechnical change (Edmondson *et al.*, 2018). In order to explicitly consider the role of policy processes in this co-evolutionary process we have drawn on Policy Feedback Theory, which focuses on how policies shape politics and the resulting effects on further policymaking. Integrating this theory of the policy process with sociotechnical transitions thinking allows us to account for multiple policy effects (resource, interpretative and institutional) on sociotechnical change and resultant feedback mechanisms (cognitive, administrative and fiscal) influencing the policy processes that underpin further policy mix change.

We have illustrated this novel analytical framework using the case of the UK zero carbon homes policy mix. This is an example where an ambitious policy target lost political support over time due to a range of policy effects and feedback mechanisms, ultimately leading to its abandonment. The example highlights that policy mixes for sustainable energy transitions should be designed to create incentives for beneficiaries to mobilise further support, while at the same time addressing a number of prevailing challenges that may undermine political support over time. Overall, we think that drawing on a range of policy process theories can enrich academic analysis and provide more adequate policy thinking about policy mixes for energy transitions.

Conclusions and policy recommendations

The research summarised in this chapter aptly demonstrates that policy mix thinking is an important analytical perspective in the context of policies to support sustainable energy transitions. This is because various policy mix conceptualisations allow scholars and policy analysts to better deal with the complexity of real-world policymaking rather than simplistic economic theoretical thinking about policy that still dominates scholarly and policy debates to some extent.

While instrument interactions matter and have been the subject of much research, this chapter has argued that there are important other issues to consider. Such issues include different policy mix rationales, processes of designing and maintaining effective policy mixes over time, the need to deliberately phase out unsustainable technologies and practices, and the important role of actors and institutions in influencing the design and implementation of policy mixes. The research summarised above provides important early insights into these issues.

Research implications

Our research identified a variety of avenues for future investigations. Given the complexity of studying policy mixes (the larger the mix analysed, the less depth and complexity can be addressed), there is only relatively little research that takes a comparative perspective. However, as our CIED research discussed above shows, comparative analysis can lead to interesting insights. It can highlight issues in a policy mix that only reveal themselves in comparison to another policy mix (in a different country or sector, for example). Such comparative work could help explain key similarities and differences, thereby potentially identifying generic (e.g. technology or sector-related factors) as well as country specific factors (e.g. national policy traditions or policy styles) (e.g. see Howlett and Ramesh, 1993).

There are also open questions about which institutional arrangements can foster the development of well-coordinated and comprehensive policy mixes and which capabilities are required for managing complex policy mixes. Few if any studies focus on how policymakers and implementing organisations can acquire or develop such competences. This development of competences is especially difficult in administrative contexts, where civil servants' work is focused on developing one specific new policy instrument in isolation and who often remain in a department only for a short time, as is the case in the UK.

In addition, the evaluation of real-world (rather than intended) policy mixes for transitions has been little addressed (Kivimaa et al., 2017). There are important questions about how to evaluate the impact of such policy mixes ex post and *ex ante*. While much research focuses on characteristics of policy mixes such as coherence or consistency as a proxy for potential success, more sophisticated methodologies are needed to analyse potential or actual effects of complex policy mixes on energy transitions. One possibility to make progress in this regard is to draw on the evaluation literature and adopt their approaches to policy mix analysis as has been proposed by Kivimaa *et al.* (2017).

One further avenue for future research is to explore how to phase policies and alter policy mixes in line with progress of energy transitions. This is an issue raised recently e.g. in Meckling *et al.* (2017) but there is not much explicit consideration of these issues in the sustainability transitions literature yet. Future research should therefore combine policy mix thinking with the work on different phases of transitions and work out which combinations of instruments are most appropriate for which phase.

Finally, future research should pay greater attention to the politics of designing policy mixes. While Rogge and Reichardt (2016) include policy processes in their framework for studying policy mixes, there is little detail on how the politics of such processes might be conceptualised. As summarised above we conducted a review of different policy process theories and their potential use in the context of studying the politics of sustainability transitions (Kern and Rogge, 2018) and have developed a novel framework based on the policy feedback literature (Edmondson et al., 2018) but much research remains to be done on this important aspect of policy mixes.

Policy recommendations

Given the broad scope of the research reported in this chapter and the varied empirical cases which our research has covered, the idea here is to draw out some broad principles for policymaking rather than providing suggestions about specific interventions or policy changes.

In line with existing work on energy efficiency policy our research on energy efficiency policy mixes strengthens the notion that there are no 'silver bullets', i.e. single instruments that can bring about energy transitions. Instead policymakers need to develop well-managed portfolios of policy goals, strategies and instruments to foster energy transitions. These policy mixes need to be continuously (re-)assessed and modified as necessary when the transition progresses and may need to be backed up with supportive changes in administrative organisations and processes.

In terms of supporting energy efficiency and energy demand reduction as a core contribution to sustainable energy transitions, it has been argued that 'efficiency first' should be a primary policy goal (Rosenow and Cowart, 2017). For policymakers this principle means that, put simply, the policy mix should prioritise incentivising investments in customer-side efficiency (including end-use energy efficiency and demand response) whenever they would cost less, or deliver more value, than investing in energy infrastructure, fuels, and supply alone. While this may sound very much like common-sense policy, unfortunately this principle is not heeded in much energy policy. For example, in the UK not a single pound of the £256 billion investment pipeline for energy infrastructure is allocated to energy efficiency improvements (Rosenow and Cowart, 2017).

Beyond such broad principles informing policy mix design, there is 'no one size fits all', ideal instrument mix, but policy mixes need to be tailored to specific goals and the (institutional and country) settings they are applied in. What is important is that this tailoring sufficiently acknowledges the existing policy mix, on top of which new policies are designed. This may require the phase out of existing policies supporting unsustainable energy production or consumption, while creating or maintaining sufficient support for innovation (Kivimaa and Kern, 2016). We have also emphasised the need for comprehensive instrument mixes, which in the case of aiming for radical energy efficiency improvements, for example, means not just to focus on near-term, relatively cost-effective technologies, but also requires incentives for costlier, more complex technologies which are needed for deep energy efficiency improvements (Rosenow et al., 2017).

Much of the emphasis on the coherence and consistency of policy mixes in the literature may suggest that in order to develop successful mixes, policymakers need to aim for a complete overhaul of existing policy arrangements or completely new policy packages. However, Howlett and Rayner argue that policy patching can also be a successful strategy, 'much in the same way as software designers issue "patches" for their operating systems and programmes in order to correct flaws or allow them to adapt to changing circumstances' (Howlett and Rayner, 2013, p. 177). Our empirical research on Finnish energy efficiency policy shows that a patching strategy can be successful so it is important for policymakers to know that developing promising policy mixes does not require 'starting from a clean slate' (Kern et al., 2017). This is a promising insight for policymakers and should encourage an honest assessment of current policy mixes for energy transitions along the lines discussed above, which can then be used to inform suitable patching strategies.

Notes

1 Various other terms such as policy portfolios, policy bundles or policy packages have been used in similar ways (see Howlett et al., 2015). However, we prefer to use the notion of policy mixes as these other terms often have a connotation of deliberate and well-designed policy mixes, which does not characterise most policy mixes in reality.
2 See www.sciencedirect.com/journal/energy-research-and-social-science/vol/33.
3 These included: IEA policies and measures databases on energy efficiency, the European Environmental Agency's database on climate change mitigation policies and measures in Europe; the IEA Sustainable Buildings Centre's Building Energy Efficiency Policies database and the ODYSSEE–MURE database.
4 All of the experts were part of the Energy Saving Policies and Energy Efficiency Obligation Schemes (ENSPOL) project, which was funded by the European Commission. The full list of involved institutions can be found on the ENSPOL project website (http://enspol.eu).

References

Abernathy, W.J. and Clark, K.B. (1985) Innovation: Mapping the winds of creative destruction. *Research Policy* 14(1): 3–22.

Atuahene-Gima, K. and Murray, J. (2004) Antecedents and outcomes of marketing strategy comprehensiveness. *Journal of Marketing* 68(4): 33–46.

Bahn-Walkowiak, B. and Wilts, H. (2017) The institutional dimension of resource efficiency in a multi-level governance system – Implications for policy mix design. *Energy Research and Social Science* 33: 163–172.

Baumgartner, F.R. and Jones, B.D. (1993) *Agendas and Instability in American Politics*. University of Chicago Press, Chicago, USA.

Bergek, A., Jacobsson, S., Carlsson, B., Lindmark, S. and Rickne, A. (2008) Analyzing the functional dynamics of technological innovation systems: A scheme of analysis. *Research Policy* 37(3): 407–429.

Braathen, N.A. (2007) Instrument mixes for environmental policy: How many stones should be used to kill a bird? *International Review of Environmental and Resource Economics* 1(2): 185–235.

Braungardt, S., Eichhammer, W., Elsland, R., Fleiter, T., Klobasa, M., Krail, M., Pfluger, B., Reuter, M., Schlomann, B., Sensfuss, F., Tariq, S., Kranzl, L., Dovidio, S. and Gentili, P. (2014) Study evaluating the current energy efficiency policy framework in the EU and providing orientation on policy options for realising the cost-effective energy efficiency/saving potential until 2020 and beyond. PWC on behalf of DG ENER. Fraunhofer ISI, TU Vienna, Karlsruhe, Vienna, Rome.

Cantner, U., Graf, H., Herrmann, J. and Kalthaus, M. (2016) Inventor networks in renewable energies: The influence of the policy mix in Germany. *Research Policy* 45(6): 1165–1184.

Christensen, C.M. (1997) *The Innovator's Dilemma*. Harvard Business Press, Harvard, USA.

David, M. (2017) Moving beyond the heuristic of creative destruction: Targeting exnovation with policy mixes for energy transitions. *Energy Research and Social Science* 33: 138–146.

del Río, P. (2010) Analysing the interactions between renewable energy promotion and energy efficiency support schemes: The impact of different instruments and design elements. *Energy Policy* 38(9): 4978–4989.

del Río, P. and Cerdá, E. (2017) The missing link: The influence of instruments and design features on the interactions between climate and renewable electricity policies. *Energy Research and Social Science* 33: 49–58.

Edmondson, D., Kern, F. and Rogge, K.S. (2018) The co-evolution of policy mixes and socio-technical systems: Towards a conceptual framework of policy mix feedback in sustainability transitions. *Research Policy* (in press).

Elzen, B., Geels, F.W., Leeuwis, C. and Van Mierlo, B. (2011) Normative contestation in transitions 'in the making': Animal welfare concerns and system innovation in pig husbandry. *Research Policy* 40: 263–275. DOI: 10.1016/j.respol.2010.09.018.

EU (2012) Directive 2012/27/EU of the European Parliament and of the Council of 25 October 2012 on Energy Efficiency. Official Journal of the European Union 315/1. Brussels.

Falcone, P.M., Lopolito, A. and Sica, E. (2017) Policy mixes towards sustainability transition in the Italian biofuel sector: Dealing with alternative crisis scenarios. *Energy Research and Social Science* 33: 105–114.

Flanagan, K., Uyarra, E. and Laranja, M. (2011) Reconceptualising the 'policy mix' for innovation. *Research Policy* 40(5): 702–713.

Geels, F.W. and Penna, C.C.R. (2015) Societal problems and industry reorientation: Elaborating the Dialectic Issue LifeCycle (DILC) model and a case study of car safety in the USA (1900–1995). *Research Policy* 44(1): 67–82.

Hajer, M.A. (1995) *The Politics of Environmental Discourse: Ecological Modernization and the Policy Process*. Clarendon Press, Oxford, UK.

Hoogma, R., Kemp, R., Schot, J. and Truffer, B. (2002) *Experimenting for Sustainable Transport: The Approach of Strategic Niche Management*. Spon Press, London, UK.

Hoppmann, J., Huenteler, J. and Girod, B. (2014) Compulsive policy-making. The evolution of the German feed-in tariff system for solar photovoltaic power. *Research Policy* 43(8): 1422–1441.

Howlett, M. and Ramesh, M. (1993) Patterns of policy instrument choice: Policy styles, policy learning and the privatization experience. *Review of Policy Research* 12(1–2): 3–24.

Howlett, M. and Rayner, J. (2006) Understanding the historical turn in the policy sciences: A critique of stochastic, narrative, path dependency and process sequencing models of policy-making over time. *Policy Sciences* 39(1): 1–18.

Howlett, M. and Rayner, J. (2007) Design principles for policy mixes: Cohesion and coherence in 'new governance arrangements'. *Policy and Society* 26(4): 1–18.

Howlett, M. and Rayner, J. (2013) Patching vs packaging in policy formulation: assessing policy portfolio design. *Politics and Governance* 1(2): 170–182.

Howlett, M., How, Y.P. and del Río, P. (2015) The parameters of policy portfolios: verticality and horizontality in design spaces and their consequences for policy mix formulation. *Environment and Planning C: Government and Policy* 33(5): 1233–1245.

IEA (2011a) Interactions of Policies for Renewable Energy and Climate. International Energy Agency, Paris, France.

IEA (2011b) Summing up the Parts, Combining Policy Instruments for Least Cost Climate Mitigation Strategies. International Energy Agency, Paris, France.

IEA (2015) World Energy Outlook Special Report 2015: Energy and Climate Change. International Energy Agency, Paris, France.

IEA (2017) Real-world Policy Packages for Sustainable Energy Transitions. International Energy Agency, Paris, France.

Jacobsson, S. and Bergek, A. (2011) Innovation system analyses and sustainability transitions: Contributions and suggestions for research. *Environmental Innovation and Societal Transitions* 1(1): 41–57.

Jacobsson, S., Bergek, A. and Sandén, B. (2017) Improving the European Commission's analytical base for designing instrument mixes in the energy sector: Market failures versus system weaknesses. *Energy Research and Social Science* 33: 11–20.

Kemp, R. and Rotmans, J. (2004) Managing the Transition to Sustainable Mobility. In: Elzen, B., Geels, F.W. and Green, K. (Eds) *System Innovation and the Transition to Sustainability*. Cheltenham: Edward Elgar, 137–167.

Kern, F. and Howlett, M. (2009) Implementing transition management as policy reforms: A case study of the Dutch energy sector. *Policy Sciences* 42(4): 391–408.

Kern, F., Kivimaa, P. and Martiskainen, M. (2017) Policy packaging or policy patching? The development of complex energy efficiency policy mixes. *Energy Research and Social Science* 23: 11–25.

Kern, F. and Rogge, K.S. (2016) The pace of governed energy transitions: Agency, international dynamics and the global Paris agreement accelerating decarbonisation processes? *Energy Research and Social Science* 22: 13–17.

Kern, F. and Rogge, K.S. (2018) Harnessing theories of the policy process for analysing the politics of sustainability transitions: A critical survey. *Environmental Innovation and Societal Transitions* 27: 102–117.

Kingdon, J.W. (1995) *Agendas, Alternatives, and Public Policies*. Longman, New York, USA.

Kivimaa, P. and Kern, F. (2016) Creative destruction or mere niche support? Innovation policy mixes for sustainability transitions. *Research Policy* 45(1): 205–217.

Kivimaa, P., Kangas, H-L. and Lazarevic, D. (2017) Client-oriented evaluation of 'creative destruction' in policy mixes: Finnish policies on building energy efficiency transition. *Energy Research and Social Science* 33: 115–127.

Lehmann, P. (2012) Justifying a policy mix for pollution control: A review of economic literature. *Journal of Economic Surveys* 26(1): 71–97.

Markard, J., Raven, R. and Truffer, B. (2012) Sustainability transitions: An emerging field of research and its prospects. *Research Policy* 41(6): 955–967.

Markard, J., Suter, M. and Ingold, K. (2016) Socio-technical transitions and policy change – Advocacy coalitions in Swiss energy policy. *Environmental Innovation and Societal Transitions* 18: 215–237.

Meckling, J., Sterner, T. and Wagner, G. (2017) Policy sequencing toward decarbonization. *Nature Energy* 2: 918–922.

Miller, C. (2008) Decisional comprehensiveness and firm performance: Towards a more complete understanding. *Journal of Behavioral Decision Making* 21(5): 598–620.

Nilsson, M. (2012) Energy Governance in the European Union: Enabling Conditions for a Low Carbon Transition? In: Verbong, G. and Lorbach, D. (Eds) *Governing the Energy Transition: Reality, Illusion, or Necessity?* Routledge, New York, USA, 296–316.

Normann, H.E. (2015) The role of politics in sustainable transitions: The rise and decline of offshore wind in Norway. *Environmental Innovation and Societal Transitions* 15: 180–193.

Pierson, P. (1993) When effect becomes cause: Policy feedback and political change. *World Politics* 45(4): 595–628.

Rehner, R. and McCauley, D. (2016) Security, justice and the energy crossroads: Assessing the implications of the nuclear phase-out in Germany. *Energy Policy* 88: 289–298.

Reichardt, K., Negro, S.O., Rogge, K.S. and Hekkert, M.P. (2016) Analyzing interdependencies between policy mixes and technological innovation systems: The case of offshore wind in Germany. *Technological Forecasting and Social Change* 106: 11–21.

Reichardt, K., Rogge, K.S. and Negro, S.O. (2017) Unpacking policy processes for addressing systemic problems in technological innovation systems: The case of offshore wind in Germany. *Renewable and Sustainable Energy Reviews* 80: 1217–1226.

Rogge, K.S. and Johnstone, P. (2017) Exploring the role of phase-out policies for low-carbon energy transitions: The case of the German Energiewende. *Energy Research and Social Science* 33: 128–137.

Rogge, K.S., Kern, F. and Howlett, M. (2017) Conceptual and empirical advances in analysing policy mixes for energy transitions. *Energy Research and Social Science* 33: 1–10.

Rogge, K.S. and Reichardt, K. (2016) Policy mixes for sustainability transitions: An extended concept and framework for analysis. *Research Policy* 45(8): 1620–1635.

Rosenbloom, D., Berton, H. and Meadowcroft, J. (2016). Framing the sun: A discursive approach to understanding multi-dimensional interactions within socio-technical transitions through the case of solar electricity in Ontario, Canada. *Research Policy*, 45L 6: 1275–1290. DOI: 10.1016/j.respol.2016.03.012.

Rosenow, J. and Cowart, R. (2017) Efficiency First: Reinventing the UK's Energy System Growing the Low-Carbon Economy, Increasing Energy Security, and Ending Fuel Poverty, The Regulatory Assistance Project. Available at: www.raponline.org/knowledge-center/efficiency-first-reinventing-uks-energy-system.

Rosenow, J., Kern, F. and Rogge, K. (2017) The need for comprehensive and well targeted instrument mixes to stimulate energy transitions: The case of energy efficiency policy. *Energy Research and Social Science* 33: 95–104.

Rotmans, J., Kemp, R. and van Asselt, M. (2001) More evolution than revolution: Transition management in public policy. *Foresight* 3(1): 15–31.

Sabatier, P.A. (1988) An advocacy coalition framework of policy change and the role of policy-oriented learning therein. *Policy Sciences* 21(2): 129–168.

Sabatier, P.A. and Weible, C.M. (2014) *Theories of the Policy Process*. Westview Press, Boulder, CO, USA.

Smith, A. and Kern, F. (2009) The transitions storyline in Dutch environmental policy. *Environmental Politics* 18(1): 78–98. DOI: 10.1080/09644010802624835.

Smith, A. and Raven, R. (2012) What is protective space? Reconsidering niches in transitions to sustainability. *Research Policy* 41(6): 1025–1036.

Sorrell, S. (2004) Understanding Barriers to Energy Efficiency. In: Sorrell, S., O'Malley, E., Schleich, J. and Scott, S. (Eds) *The Economics of Energy Efficiency – Barriers to Cost-Effective Investment*. Edward Elgar, Cheltenham, UK, 25–94.

Sorrell, S. and Sijm, J. (2003) Carbon trading in the policy mix. *Oxford Review of Economic Policy* 19(3): 420–437.

Sovacool, B.K. (2009) The importance of comprehensiveness in renewable electricity and energy-efficiency policy. *Energy Policy* 37(4): 15–29.

Spyridaki, N.A. and Flamos, A. (2014) A paper trail of evaluation approaches to energy and climate policy interactions. *Renewable and Sustainable Energy Reviews* 40: 1090–1107.

Suurs, R. and Hekkert, P (2009) Cumulative causation in the formation of a technological innovation system: The case of biofuels in the Netherlands. *Technological Forecasting and Social Change* 76(8): 1003–1020.

Walls, M. and Palmer, K. (2001) Upstream pollution, downstream waste disposal, and the design of comprehensive environmental policies. *Journal of Environmental Economics and Management* 41(1): 94–108.

Weber, K.M. and Rohracher, H. (2012) Legitimizing research, technology and innovation policies for transformative change: Combining insights from innovation systems and multi-level perspective in a comprehensive 'failures' framework. *Research Policy* 41(6): 1037–1047.

13 Managing energy and climate transitions in theory and practice

A critical systematic review of Strategic Niche Management

Kirsten E.H. Jenkins and Benjamin K. Sovacool

Introduction

Ambitious goals for reducing carbon emissions require a rapid and extensive deployment of low-carbon technologies throughout the economy, with far-reaching implications for infrastructures, institutions, social practices and cultural norms. As the International Renewable Energy Agency (IREA) and International Energy Agency (IEA) (2017) recently noted, meeting the goals enshrined in the Paris Agreement – limiting global temperature rise to below 2°C above pre-industrial levels – demands we reduce the carbon dioxide intensity of the global economy by 85 per cent in 35 years. This corresponds to an average reduction of energy-related carbon dioxide emissions of about 2.6 per cent per year, or 0.6 Gigatons per year (IREA and IEA, 2017).

Meeting such targets will necessitate low-carbon transitions across multiple sociotechnical domains, namely electricity and heat, industry and buildings, forestry and agriculture, and transport, to name a few (Geels *et al.*, 2017). Yet, so far, progress on global energy and climate policy has been phlegmatic, and the pace of change is set to become ever more lethargic in the wake of the Trump Administration's plans to reinvest in carbon-intensive forms of energy such as tar-sands, oil and natural gas, and coal. Admittedly, the rate and scale of this transformative change has few historical precedents and represents a major policy challenge (Sovacool, 2016).

Transitions frameworks for understanding the pathways by which these changes occur have emerged in response to universal and localised energy challenges, including the coupled threats of climate change, fossil fuel depletion and fuel poverty. This chapter focuses on new technological innovation, the destabilisation of dominant energy regimes and the reframing of transitions goals through one such framework: Strategic Niche Management (SNM) (Markard *et al.*, 2012; Schot and Steinmuller, 2016). SNM promotes 'the reflexive management of real-world experiments in the form of pilot and demonstration projects, in which new sociotechnical configurations can grow and conditions for their 'up-scaling' can be elaborated"' (van den Bergh *et al.*, 2011, p. 13). In essence, SNM offers a means of learning about and enhancing the development and diffusion of new technologies with the aim of meeting low-carbon and low-energy goals.

Since the articulation and development of the SNM field by the end of the 1990s (Kemp et al., 1998; Weber et al., 1999), applications have increased steadily, gaining particular traction with energy scholars. Articles have been published with regards to electricity production and use with a focus on biomass (Raven, 2005; van der Laak et al., 2007; Verbong et al., 2010), zero energy buildings (Jain et al., 2016; Martiskainen and Kivimaa, 2018), clean vehicles (Hoogma et al., 2002; Sushandoyo and Magnusson, 2014) and the role of actors (Caniëls and Romijn, 2008; Lovell, 2007) among others. As a further example, Sovacool (2017) surveyed social science theorists about conceptual approaches for transitions in the transport sector to electric mobility, where both SNM and sociotechnical transitions theory were prominently discussed.

In this chapter, we provide an updated, confirmatory analysis of the ten-year literature review of the SNM literature provided by Schot and Geels (2008). Moreover, we present a synthesis of lessons learnt for both academic studies and policymakers working towards low-energy transitions. Specifically, we ask: what insights do 15 years of SNM literature offer for how to manage, or accelerate, low-carbon energy transitions? Arranz (2017) notes in a meta-survey of the sociotechnical transitions field the need for more refined understanding of what it takes to drive energy transitions – a progress typically undertaken by looking at past transitions (Grubler et al., 2016). This chapter makes another important step towards this goal through its focus on forward-looking policy recommendations.

We begin with a brief overview of the SNM heuristic, necessary to contextualise the results that follow. We then present the methodology behind our systematic review and content analysis covering 15 years of SNM literature (between 2002 and 2016) published across 9 databases, as well as 22 newsletters developed by the Sustainability Transitions Research Network (STRN). This approach allows for the identification of emergent research themes, as well as data about authors, their methods, and case studies in addition to analytical strategies and proselytised policy recommendations.

Theorising sociotechnical transitions: a brief summary of Strategic Niche Management

The development of the SNM theory coincided and coevolved with that of the multi-level perspective (MLP) model on sociotechnical systems (introduced in Chapter 2), and arguably appears a subset of this larger theory. Much of the SNM literature concerns the early adoption of new technologies that have the potential to contribute to sustainable development goals, with the assumption that innovation journeys can be facilitated by creating and supporting technological niches, protected spaces that allow experimentation (Schot and Geels, 2008). This, in theory, supports new technological pathways capable of penetrating the prevailing regime, destabilising or replacing unsustainable technologies in the process. In focusing on developments at the niche level, SNM presents 'a necessary and reflexive component of intentional transformation processes of regimes', where actors can push new, sustainable technologies on to the market (Kemp et al., 1998, p. 185). Table 13.1 provides an overview of the concept's main features.

Table 13.1 Strategic Niche Management guidelines and potential dilemmas

Policy area	Policy guidelines and potential dilemmas
Expectations, visions	• Be flexible, engage in iterative visioning exercises; adjust visions to circumstances and take advantage of windows of opportunity • Be persistent, stick to the vision, persist when the going gets tough
Learning	• Create variety to facilitate broad learning • Too much variety dilutes resources and prevents accumulation. It also creates uncertainty and may delay choices/commitments (by consumers, policymakers)
Network	• Work with incumbent actors, who have many resources, competence and 'mass'. Try to change their agenda, visions • Work with outsiders, who think 'out of the box' and have new ideas. Avoid incumbents have too many vested interests and will hinder progress
Protection	• Protection is needed to enable nurturing of niche-innovations • Do not protect too long and too much. This might lead to limited exposure to selection pressures
Niche–regime interaction	• Wait for cracks in the regime, and vigorously stimulate niche-innovations. Until such windows of opportunity arise, niches should be nurtured to facilitate stabilisation • Use niche experiences to influence perceptions of regime actors and actively create cracks in the regime

Source: adapted from Schot and Geels (2008).

While SNM was initially created as a management tool, in practice, it has also been used as an analytical one. Given that SNM is planned, focused and intended to direct sociotechnical change, it follows that studies investigating or utilising the model should engage with statements on (1) what we are transitioning towards, and (2) how it is possible to achieve this. To this end, we seek to summarise any emergent recommendations from 15 years of research.

Research methods: a systematic review and content analysis

To collect data for our study, a systematic and extensive search was conducted for peer-reviewed academic, energy-related articles on SNM published between 2002 and 2016, in addition to a complete search of the newsletters of the STRN. We acknowledge from the offset that there are earlier publications in the SNM field not captured by this data range (indeed, some of the most cited emerging in the late 1990s). Nevertheless, we focus on the last 15 years in order to provide a state-of-the-art summary of the most recent developments in the field, building on and corroborating the analysis done by Schot and Geels (2008). Furthermore, we acknowledge that our energy-related sample excludes transitions in other sectors. Yet, given the book is concerned with energy transitions specifically, this segregation is as useful as it is necessary.

For research articles, only those published between 1 January 2002 and 31 December 2016 were collected. To identify relevant articles, the authors searched for the paired terms of 'Strategic Niche Management' and 'energy', 'SNM' and 'electricity', 'SNM' and 'buildings', 'SNM' and 'transport' and 'SNM' and 'vehicles' within three fields, the article title, abstract and keywords. These categories were inclusive, meaning that a single article could not be counted multiple times in different categories, i.e. if they appeared in 'energy' *and* 'electricity' they would only be coded once. We coded for 'energy' first, meaning that the majority of papers were allocated here – later categories then allowed us to capture ones that would otherwise have been excluded from the sample. These searches were undertaken across nine article databases, resulting in the following sample from each: ScienceDirect (31 papers), JSTOR (0), Project Muse (0), Hein Online (0), SpringerLink (0), Taylor and Francis Online (7), Wiley Online (1), Sage Journals (3) and Annual Reviews (0), selecting only the articles that were peer reviewed, full-length and written in English.

For the STRN newsletter, we collated all editions published from its establishment in 2011, to the end of 2016 – the latest publication at the time of data collection – and, using the same sampling search criteria as above, sampled from the articles listed in their 'publications' section. Given that these newsletters are designed to explicitly engage with the SNM literature, it was not necessary to select articles based on three fields outlined above, the title, abstract and keywords. This led to the identification of one additional article.

In total, we generated a population of 45 papers. To analyse these resources, we used a content analysis methodology similar to Sovacool (2014). In keeping with a contents analysis approach, we primarily focused on quantitative forms of assessment. In all but two instances (see sections headed 'Topics' and 'Policy recommendations'), categories of analysis were identified before coding began, allowing us to record the articles that met our criteria across those categories only, resulting in a targeted sample that focused on our key questions.

We coded the contents of articles according to ten main categories of analysis; author discipline, author region, author gender, method, case study, topic, attitude towards the speed of transitions, contribution type, analytical strategy and policy recommendations. Each category contained a number of more specific codes, which are outlined in Table 13.2. To determine the coding outcome, coders read the title, abstract and article keywords (when available), before searching the rest of the article for key terms and phases. We recognise that there is some subjectivity inherent in this process. Across all coding categories, basic statistics were then conducted to calculate percentages, frequencies and distribution across years.

Table 13.2 Content analysis coding framework

Category	Sub-category
Author discipline	Science & Engineering; Economics & Statistics; Social Science; Arts & Humanities; Interdisciplinary; Other
Author region	Africa; Asia–Pacific (including Australia and New Zealand); Europe (including Russia and Turkey); Latin America and Caribbean; Middle East; North America
Then each specific country
Author gender	Male; Female; Indeterminate
Method	Experimental; Surveys; Modelling; Qualitative; Literature Review; None; Mixed Method
Case study	Geographical Case Study; Technological Case Study; If so, comparative?
Topic	Solar; Biofuel; Wave; Wind; Smart Grids; Community Energy; Energy Efficiency; Zero Energy Buildings; LEDs; Electric Vehicles; Incumbents; Other; Theoretical
Attitude to speed of transition	Negative; Neutral; Optimistic
Analytical strategy	Agency; Structure; Meaning
Policy recommendations	Present; Absent
Financial Support; Regulatory Support; Policy Mixes; Intermediaries Resourcing; Sectoral Diversity; Diverse Performance Targets; Mutual Learning; Flexible Institutional Structure; Brokering and Partnership Management; Private Sector Empowerment; Evaluation and Feedback; User Training and Awareness; Standardisation and Licensing

Source: the authors.

Author demographics

We began the content analysis by looking at three categories relating to the demographics of authors: disciplinary affiliation, location and gender. In all categories, we coded for each individual author, not just the lead author. This meant that a paper could receive numerous counts for the same category e.g. two 'female' and one 'male'.

For *disciplinary affiliation*, we coded the affiliation listed for all paper authors and classified those based on the categories used in Scopus, along with the categories of 'not listed' or 'other'. As authors can list multiple affiliations these categories were inclusive. This meant that an author working under 'management' and 'engineering' would be coded for both 'management' and 'engineering'. Where multiple authors on the paper recorded the same affiliation, it was only coded once. A paper was coded as 'interdisciplinary' if two or more of the authors listed different disciplinary affiliations, if the affiliation itself mentioned more than one discipline e.g. 'Department of Management and Engineering', or if it said 'interdisciplinary' in its title.

For *author region*, we coded each country listed on the paper. To get an accurate sense of geographical bias, if any, we coded for all paper authors. Where authors listed multiple country affiliations, each one was scored.

Finally, for *author gender*, authors were coded into 'male', 'female', and 'indeterminate' as some authors only used initials or had names common to both genders.

Methods

For paper *method*, we coded for seven categories: 'experimental', 'surveys', 'modelling', 'qualitative', 'literature review', 'none' and 'mixed method'. Articles were only coded for each method listed, meaning they could achieve more than one score. Where more than one method was present, they were determined to be 'mixed method' and further notes were taken. Examples include the presence of semi-structured interviews *and* documentary analysis, or a documentary analysis, field study *and* participatory observation.

Case studies

We coded for whether the articles used geographical case studies (at any scale), technological case studies, or both, and whether these case studies were comparative. Studies were coded as comparative whenever they compared two or more geographical areas or technologies, with further analysis undertaken as to the depth and content of the case study examination. This allowed us to record the scope of the research articles. For papers using geographical case studies, we then coded for which countries they were studying.

Topics

For this category, our purpose was to discern the general topic of research, rather than to determine the exact nuances of the papers. We began by recording short notes on the topic of the article – derived from the title, keywords and abstract, where possible. This included the terms 'incumbents', 'community energy' and 'electric vehicles' for example, as well as a 'theoretical' category. We then inductively built a list of 15 final topics. From these notes, 15 topical themes emerged as summarised in Table 13.2. Each paper was then allocated to all appropriate categories. Papers could be coded in multiple categories i.e. if they referred to both 'smart grids' and 'incumbents'.

Pace of transitions

We were also interested in attitudes towards the speed of energy transitions, and whether the articles assessed whether it would, could, or should be a fast or a slow process. Articles were coded as 'positive', 'negative' or 'neutral'. Papers were determined to be 'positive' either if they reported potential transitions within ten years or used the term related to a quicker pace, e.g. 'fast', 'rapid', 'quick' or 'accelerated'. 'Neutral' papers either referenced uncertainty over the pace of transitions or gave no comment, including those that did not offer any estimated time in years. Papers were determined to be 'negative' if they stated it would take longer than ten years or used terms such as 'slow', 'long' and 'gradual'.

Analytical strategy

For *analytical strategy*, we categorised articles according to whether the publications tended to centre their analytical strategy on 'agency', 'structure' or 'meaning'; in keeping with the classic social theory triangle of agency, structure and meaning – or the three 'I's of interests, institutions and ideas – that often guide the analysis of analytic strategy and empirical focus (Sovacool and Hess, 2017). We use the term 'centre' to acknowledge that each paper may involve elements of multiple types. Papers coded to the category of 'agency' prioritised the agency of people and their strategies, covering a range of actors, from individuals, organisations and collective groups. 'Structure' refers to macrosocial, infrastructural and institutional hardware; and 'meaning' refers to the cognitive, discursive and normative systems that orient action. For meaning, this implies a focus on language, symbolism, narratives, performativity, and how technologies co-construct and negotiate meaning for human subjects. Most papers were coded to the one category representing their main focus. However, where they explicitly compared two or three theoretical approaches, they were coded in both.

Policy recommendations

Finally, we coded for *policy recommendations* by recording information on the suggested recommendations for users, decision-makers, planners, policymakers and regulators in each article. A recommendation was determined to be 'non-academic' when it provided advice or explained lessons for an external stakeholder group, including policymakers and business groups, for example. Coders read the discussion and conclusion and determined whether non-academic recommendations were either 'present' or 'absent'. When they were present, the relevant information was extracted and then analysed inductively.

Results: unveiling 15 years of Strategic Niche Management research on energy transitions

This section of the chapter presents the results of the content analysis, following the same structure outlined above.

Author demographics

Across all articles analysed in the sample, a total of 96 author affiliations were listed, covering a range of disciplines. As panel A of Figure 13.2 shows, authors were strongly associated with social sciences and management disciplines, which made up a total of 74 per cent of the overall sample. The least represented disciples were life sciences and medicine and engineering and technology, which may be unsurprising. Of the papers analysed, only 12 were identified as having interdisciplinary authorship meaning that despite an increasing shift in academic pedagogy towards interdisciplinary research, many authors continue to work in the confines of their own disciplines.

Authors reported affiliations with all global regions, although panel B of Figure 13.1 shows that there was a heavy bias towards European contributions, which made up 82 per cent of the sample. Of those from Europe, 27 per cent were from Dutch authors and 26 per cent were from UK authors. We acknowledge that this is likely related to the availability of research funds in particular countries (as well as, arguably, the origin of the transitions concept itself). It was also possible here to identify recurring authors – (eight papers), Smith (five papers), Seyfang (four), Lovio (four), Hielscher (three), Verbong (three) and Hargreaves (three). This shows not only a geographical limit to the diffusion of this concept, but also an authorial limit. Particularly underrepresented global regions include the Africa (1 per cent), Middle East (2 per cent) and North America (2 per cent), although we do note that our sample only selected papers written in English from major databases, which may not be readily accessible in some global regions.

For author gender, shown the bottom panel of Figure 13.1, while male contributions did dominate (65 per cent), female authors were fairly well represented at 34 per cent of the total sample. Only 1 per cent of author genders could not be identified. We identify this gender balance as being positive. Sovacool (2014) recognises in his content analysis of 9,549 papers published in the social science research in the energy field, only 15.7 per cent could be identified as female. Although our sample is evidentially significantly smaller, our analysis shows that the SNM literature is currently more gender progressive.

(A) Author discipline

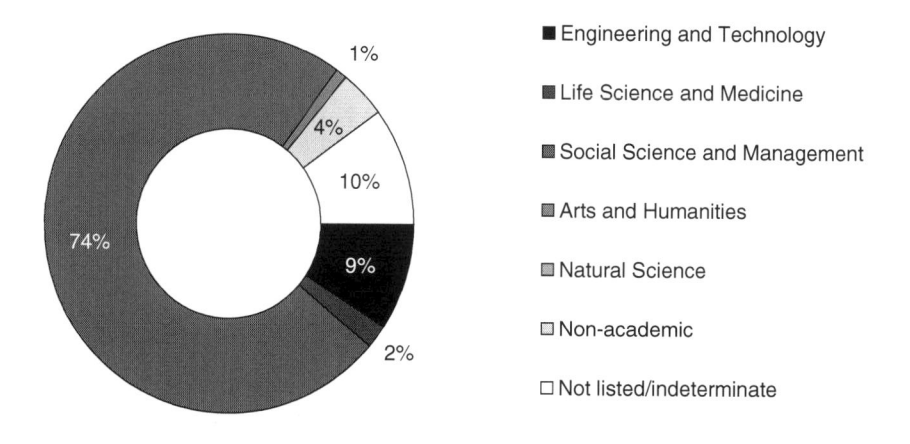

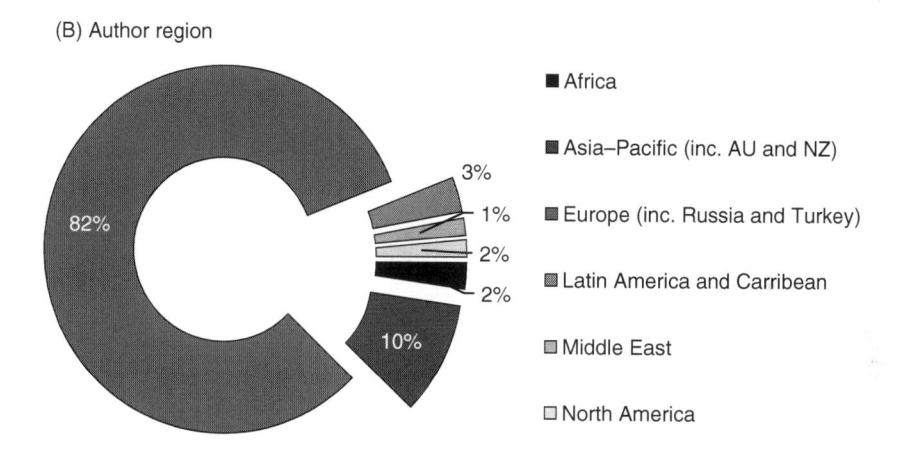

- Engineering and Technology
- Life Science and Medicine
- Social Science and Management
- Arts and Humanities
- Natural Science
- Non-academic
- Not listed/indeterminate

74%, 10%, 9%, 4%, 2%, 1%

(B) Author region

- Africa
- Asia–Pacific (inc. AU and NZ)
- Europe (inc. Russia and Turkey)
- Latin America and Carribean
- Middle East
- North America

82%, 10%, 3%, 2%, 2%, 1%

(C) Author gender

Gender	Total (%)
Male	65
Female	34
Indeterminate	1

Figure 13.1 Strategic Niche Management author demographics (n = 100).
Source: the authors.

Methods

For methods, shown in the top panel of Figure 13.2, the most favoured approach was secondary data analysis (44 per cent) followed closely by qualitative (43 per cent). The comparative lack of survey (5 per cent), modelling methodologies (3 per cent) and experimental (0 per cent) methodologies shows new avenues for methodological expansion. Few studies stated no method (5 per cent). Of the total papers coded, 35 per cent were identified as being mixed methods studies, primarily relying on a combination of semi-structured interviews and documentary analysis, although with a number of papers also utilised participatory fieldwork including fieldtrips, meeting participation and participant observation. There was methodological diversity throughout the years.

The finding that the top methods employed by SNM researchers are qualitative and dependent on secondary data sources has at least two implications. First, and positively, as qualitative analysis is typically used to explore the socially constructed nature of a phenomenon, the predominance of these methods suggests a focus on the social elements of energy transitions. Whereas Sovacool (2014) identified a need for energy studies research to expand methodologically towards socially sensitive approaches – including the utilisation of more research interviews, field research, focus groups and other human-centred methods of data collection – the SNM literature is, therefore, already stronger in this regard.

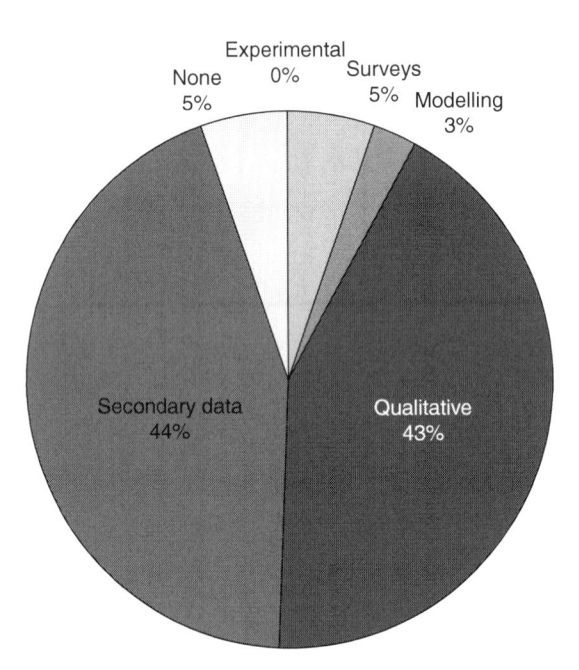

Figure 13.2 Strategic Niche Management article methods (n = 45).
Source: the authors.

Second, as both a positive and a negative, qualitative analyses also give findings that are context specific. Positively, research conclusions and policy recommendations independent of context can be meaningless, thereby context enriches results. Negatively, outside of the presence of multiple cases reinforcing the same results or comparative studies, this hinders the ability for widespread application of the findings and the ability to draw policymaking principles from them. Further, without transparent documentation of the processes undertaken, qualitative analyses can be hard to replicate.

As a result, and drawing on the relative under-emphasis of experimental, survey and modelling methodologies in this sample, we argue for both a continuation of socially oriented methodologies and, crucially, increased attention to modelling techniques and approaches, which may be more testable to statistics, and falsifiable and replicable. The result is findings that may be more readily applicable in practice.

Indeed, a preference for qualitative or literature-based methods ignores experimental designs that have shown great promise across the fields of behavioural science, psychology, and applied sociology (Sorrell 2007; Sovacool and Hess 2017), especially when they investigate scales of social action and actors' roles (Stern et al., 2016). Raven (2005) stated that

> SNM can be used for improving the design of experiments, for evaluating policies in the past, for using SNM as part of scenario development, or for designing future policies on niche management. However, SNM has not been used as such in practice, but mainly as a research tool. The policy claims that are often made by SNM researchers still remain a promise; SNM needs real-life experimentation in society.

This is a promise of SNM that remains largely unfilled.

Case studies

Papers contained a range of both geographical and technological case studies. In the case of geographical case studies, these were not well distributed globally. Few studies utilised examples or analysis of countries or regions in the Global South. Further, most publications were nationally focused, with limited investigations into international commodity systems or international intermediaries such as European Union policymakers. Across all papers analysed, only 16 per cent utilised comparative case studies, whether it was two technologies, two countries or several technologies across different countries. We take this to represent limited engagement with 'lessons learnt' and their application to different international and technological contexts.

There was no clear trend in the countries studied through time. Nonetheless, there was a predominance of SNM scholarship both originating from and using case studies of the United Kingdom and the Netherlands. This indicates a relatively limited scope of global application, empirically supporting the statement from Caniëls and Romijn (2008, p. 257) that 'the main preoccupation of the SNM researchers has clearly been on the initiation and management of (individual) experiments', with little consideration for the next stages of market establishment and beyond. Moreover, our finding further reinforces the conclusion of Hoogma et al.'s (2002) book, which has led to substantial research on the local-global distinction between experiments and niches (see Geels and Raven, 2006) as well as spatially informed research on cross-local/transnational dynamics (see Sengers and Raven, 2015).

Indeed, this points to a continued failure to better understand the conditions under which niches, or mainstream alternatives, can truly break through into the wider system (Arranz, 2017). An implication is the need for a greater emphasis on knowledge transfer within the SNM research community, both within academic circles and outside. While innovation niches may first emerge within one country or area – the place where the technology is first designed, piloted or implemented, for example – a technology will not 'break through' into the regime without the supporting impetus to do so. This comes, in part, from widespread geographical diffusion and embedding consistent with a technology or social innovation entering the mainstream marketplace.

Topics

An article's topic reveals what is considered to be a 'niche' technology or process. Overall, our analysis demonstrated a slight favour for the reporting of energy production niches (43 per cent) as opposed to energy consumption (39 per cent), actors (including intermediaries) (9 per cent) or other (9 per cent). The category 'other' included theoretical contributions and the ideas of Citizen Participation Initiatives (CPI), donor interventions, product–service systems and alternative technology movements, among others. Generally speaking, niche technologies were considered across the full range of the energy system and at both commercial and domestic scales. Although our chapter covers a short time span, Figure 13.3 shows some changes in the popularity of different niche topics. Alongside a general increase in SNM publications throughout the years, there was a clear increase in research on intermediaries, for instance, driven mostly by UK academics including, notably, papers by Seyfang et al. (2014), Hamilton et al. (2014) and Smith et al. (2016).

Moreover, our qualitative coding process did reveal three other topics that occurred with frequency: financial support, mutual learning and brokering. These topics reaffirm and update similar findings and 'conceptual categories' presented by Schot and Geels (2008), where 'learning' and 'networking' were included as key recommendations (although to make the meaning of these categories more transparent, we name them slightly differently). To be specific, 'protection' becomes 'financial support' (including market protection through subsidies and regulation), 'learning' becomes 'mutual learning' and 'networks' becomes 'brokering and partnership management'. Surprisingly given the origins of the literature, there was no clear category relating to expectations/visions.

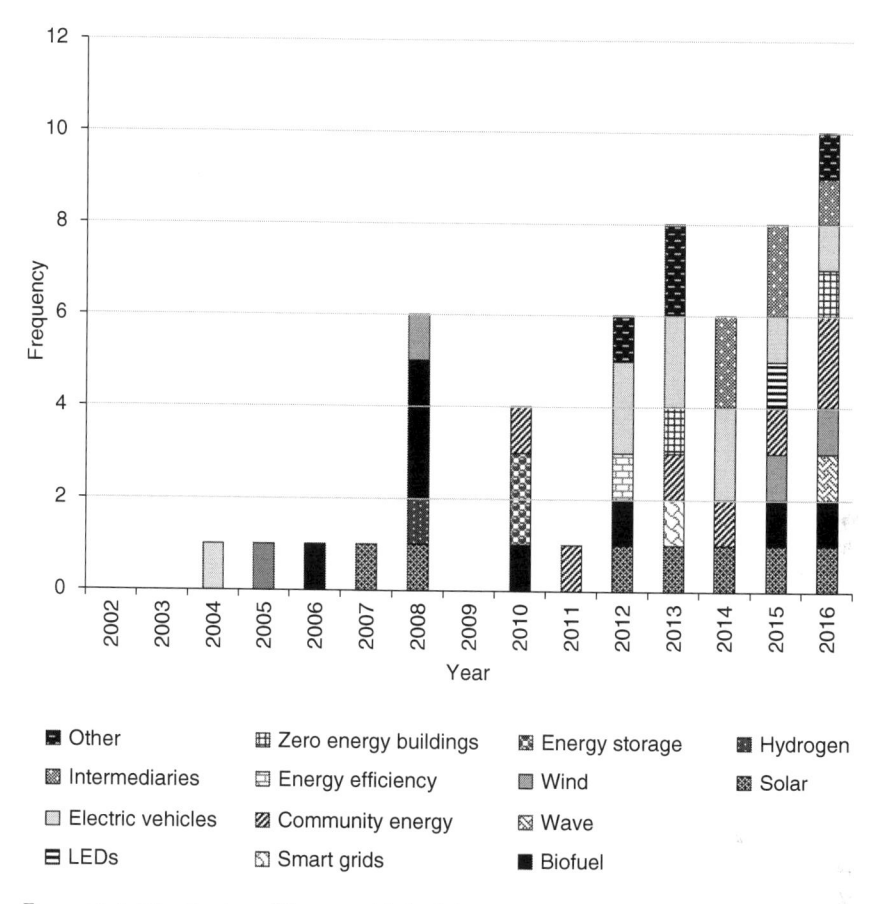

Figure 13.3 Distribution of Strategic Niche Management cases by technology (n = 24).
Source: the authors.

Sixteen of the papers referenced the need for what we term *financial support* of certain elements of niche innovation. These recommendations manifested as either funding particular intermediary groups or as providing subsidies for niche technologies or taxing others. As an illustration, in the context of smart grid projects, Verbong *et al.* (2013, p. 123) state 'users are often regarded as a potential barrier to smart grids deployment and financial incentives the best instrument to persuade or seduce the users'. In keeping, Kamp and Forn (2016) referenced that in their case of Ethiopian biogas, entrepreneurial activities were hindered by inadequate financing possibilities. Kamp and Vanheule (2015) explain that improving finance mechanisms will enable end-users to purchase small wind turbines. In all cases, appropriate financing was seen to provide opportunity to remove a barrier and to increase niche diffusion. Steinhilber *et al.* (2013, p. 537) summarise such recommendations particularly effectively as they write 'governments must therefore find the right mix of regulatory pressure and funding options corresponding to the current condition of its national industries and markets, to make innovation attractive for both the supply and demand side'. Thus, in order to manage energy and climate transitions, niches must be appropriately supported fiscally.

Fifteen papers referred to the need for *mutual learning*, a category that contained notions of face-to-face learning, aggregated knowledge and knowledge exchange, emphasising the social embedding of energy technologies. Seyfang et al. (2014) stress the importance of learning throughout their paper, explaining that in order to strengthen a UK grassroots energy niche, a group of intermediary organisations is needed that has the capacity to consolidate and aggregate the learning and experiences of local projects with a view to repackaging them for implementation elsewhere. This includes learning-by-doing and pro-active learning interactions. As recognition that a community energy niche is heavily grounded in civil society and community engagement, this draws on the social, human and organisational capital alongside and in complement to finance, natural or manufactured capital. Taking a policy focus towards mutual learning, Browne et al. (2012, p. 149), referencing Schwoon (2008), suggest that 'policy-makers should also focus on "technological learning" or "learning-by-doing", which can lead to substantial cost reductions and result in "early mover advantage"'.

Twelve papers then referred to the topics of *brokering and partnership management*. Building on the category of mutual learning above, this considered the role of *particular groups* in directing niche implementation – i.e. who is able to share knowledge. We base the title of this category on the paper by Hargreaves et al. (2013, p. 878), who relay that ideas of 'brokering and partnership management' exist as growing recognition that community energy intermediaries 'can no longer focus solely on internally building local community energy projects but must actively try to work beyond the community energy sector – brokering partnerships and engaging in lobbying activities – to try to shape wider contexts'. This idea is supported in other papers. Kamp and Vanheule (2015, p. 479) add that 'upscaling can be enabled by network expansion through the development of key partnerships with local authorities, financing institutions and local NGOs'. Hamilton et al. (2014) refer to the body responsible for this as a 'moderator' who can convene participatory forums, encourage learning processes and provide leadership, capacity and institutional support with a view to facilitating long-term goals. Importantly, this idea builds on the role of 'intermediation', where certain groups are able to consolidate, grow and diffuse niche innovations. According to Seyfang et al. (2014, p. 40), better-resourced intermediary groups 'could take the initiative in offering resources to new projects, transferring lessons from local projects, liaising with energy utilities and policymakers, and developing standardized models for easier replication'.

Pace of transitions

In terms of the temporal pace of transitions, the majority of papers either did not give an estimated timespan (47 per cent) or gave a pessimistic or negative view (47 per cent) implying that it would either take more than ten years or emphasising qualitatively that transitions processes were long, slow, and cumbersome. Negative statements on timespan typically occurred in reference to economic, institutional and cultural barriers to niche development.

In the few cases where a positive timespan was given (6 per cent), it was typically in relation to electric vehicles niches and came with a caveat – that diffusion may appear fast but dispersed, and therefore not necessarily holistic or efficacious. For example, Bakker et al. (2015) exhibit that while quick progress was made for electric vehicle recharging infrastructure, several market movers made different types of electric vehicle plugs, flooding the market and inhibiting the development of a global standard. Thus, 'fast' may not mean effective for the related niche development of electric vehicles. Sushandoyo and Magnusson (2014) illustrate potentially uneven transitions paces as they explored a 20-year timeline for the development of hybrid buses. Their results demonstrated rapid diffusion only towards the end of this range. This shows different pace along with design, diffusion and stabilisation phases of innovation making 'take-off' and the pace of transitions hard to predict.

Such pessimistic framing of transitions does go against some recent empirical evidence highlighting numerous 'fast' transitions that have occurred in both national energy supply and the diffusion of end-use devices (Sovacool, 2016). We note, however, that while market diffusion may sometimes occur quickly, it often only results after decades of pre-development. Rotmans et al. (2001) support this idea as they note that transitions may appear to be quick, even when their pre-development is long. This phase – also known as the valley of death – is, in many ways, what the SNM literature has been developed to understand. Nevertheless, with conscious of the real path dependencies and lock-ins in incumbent regimes, we refer here, to the need for positivity and practically oriented forward thinking about the potential of energy transitions.

Analytical strategy

When coding for analytical strategy, there was a slight dominance of papers in the 'agency' category (55 per cent), followed by 'structure' (36 per cent) and 'meaning' (9 per cent). While one paper did engage with all three approaches (Caniëls and Romijn, 2008), this was a literature review. Reflecting this, papers discussed a range of actors, from individuals and organisations to collective groups. This included, most notably, a strong focus on community energy groups and intermediaries as drivers and enablers of energy transitions. Given the planned nature of SNM outlined above, and indeed, the focus on 'management' within the approach in general, this focus on agency was determined as necessary and unsurprising as we consider who manages transitions or is affected by them. Nonetheless, analytical strategies emphasising agency can assume that people are atomistic agents whose action can be explained without deep consideration of structure (Jackson, 2005). Indeed, Mouzelis (1995) warns that the micro-turn in social theory towards agency has led to an almost complete neglect of asking questions about bigger entities, reification, or the structural or functional attributes of larger systems.

In our sample, the almost complete emphasis within the SNM literature on only agency and structure – and not meaning – implies a failure to appreciate the power of language, symbolism, narratives, performativity, rhetorical visions, and how technologies can co-construct and negotiate meaning for human subjects (although this was touched upon briefly in some instances e.g. through Verbong et al.'s (2008) statements on expectations and visions). This absence of papers considering the meaning category will be surprising to some, given the potential argument that SNM was built upon science and technology studies with a particular emphasis on the dynamics of expectations as one way of giving meaning to the (future) world.

In this vein, we note that it may be possible to analyse and explore how (and why) the SNM research has been guided to an overemphasis of agency and structure at the expense of meaning. Indeed, it seems fruitful for the SNM community to continue to engage with relational theories that emphasise agency, structure and meaning together. Such approaches may emphasise social relations and interactions, but they also highlight the webs of social structure and meaning in which actors are suspended and which they change through their action (Geels, 2009; Rutherford and Coutard, 2014). We say 'continue to engage' as acknowledgement that although relational perspectives on niche development can be developed in more detail, they do already exist.

At least four papers in the sample seemed to recognise the value of a processual approach, or at least combined different analytical strategies. Hatzl et al.'s (2016, p. 58) paper explained the typical distinction between (a) grassroots social innovation and (b) market-based technological innovations, representing an 'agency' vs. 'structure' split. In their work they argued that CPIs for niche developments, represented 'a continuum from market-based to grassroots characteristics in order to facilitate a more detailed picture of niche development'. Here, social arrangements such as CPIs are both ideologically motivated but also show market-based characteristics such as profit seeking.

Policy recommendations

Across the entire sample of SNM literature, 56 per cent of papers contained explicit policy recommendations, leaving 44 per cent without non-academic impact statements. In terms of type and scope of recommendations, financial support (n = 16); regulatory support (n = 4); policy mixes (n = 4); intermediaries resourcing (n = 3); sectoral diversity (n = 4); diverse performance targets (n = 3); mutual learning (n = 15); flexible institutional structure (n = 3); brokering and partnership management (n = 12); private sector empowerment (n = 3), evaluation and feedback (n = 4), user training and awareness (n = 1) and standardisation and licensing (n = 3) were suggested repeatedly. Table 13.3 provides one indicative recommendation for each of the categories listed above. These ideas are then further explored in the conclusion.

Table 13.3 Indicative policy recommendations by analytical category

	Indicative recommendation
Financial support	'Financial support for these dedicated sector-development organizations and networks is critical to help the sector coalesce into an effective niche' (Seyfang et al., 2014, p. 40)
Regulatory support	'Consideration could also be given to whether policies require more protection than already provided by the regulatory framework' (Ieromonachou et al., 2004, p. 78)
Policy mixes	'Policy-makers have a range of options and should … ensure a consistent mix of policy and regulatory signals, which offer long-term certainty' (Browne et al., 2012, p. 150)
Intermediaries resourcing	'Identifying and finding ways to involve relevant overlooked group of middle actors could contribute to the development of strategic niche management efforts, although resourcing this would be necessary' (Hamilton et al., 2014, p. 473)
Sectorial diversity	'The transition to a low-carbon society is imperative to climate change mitigation and requires cross-sectoral action at multiple levels' (Hamilton et al., 2014, p. 463)
Diverse performance targets	'Another important issue is related to creating diversity among experiments, which is considered crucial because can led to learning about different designs in different use environments' (Ceschin, 2013, p. 85)
Mutual learning	'This might demand greater resource input to face-to-face mutual learning, rather than attempts to codify and standardise action on the ground' (Seyfang et al., 2014, p. 42)
Flexible institutional structure	'If community energy in the UK is to contribute to a shifting energy mix, … it requires imaginative policy support, … this may require flexible institutional infrastructure' (Seyfang et al., 2014, p. 42)
Brokering and partnership management	'Niche intermediaries can help mitigate these intrinsic & diffusion challenges and accelerate the transition of community energy as a socio-technical innovation by … brokering and managing partnerships between projects and external actors' (Klein and Coffey, 2016, p. 877)
Private sector empowerment	'Another important aspect is the empowerment and inclusion of the private sector by rethinking the way the technology is disseminated' (Kamp and Forn, 2016, p. 486)
Evaluation and feedback	'The central government therefore should strengthen the evaluation component of the [technology] demonstration programs' (Xue et al., 2016, p. 16)
User training and awareness	'To pay special attention to training and awareness of masons and users in the rural areas especially adapted to circumstances with high illiteracy rates and many different mother tongues' (Kamp and Forn, 2016, p. 487)
Standardisation and licensing	'Policy-makers should endeavour to remain "technology-agnostic" and introduce standards or taxation measures that do not incentivise any particular fuel or technology but set the appropriate conditions for consumers and investors' (Brown et al., 2012, p. 150)

Source: the authors.

Interestingly, recommendations were given for both government policymakers, local authorities and specific technology user groups, including industry members and intermediaries. Kamp and Vanheule (2015, p. 497) stated to this end that 'niche upscaling is not solely in the hands of the government. Rather than standing aside, technology supplies, research institutes and NGOs can enable change themselves with targeted and joint efforts'. In this regard, the range of non-academic stakeholders deemed to be of relevance was wide.

Moreover, where recommendations were present, they contained a diverse array of suggestions, ranging from both directly implementable options to broader policymaking principles (and even, in one instance, recommendations to avoid biofuel crop monoculture (Eijck and Romijn, 2008)). As illustrated in Table 13.3, this includes a distinction between technocratic solutions – e.g. technology standardisation and licensing (e.g. Kamp and Vanheule, 2015) – and socially oriented user-based engagement (e.g. Hargreaves et al., 2013), where Raven et al. (2008, p. 475) identified that ultimately, 'ready-made solutions cannot be dropped into a context without local negotiations'. Fitting the socially oriented recommendations category, Seyfang et al. (2014, p. 42) advocated for 'imaginative policy' support before going on to suggest what this might look like, including steers towards flexible institutional infrastructure. Browne et al. (2012, p. 150) outlined a clear series of recommendations that crossed these two categories, where they advised that policymakers should consider:

1   developing a transition strategy and engaging in scenario planning with industry stakeholders;
2   identifying potential 'lead adopters' and develop a strategy for SNM;
3   developing stakeholder partnerships with industry and consumer groups;
4   promoting the adoption of new sociotechnical regimes through awareness campaigns and education;
5   changing the taxation structure to tax negative externalities and create positive incentives through excise relief and subsidies;
6   providing long-term certainty through a constant mix of policy and regulatory signals.

A further strong example was provided by Tsoutsos and Stamboulis (2005) who, in a standalone section, suggest three key policy aims: (1) the development of focused learning mechanisms, (2) the encouragement of new types of players and 3) flexible financing mechanisms, adapted to the characteristics of individual applications and environmentally consistent academic evaluation.

Some studies introduced a 'mutual learning' category to encompass suggestions for creating individual groups responsible for knowledge sharing or increased education across the full range of relevant stakeholders (e.g. Kamp and Vanheule, 2015). Sushandoyo and Magnusson (2012) argue that this should include provision of information on internal synergies, scale economies and projections of future sales and production volumes in the case of developing technologies. Verbong et al. (2008) identified the need for clearly expressed expectations and visions at the beginning of and during niche transformations.

Admittedly, although our analysis does not reveal the degree to which SNM findings and approaches are integrated in practice, it does illustrate the range of potential options for doing so. This diversity of potential policy options can be perceived as strength as it illustrates a range of different tools that can be applied in different contexts, plus the ability to tailor solutions. However, more work is needed that looks at exactly *how, if and where* policy lessons have been applied as attempts to and lessons from translating them into actual politics have not been well documented. Stemming from our analysis, we suggest that one fruitful avenue would be to look at the politics behind the policies, creating coalitions of stakeholders, so that low-carbon transitions occur.

Conclusions and policy recommendations

Our systematic review of the literature has exposed both trends and gaps in the SNM literature and importantly, summarised key recommendation for transitions management. With this in mind, we offer four key findings and conclusions.

First, there is an intellectual diversity within the transitions management community that we laud. Our review has demonstrated some positive demographic trends within authors. We have found contributions from across many academic disciplines, a fair balance of male and female contributors to the literature, a diversity of methods applied variously and analytical strategies that focus on the social integration and fostering of innovations, including concerns for energy groups and intermediaries. The recent SNM literature engaged with a range of geographical and technological cases, with studies primarily focusing on the UK and the Netherlands, but also with less typical case studies of China, India, Ethiopia and Malaysia. Cumulatively across the sample years, the top three technological case studies included solar energy, biofuels and electric vehicles.

Second, we identify positive qualitative or topical trends within the community. There is a commitment to technological agnosticism, illustrated by a range of technological and geographical cases used within papers. There also appears to be a ready commitment to policy engagement, in that that a majority of publications analysed (56 per cent) provide a series of non-academic recommendations stemming from empirical and theoretical work. These recommendations were targeted at government officials, industry members and intermediaries, among others, showing wide application and relevance. The most frequently reoccurring recommendations or topics focused on (1) the appropriate financing of niche innovations, (2) mutual learning between stakeholders and across niches, and (3) brokering and partnership in order to strengthen niche development. As a key contribution of this chapter in line with the policy focus of the book it sits within, focusing on these areas increases our ability to manage energy and climate transitions.

Third, there were also worrying demographic trends related to geographic and methodological bias. There was a heavy bias towards European authorship, which contributed 82 per cent of the sample (with 26 per cent from UK authors and 27 per cent from Dutch authors). In addition, geographic case studies are not well distributed globally with few studies in the Global South and most publications taking on national case studies, with limited investigations into international commodity systems or international intermediaries. Further, while positive in some regards, the predominance of qualitative and secondary data methodologies has downsides as well. This methodical focus leads to results that may be hard to replicate and, in many cases, and context specific, with knock-ons for the ability for widespread application of the findings and the ability to draw policymaking principles from them. Experimental research designs have also been largely avoided by the SNM community.

On the grounds of these failings, we make recommendations for future research *and* policy-oriented commitments. It follows that our recommendation is to improve this negative demographic trend is for greater emphasis to be placed not only on increasing the breadth of case studies but also on comparing them. Doing so could aid the identification of more generalisable or scalable lessons or policy principles. Moreover, we appeal for broader methods, capable of capturing a broader range of perspectives on the challenges we face and most pressingly, their solutions.

Finally, and also critically, there was an apparent inability to draw comparative lessons either across technologies or countries within the SNM literature, and in some cases, where they were absent, engage with non-academic impact statements. These failures have at least three consequences. First, a failure to translate information between countries may leave niches isolated, preventing mainstream adoption. Second, a failure to compare technologies restricts lesson learning, therefore increasing the probability of negative mistakes. As demonstrated by Bakker *et al.* (2015), it also inhibits the development of a global standard technology that, through pervasive uptake, can challenge regime technologies. Third, a failure to engage with non-academic impact statements represents, in essence, a failure to maximise the discursive potential of academic research. It follows that we call for more synthesis across the literature and further attempts to take academic discussion towards practical implementation. We have identified a conservative and pessimistic approach to envisioning the timing, temporal pace or accelerative potential of low-carbon energy transitions within the SNM literature as well. We wonder if the very hesitancy from SNM theorists to validate the notion of expedient transitions, and the continued dominance of techno-economic analyses rooted in modelling, contributes in part to the very 'lock-in' or 'path dependency' they critique. The theorists endow the fossil fuel regime with perhaps more agency than it actually has or need have. The point being that the very discourse we in the academy utilise to frame and engage on energy and carbon transitions can distort and even reinforce trends in that very empirical space.

References

Arranz, A.M. (2017) Lessons from the past for sustainability transitions? A meta-analysis of socio-technical studies. *Global Environmental Change* 44: 125–143.

Bakker, S., Leguijt, P. and van Lente, H. (2015) Niche accumulation and standardization – the case of electric vehicle recharging plugs. *Journal of Cleaner Production* 94: 155–164.

Bergek, A., Jacobsson, S., Carlsson, B., Lindmark, S. and Rickne, A. (2008) Analysing the functional dynamics of technological innovation systems: A scheme of analysis. *Research Policy* 37(3): 407–429.

Browne, D., O'Mahony, M. and Caulfield, B. (2012) How should barriers to alternative fuels and vehicles be classified and potential policies to promote innovative technologies be evaluated. *Journal of Cleaner Production* 35: 140–151.

Caniëls, M.C.J. and Romijn, H.A. (2008) Actor networks in strategic niche management: insights from social network theory. *Futures* 40(7): 613–629.

Ceschin, F. (2013) Critical factors for implementing and diffusing sustainable product-Service systems: Insights from innovation studies and companies' experiences. *Journal of Cleaner Production* 45: 74–88.

Eijck, van J. and Romijn, H. (2007) Prospects for Jatropha biofuels in Tanzania: An analysis with Strategic Niche Management. *Energy Policy* 36(1): 311–325.

Geels, F.W. (2009) Foundational ontologies and multi-paradigm analysis, applied to the socio-technical transition from mixed farming to intensive pig husbandry (1930–1980). *Technology Analysis and Strategic Management* 21(7): 805–832.

Geels, F.W. (2014) Regime resistance against low-carbon transitions: Introducing politics and power into the multi-level perspective. *Theory, Culture and Society* 31(5): 21–40.

Geels, F.W. and Raven, R. (2006) Non-linearity and expectations in niche-development trajectories: Ups and downs in Dutch biogas development (1973–2003). *Technology Analysis and Strategic Management* 18(3–4): 375–392.

Geels, F.W., Sovacool, B.K., Schwanen, T., Sorrell, S. (2017) Sociotechnical transitions for deep decarbonisation. *Science* 357(6357): 1242–1244.

Grubler, A., Wilson, C. and Nemet, G. (2016) Apples, oranges, and consistent comparisons of the temporal dynamics of energy transitions. *Energy Research and Social Science* 22: 18–25.

Hamilton, J., Mayne, R., Parag, Y. and Bergman, N. (2014) Scaling up local carbon action: The role of partnerships, networks and policy. *Carbon Management* 5(4): 463–476.

Hargreaves, T., Hielscher, S., Seyfang, G. and Smith, A. (2013) Grassroot innovations in community energy: The role of intermediaries in niche development. *Global Environmental Change* 23(5): 868–880.

Hatzl, S., Seebauer, S., Fleiß, E. and Posch, A. (2016) Market-based vs. grassroots citizen participation initiatives in photovoltaics: A qualitative comparison of niche development. *Futures* 78–79: 57–70.

Hoogma, R., Kemp, R., Schot, J. and Truffer, B. (2002) *Experimenting for Sustainable Transport. The Approach of Strategic Niche Management*. Routledge, London, UK.

Ieromonachou, P., Potter, S. and Enoch, M. (2004) Adapting strategic niche management for evaluating radical transport policies-the chase of Durham Road access charging scheme. *International Journal of Transport Management* 2(2): 75–87.

International Renewable Energy Agency and International Energy Agency (2017) Perspectives for the Energy Transition: Investment Needs for a Low-carbon Energy System. Available at: www.irena.org/menu/index.aspx?mnu=Subcat&PriMenuID=36&CatID=141&SubcatID=3828.

Jackson, T. (2005) Motivating Sustainable Consumption: A Review of Evidence on Consumer Behaviour and Behavioural Change. A Report to the Sustainable Development Research Network. Sustainable Development Research Network, Surrey, UK. Available at: www.sustainablelifestyles.ac.uk/sites/default/files/motivating_sc_final.pdf.

Jain, M., Hoppe, T. and Bressers, H. (2016) Analyzing sectoral niche formation: The case of net-zero energy buildings in India. *Environmental Innovation and Societal Transitions* 25: 47–63.

Kamp, L.M. and Forn, E.B. (2016) Ethiopia's emerging domestic biogas sector: Current status, bottlenecks and drivers. *Renewable and Sustainable Energy Reviews* 60: 475–488.

Kamp, L.M. and Vanheule, L.F.I. (2015) Review of small wind turbine sector in Kenya: Status and bottlenecks for growth. *Renewable and Sustainable Energy Reviews* 49: 470–480.

Kemp, R., Schot, J. and Hoogma, R. (1998) Regime shifts to sustainability through processes of niche formation: The approach of strategic niche management. *Technology Analysis and Strategic Management* 10(2): 175–198.

Klein, S.J.W. and Coffey, S. (2016) Building a sustainable energy future, one community at a time. *Renewable and Sustainable Energy Reviews* 60: 867–880.

Lovell, H. (2007) The governance of innovation in socio-technical systems: The difficulties of strategic niche management in practice. *Science and Public Policy* 34(1): 35–44.

Markard, J., Raven, R. and Truffer, B. (2012) Sustainability transitions: An emerging field of research and its prospects. *Research Policy* 41(6): 955–967.

Martiskainen, M. and Kivimaa, P. (2018) Creating innovative zero carbon homes in the United Kingdom – intermediaries and champions in building projects. *Environmental Innovation and Societal Transitions* 26: 15–31.

Mouzelis, N. (1995) *Sociological Theory: What Went Wrong? Diagnoses and Remedies.* Routledge, London, UK.

Raven, R.P.J.M. (2005) Strategic niche management for biomass: A comparative study on the experimental introduction of bioenergy technologies in the Netherlands and Denmark. PhD Thesis. Technische Universiteit Eindhoven.

Raven, R.P.J.M., Heiskanen, E., Lovio, R., Hodson, M. and Brohmann, B. (2008) The contribution of local experiments and negotiation processes to field-level learning in emerging (niche) technologies: Meta-analysis of 27 new energy projects in Europe. *Bulletin of Science, Technology and Society* 28(6): 464–477.

Rotmans, J., Kemp, R. and Van Asselt, M. (2001) More evolution than revolution: Transition management in public policy. *Foresight* 3: 15–31.

Rutherford, J. and Coutard, O. (2014) Urban energy transitions: Places, processes and politics of socio-technical change. *Urban Studies* 51(7): 1353–1377.

Schot, J.W. and Geels, F.W. (2008) Strategic niche management and sustainable innovation journeys: Theory, findings, research agenda, and policy. *Technology Analysis and Strategic Management* 20(5): 537–554.

Schwoon, M. (2008) Learning by doing, learning spillovers and the diffusion of fuel cell vehicles. *Simulation Modelling Practice and Theory* 16(9): 1463–1476.

Sengers, F. and Raven, R. (2015) Towards a spatial perspective on niche development: The case of bus rapid transit. *Environmental Innovation and Societal Transitions* 17: 166–182.

Seyfang, G., Hielscher, S., Hargreaves, T., Martiskainen, M. and Smith, A. (2014) A grassroots sustainable energy niche? Reflections on community energy in the UK. *Environmental Innovation and Societal Transitions* 13: 21–44.

Smith, A., Hargreaves, T., Hielscher, S., Martiskainen, M. and Seyfang, G. (2016) Making the most of community energies: Three perspectives on grassroots innovation. *Environment and Planning A* 48(2): 407–432.

Sorrell, S. (2007) Improving the evidence base for energy policy: The role of systematic reviews, *Energy Policy* 35(3): 1858-1871.

Sovacool, B.K. (2014) What are we doing here? Analysing 15 years of energy scholarship and proposing a social science research agenda. *Energy Research and Social Science* 1: 1–29.

Sovacool, B.K. (2016) How long will it take? Conceptualizing the temporal dynamics of energy transitions, *Energy Research & Social Science* 13: 202–215.

Sovacool, B.K. (2017) Experts, theories, and electric mobility transitions: Toward an integrated conceptual framework for the adoption of electric vehicles. *Energy Research and Social Science* 27: 78–95.

Sovacool, B.K. and Hess, D.J. (2017) Ordering theories: Typologies and conceptual frameworks for sociotechnical change. *Social Studies of Science* 7(5): 703–750. DOI: 10.1177/0306312717709363.

Steinhilber, S., Wells, P. and Thankappan, S. (2013) Socio-technical inertia: Understanding the barriers to electric vehicles. *Energy Policy* 60: 531–539.

Stern, P.C., Sovacool, B.K. and Dietz, T. (2016) Towards a science of climate and energy choices. *Nature Climate Change* 6: 547–555.

Sushandoyo, D. and Magnusson, T. (2014) Strategic niche management from a business perspective: Taking cleaner vehicle technologies from prototype to series production, *Journal of Cleaner Production* 74(1): 17–26.

Tsoutsos, T.D. and Stamboulis, Y.A. (2005) The sustainable diffusion of renewable energy technologies as an example of an innovation-focused policy. *Technovation* 25(7): 753–761.

Van den Bergh, J.C.J.M., Truffer, B. and Kallis, G. (2011) Environmental innovation and societal transitions: Introduction and overview. *Environmental Innovations and Societal Transitions* 1(1): 1–23.

Van der Laak, W.W.M, Raven, R.P.J.M. and Verbong, G.P.J. (2007) Strategic niche management for biofuels: Analysing past experiments for developing new biofuel policies. *Energy Policy* 35(6): 3213–3255.

Verbong, G.P.J., Beemsterboer, S. and Senger, F. (2013) Smart grids or smart users? Involving users in developing a low carbon electricity economy. *Energy Policy* 52: 117–125.

Verbong, G.P.J., Christiaens, W., Raven, R. and Balkema, A. (2010) Strategic niche management in an unstable regime: Biomass gasification in India. *Environmental Science and Policy* 12(4): 272–281.

Verbong, G., Geels, F.W. and Raven, R. (2008) Multi-niche analysis of dynamics and policies in Dutch renewable energy innovation journeys (1970–2006): Hype-cycles, closed networks and technology-focused learning. *Technology Analysis and Strategic Management* 20(5): 555–573.

Weber, M., Hoogma, R. Lane, B. and Schot, J. (1999) Experimenting with Sustainable Transport Innovations: A Workbook for Strategic Niche Management. University of Twente, Seville/Enschede.

Xue, Y., You, J., Liang, X. and Liu, H-C. (2016). Adopting strategic niche management to evaluate EV demonstration projects in China. *Sustainability* 8(2): 142. DOI: 10.3390/su8020142.

Part V
Conclusion

14 Conclusion

Towards systematic reductions in energy demand

Kirsten E.H. Jenkins, Debbie Hopkins and Cameron Roberts

Introduction

Transitions in Energy Efficiency and Demand began by outlining the challenge put forth by the Paris Agreement – the aspirational target of keeping the increase in global average temperature to 1.5°C above pre-industrial levels, and the firm target of achieving an increase well below 2°C. At the time of writing this book, media headlines suggest that this target is slipping through our grasp and the scientific evidence warns of failing targets (Climate Action Europe, 2018; Rogelj *et al.*, 2016). Indeed, to have a reasonable chance of reaching these goals, emissions must peak by 2020 and fall by more than 70 per cent in the next 35 years (Cooper and Hammond, 2018; Geels *et al.*, 2018); a formidable and unprecedented challenge that requires radical and far-reaching transformations of the whole energy system, including significant increases in energy efficiency and considerable reductions in energy demand.

In the introduction, we noted too, that the process by which we tackle the issue of energy demand defies any simple solution: No single policy or innovation is likely to make a notable impact. Thus, the means of reducing carbon emissions are various, ranging from the increased efficiency of existing energy-using devices, to the development of entirely new systems, and complex, relying on elaborate combinations of technology, policy, social practices, infrastructure, and culture to succeed. There are also many stumbling blocks – both historically embedded and contemporary – that might prevent rapid and consistent progress. Examples of this include the complexity of energy demand; the need for large-scale, rapid change; growing demand for energy; societal disinclination to change; the insufficiency of market mechanisms, and the plethora of economic barriers. Against this somewhat pessimistic backdrop, this chapter offers overarching insights gathered from the book's various and wide-ranging case studies, to end with a cautiously hopeful, and optimistic path forward.

Through a common commitment to a sociotechnical approach, the chapters presented in this book have sought to use a wide range of social science perspectives to tackle the complexity of the energy demand reduction challenge. Our aim was to do this for an academic audience, while also having in mind the needs of various decision-makers, such as policymakers, entrepreneurs, engineers, activists, researchers and others involved with the areas of energy efficiency and energy demand. The result has been a diverse range of chapters that examine how new low-energy innovations emerge and diffuse, and how this process is shaped by market forces, government policy, social interactions and cultural norms, as well as the complex interactions between all of these – and other – factors. The innovations examined in this book include new technologies, energy systems, business models and behaviours, as well as combinations of these.

This chapter proceeds as follows: First, we restate the importance and utility of the sociotechnical approach and the framework of innovation emergence, diffusion and impact first outlined in Chapter 2. Second, we discuss the importance of this approach for energy demand reduction in the UK context. Third, we outline what this means for countries beyond the UK. Finally, we introduce a series of policy principles that set the scene for our overall conclusion, an agenda for ongoing research and policy action into the systematic reduction of energy demand.

This conclusion necessarily takes a big-picture perspective. Because energy demand is such a complex phenomenon, running through virtually every aspect of human society in one way or another, and influencing (and being influenced by) so many different sociotechnical systems, there can be no one-size-fits-all solution. Thus, we note from the offset that given the wide breadth of topics covered throughout this book, only each chapter can give specific, case-relevant recommendations. Rather than duplicating these, this conclusion aims to offer general principles and heuristics.

Restating the sociotechnical approach

The most obvious takeaway from this volume, especially for a non-academic reader, should be the value of a sociotechnical perspective. This has, first and foremost, offered a novel framework for considering not only the technical aspects of the demand reduction challenge, but also its social, political, economic and cultural complexities. More specifically, it has drawn together a series of theoretical advances that guide the way we can and should consider such challenges, and, most pressingly, their solutions. For academics who may already be familiar with the sociotechnical approach, the chapters in this volume suggest fruitful new areas for scholarship that applies sociotechnical theory to the challenge of energy demand reduction. Beyond academia, this approach presents a useful way to imagine and understand the challenge of demand reduction and wider societal implications.

Transitions in Energy Efficiency and Demand contains many examples of the value of a sociotechnical approach. Chapter 9 (Figus *et al.*) investigates energy-saving innovations and economy-wide rebound effects and demonstrates the benefits for aggregate social welfare that can be achieved though improvements in the energy efficiency of domestic boilers, for instance. In contrast, however, Chapter 6 (Jenkins *et al.*) uses issues in the rollout of smart meters in the UK to show what happens if people are forgotten throughout the transitions process, cautioning that some social groups may become *more* vulnerable or marginalised. Similarly, Chapter 5 (Hopkins and Schwanen) points to the selectively managed experimentation with vehicle automation that involves some actors, but excludes others, perpetuating existing power dynamics. Finally, Chapter 11 (Webb) illustrates that effective policy towards energy efficiency in UK buildings may be hampered by political reliance on classical economic theories and short-term price metrics, and a failure to link up policy strategies; a uniquely social failing. In each case, the sociotechnical approach reveals the actors we need to engage with if we are to reach climate change and emissions reduction targets in a socially-just way. These insights also restate the importance of both technologically radical *and* socially radical change, or as is the ultimate aim, a systematically radical combination of both (Dahlin and Behrens, 2005).

In various places, our chapters also show that the challenge of reducing energy demand is much more complex than simply relying on market mechanisms to incentivise people to invest in the 'right' innovations. This fairly blunt approach is likely to result in unintended consequences, which could either undermine the transition, or make it harmful for vulnerable groups. Jenkins and Sovacool (Chapter 13) provide perhaps the clearest set of practical lessons that can be taken from a sociotechnical approach (at least as far as niches are concerned), when they observe that the existing literature on Strategic Niche Management emphasises the critical roles of niche experiment financing, mutual learning, and brokering partnerships for successful niche development.

A second key point revealed by our sociotechnical approach is that of *co-evolution* between different elements of a sociotechnical system (Geels, 2004). Linkages may emerge between the evolution of technologies and users, or between technology, industry structure and policy institutions (Geels, 2005). For instance, the linkages between technology, industry structure and policy institutions are shown in Brockway *et al.*'s (Chapter 8) exploration of exergy economics, and Webb (Chapter 11) identifies the inevitable interconnection between energy and political values and beliefs. These chapters reinforce our point that energy demand must be viewed as a holistic problem, and cannot be reduced to any one factor, viewed in a vacuum, nor addressed with any one solution such as a new technology or a new regulation governing energy companies. In this vein, Kern *et al.*'s work on policy mixes (Chapter 12) provides critical insights on the co-evolution of policy and sociotechnical change which calls for more explicit attention to policy processes, and which in turn may enable a better understanding of the dynamic nature and causal links between the two.

Beyond these two overarching lessons, we further identify merit in the specific sociotechnical themes of emergence, diffusion and impact, initially introduced by Geels *et al.* (Chapter 2). Although these themes overlap and are non-linear, they provide a process-oriented framework that explores how low-energy innovations develop and become established; how they achieve widespread adoption, and, crucially, how low-energy innovations ultimately impact energy demand.

Under the theme of emergence, the chapters reveal, in accordance with Geels *et al.* (2018), that we must identify the techno-economic, finance and investment, cognitive (contrasting views and perceptions around consumer preference, for example) and social (including instabilities within networks of actors) uncertainties that limit the emergence of new innovations. Doing so is especially important considering that the aforementioned co-evolution processes will inevitably create further obstacles for sustainable alternatives, which are already saddled with the teething problems facing all new technologies (e.g. smart meters) (Mokyr, 2010; Unruh, 2000).

For the theme of diffusion, gaining endogenous momentum behind innovations, and understanding how these innovations can become embedded within policy, social, business and user environments is central. Finally, in terms of impact, and with the acknowledgement that it is extremely difficult to do so, we must seek to better understand the influence of incremental innovations, such as loft and cavity wall insulation; explore rebound effects; analyse impact scenarios, and develop modelling tools for systematic sociotechnical transitions (Geels *et al.*, 2018). Taken together, the detailed consideration of emergence, diffusion and impact provided in this book provides opportunities for far-reaching transformations. From here and working towards practical utility of our research and findings, we now consider what the sociotechnical approach means for the UK, for international audiences, and for policy practice.

Reflections on the UK context

To date, the UK *has* made significant progress in decreasing energy demand through both technological innovations and the offshoring of manufacturing (Hardt *et al.*, 2018), yet as the chapters in this book demonstrate, much more can be done. Despite the recognition from the UK government that reducing demand is a more cost-effective approach to reaching national climate goals than building additional capacity, all of the so-called 'low-hanging fruit'[1] have been plucked and at the time of writing, energy efficiency and demand policy in the UK is somewhat confused. In fact, without further progress, the Committee on Climate Change (2016, 2018) warns that UK policies will fall well short of the fifth carbon budget – a legal emissions restriction that forms part of a long-term target of reducing greenhouse gas emissions to 80 per cent below 1990 levels by 2015.

Taken together, the case studies presented in this book have shown that UK progress on energy demand transitions is something of a mixed bag. Some innovations – including electric vehicles (EVs), automated vehicles (AVs, also known as 'driverless', 'self-driving' or 'autonomous' vehicles), and smart meters – certainly still appear to be in the emergence phase. EVs for personal mobility (Chapter 4), for instance, are constrained by the lack of simultaneous development of energy storage of mobile power supplies. The emergence of AVs (Chapter 5) is constrained in the experimentation phase by limited visions of the 'real world', with fewer opportunities for surprises and second-order learnings. Likewise, smart meters (Chapter 6) are still undergoing technological development and face the constraint of resistant and reluctant consumers.

Progress is slow at best as policymakers appear to reluctantly and tentatively commit to change. This is perhaps most dramatically illustrated by Bergman's work on EVs (Chapter 4), which shows that policymakers' visions of how change will occur still uphold the dominant regime structure of the car industry, which is based on the continued prominence of conventional vehicles in the medium term. Also, in the transport sector, Hopkins and Schwanen (2018a, 2018b; Chapter 5) show how a technological solutionist discourse prevails in responding to the environmental externalities of transport. This is despite the fact that transport accounts for approximately 25 per cent of the UK's CO_2 emissions (with two-thirds of that coming from cars and vans) and therefore offers a significant option for large-scale energy demand reduction (Chapter 4; Chapter 5; Committee on Climate Change, 2014). It seems that British policymakers are uncomfortable even *imagining* radical change in transport-related energy demand, much less implementing it.

The two UK cases discussed by Roberts and Geels (Chapter 10) on historical transitions show that this somewhat reticent approach to embracing radical sociotechnical change is not new to the UK. The transition to road transport in the UK was only given policy support after the British road transport system was already widely established. Policymakers dramatically accelerated the transition to modern agriculture in response to a wartime food shortage, but in that case, they acted only in response to an existential threat combined with a decade of lobbying from farmers. Moreover, their actions mainly consisted of developing a domestic system that had been in place overseas for since the late nineteenth century. If this pattern continues in the era of climate change, the UK might thus be limited to being a follower, rather than a leader, in low-carbon transitions.

A further case-in-point are current approaches to energy efficiency improvements in buildings, which account for approximately one-quarter of UK carbon emissions and therefore, present considerable potential for further savings and improvements (Clarke et al., 2008; Rosenow et al., 2018). Chapters by Brown et al. (Chapter 7) and Webb (Chapter 11) identify failures with policy initiatives such as the Green Deal, despite the fact that the Committee on Climate Change (2015, 2016) estimates that there is cost-effective potential to reduce direct emissions from all buildings by a third by 2030, and to achieve near-zero emissions by 2015. If proactive policies are implemented and support restored, this retrofit rollout alone could add approximately £25.3 billion of value added to Britain's GDP (Guertler and Rosenow, 2016). In this regard, UK decision-makers must not only imagine positive change, but also heed the positive business case for it.

Constructively, the chapters on diffusion and impact highlight a number of key findings towards averting and redressing this somewhat unsupportive policy trend and provide evidence of what successful policymaking might look like. For instance, Kern et al. (Chapter 12) demonstrate that to be effective, UK policymakers need to develop well-managed portfolios of policy goals, strategies and instruments, the mix of which needs to be (re-)assessed and modified as necessary. These policies, they argue, should focus on 'efficiency first' as a policy goal which goes against the current strategy, where, to date, none of the £256 billion investment pipeline for energy infrastructure has been allocated to energy efficiency improvements (Rosenow and Cowart, 2017).

Following a case study of stalled progress on energy efficiency policy for UK buildings, Webb (Chapter 11) reinforces our earlier point that British policy mindsets must shift away from classical economics' insistence on the efficiency of markets towards a valuation framework structured around societal responsibility for welfare and sustainable prosperity. Webb adds that to reach energy efficiency goals, the British government may even require departmental reform, as energy-saving is relevant to multiple ministries, none of which are focused on the issue (particularly in the case of low-carbon buildings). British policymakers, activists, researchers, and others trying to influence change, should more carefully look at what kinds of broader sociotechnical developments can *enable* greater political will to accelerate transitions to sustainability. Thus, UK energy demand and energy efficiency policy requires a consistent, front-and-centre seat at a number of interlinked tables.

Lessons for other countries

Roberts and Geels (Chapter 10) warn that despite being instructive, policymakers should be cautious when copying lessons from foreign examples given their different political, economic, cultural and technological particularities. Nonetheless, the primarily British case studies discussed in this book certainly have relevance beyond the British context and suggest a wider range of broadly applicable lessons.

First, we stress the importance of *visions and expectations*. While these might seem rhetorical and ultimately not constitutive of actual on-the-ground change, the evidence in this book shows that they can have powerful performative impacts not just on the uptake of energy-saving innovations, but also on the effects that these innovations have once adopted. Roberts and Geels (Chapter 10) show how visions of a motorised future in the UK; of clean, efficient, natural gas in the Netherlands; and of a cooperatively-run, sustainable heating system in Denmark, proved decisive in shaping choices by both policymakers and private actors that allowed the deliberate acceleration of sociotechnical transitions. Bergman (Chapter 4), on the other hand, demonstrates how visions and expectations not only help determine whether low-energy innovations diffuse widely, but also what form they will take after doing so. This is important given that during the emergence phase, any innovation has several ways in which it can be used in practice. EVs can simply result in the same patterns of automobility, for instance, or they can result in completely new kinds of travel patterns that have a much greater effect on energy demand. Where the UK has arguably failed in this regard, other countries can succeed.

The second international lesson relates to the importance of *incumbent power and resistance*, which appears, one way or another, in virtually all chapters of this book. Drawing on Hughes (1987), Chapter 2 (Geels *et al.*) identifies sources of so-called 'lock-ins', including sunk investments in skills, factories and infrastructures, for example, as well as economies of scale and the momentum of established rules and institutions, each of which restricts opportunities for change. Others refer to incumbent business models and industry groups. In their work on the emergence of AVs, Hopkins and Schwanen (Chapter 5) contribute to understandings of the politics and power-laden nature of urban experimentation, with reproductions of the status quo. Roberts and Geels (Chapter 10) go into detail illustrating how overcoming incumbent resistance is not just a matter of fighting it directly; sometimes, such as in the case of Dutch natural gas, incumbent actors can be bought out, or even co-opted to become active partners in the transition. If enough incumbents can be brought onside in this way, then their power acts in favour of energy transitions rather than against them, and therefore, may create the necessary pre-conditions for radical change to happen. These incumbents are not just policy and industry elites, but also users. Admittedly it is strange to think of users (who, in the case of energy demand, are essentially the various and multiple general publics), as incumbents. Like industrial and political actors, however, they have entrenched interests, practices, and preferences that they are reluctant to change. They, too, might have to be bought out for radical change to occur.

A third common point is that of *technological and policy mixes*. This book has illustrated the paramount importance of looking both at the interaction between different kinds of interventions and innovations, and at the broader sociotechnical effects that can have. Kern *et al.*'s research on policy mixes (Chapter 12) has some obvious lessons for this as it applies to policy measures that take account of existing complex policy mixes, and which simultaneously take advantage of existing innovations while developing new, more radical ones for the future. This point is articulated by Brown *et al.* (Chapter 7), who recommend a mix of standards and regulations, financial measures, new institutions, and intermediaries to address the problem. Policymakers should be fully aware of all the policy and technological mixes they are dealing with before intervening to change something. They can also look for pre-existing systems or policies they can build from, rather than trying to create radical change from scratch.

The final insight has to do with *users and practices*. Users impose normative conditions on transitions, as discussed by Jenkins and Martiskainen (Chapter 3) and Jenkins *et al.* (Chapter 6). They therefore must play a fundamental role in the innovation process not only for the sake of moral considerations such as a commitment to democracy, but also because transition policies that succeed at the expense of some vulnerable element of society are likely to be politically unstable. Users should therefore be active participants in energy transitions, rather than passive beneficiaries, a finding reiterated by Hopkins and Schwanen (Chapter 5). Ultimately, the various publics have the biggest single role (albeit a collective, often unguided, and sometimes unconscious one) in actually enacting change. As Figus *et al.* (Chapter 9) show, the ultimate impact of rebound effects depends to a large extent on the specific circumstances of particular user groups and how they compensate for energy savings in one area with increased energy use in other areas.

Taken together, these lessons may seem somewhat abstract, but alongside the specific recommendations and applications presented in each chapter – which draw on material from New Zealand, Denmark, Japan and Finland, among others – they provide important practical recommendations and cautionary tales.

Recommended policy principles

While the primary value of *Transitions in Energy Efficiency and Demand* lies with its sociotechnical approach and the range of substantive insights and conceptual contributions developed throughout each chapter, we also offer a secondary benefit: nine promising policy principles for accelerating high efficiency, low demand change. Developing the points made above, we now discuss each of these in turn.

First, policymakers should, whenever possible, be *ambitious, inclusive, and challenging* when setting their visions, roadmaps, plans and other devices for orienting a transition in energy demand. While incumbent actors might have the easiest claim to expertise on the future possibilities that exist, this comes with a bias that might lead to reproduction of existing patterns of energy demand, rather than wholesale change. Chapter 2, for instance, demands a broader view of the process, which takes into account learning and experimentation, the multiple conditions necessary for systemic change, and the coalitions of interest that can block or support emerging niche innovations (Geels *et al.*, Chapter 2). Bergman (Chapter 4), recommends that policymaker's engagement with visions includes a larger variety of futures, scenarios of disruption and failures to meet emissions reductions and other targets. These, he argues, could be commissioned from a wider variety of actors, including outsiders and niche players who can challenge, rather than support, dominant visions.

Second, policymakers should *avoid looking for single, silver bullet technological or policy interventions and move towards policy and technological mixes.* This necessitates an embrace of complex, multi-faceted approaches that include targeted regulations, subsidies, public relations campaigns, and other strategies that take account of (and, when possible, augment) existing policy and technological mixes. Critically, this includes a move beyond a sole focus on market mechanisms or drop-in technological fixes (e.g. Chapter 5; Hopkins and Schwanen). As Geels *et al.* (Chapter 2) show, this forces us to look beyond carbon pricing as a policy panacea. Research on policy mixes provides a particularly promising avenue here: Brown *et al.* (Chapter 7) develop an especially nuanced and tangible set of recommendations in keeping with this notion that includes a mix of regulations, financing and incentives along with the establishment of new institutions and the recognition of energy efficiency as a strategic infrastructure priority.

Third, *policies should aim to support present-day incremental change, while also building towards radical change in the future*. This applies to all aspects of developing efficiencies in energy demand, including technologies, networks, business models, regulatory structures and user practices, and comes as acknowledgement of both the dramatic change required, and of the cumulative effect of small steps to get there. It further suggests a rethinking of the radical/incremental dichotomy that is so prevalent in discussions about climate policy. While radical policy changes to deliberately accelerate transitions should be the ultimate goal, incremental changes should be seen not as an inferior alternative to these, but as near-term facilitators of the more aggressive cuts to energy demand. This becomes especially important when you consider that the more individuals that are successfully engaged and take on energy efficiency schemes, the greater the potential success of transition pathways.

Fourth, *users should be considered as a critical component in any process of change*. The ultimate impact of transition policies on users should be socially just, and supportive of practices that already exist at the user level. Wherever possible, policymakers should try and build on practices that users are already demonstrating or needs that they are already articulating. Jenkins and Martiskainen (Chapter 3) note, to this end, that throughout the transition process, governments and business must identify those who may be vulnerable and then both ascertain and make provision for them through targeted subsidies, exemptions and efficiency measures (e.g. in energy efficiency policy). As one very tangible option, this may take the form of a funded Energy Cafés that acts as a triage service, bringing together local authorities, health workers, community organisations and individuals in a trusted setting, providing advice and ensuring that energy needs are met. The risk of not doing this is that we fail particular social groups through insufficient consumer engagement, as is warned by Jenkins et al. in their exploration of the UK smart meter rollout (Chapter 6).

Fifth, *transitions should have a clear normative goal*. Innovations do not just take the form of new technologies, but can also be social or procedural in nature, and this makes them inherently normative affairs. The energy justice framework introduced in Chapter 3 suggests that transitions in energy demand should occur in a way that ameliorates, rather than exacerbates, energy poverty. In order to fulfil this aim, it thus becomes paramount that we engage with a wide range of both practical and normative voices. This will require the British government, for one, to address its current tardiness (shown by Webb in Chapter 11) in responding to the voices of researchers, advisory bodies and lobbyists, even when they highlight the benefit of alternative transition pathways (or indeed, caution their failure, as was the case with the Green Deal (Mallaburn and Eyre, 2014)).

Sixth, *transition policies should, where possible, act on technologies that already exist*. Radical innovations are useful, but they take time to develop and up-scale. As Geels and Roberts (Chapter 10) suggest, focusing on technologies that are already well established in other contexts means that policymakers are simply acting to consolidate, and, perhaps, to accelerate, transitions that are already well underway, and can also benefit from previously developed technical performance, user communities, and business networks. Knowing their strengths, these can be developed using innovation policies in advance of deliberate acceleration e.g. promoting them as public goods, avoiding monopolies or coercive policies that will aggrieve users.

Seventh, *transitions will always have unintended consequences, and wherever possible, these have to be acknowledged, anticipated and managed*. This is particularly important in energy demand, as energy-saving innovations have a tendency to produce rebound effects, which reduce their ultimate effectiveness. It is also key in the context of Brexit, where certain types of outcomes may be difficult or impossible to anticipate (Geels *et al*., Chapter 2). Bergman (Chapter 4) shows how this could be achieved by commissioning visioning documents from a larger variety of actors, including outsiders and niche players, who can challenge, rather than support, the views of incumbents. This, he writes, would enable more scope and choice for policymakers to meet policy goals and targets, and leave us better prepared for foreseeable and unforeseeable changes to transport in the future; a lesson that undoubtedly extends beyond his case study of EVs. It seems important, too, to acknowledge small, often knock-on consequences.

To this end, Figus *et al*. (Chapter 9) warn that a focus on rebounds highlights the failure to achieve the technologically feasible energy use reductions and neglects the wider range of economic and social impacts that energy efficiency improvements can deliver beyond climate change alone. These include, the co-benefits of household energy efficiency stimulating the UK's economy, leading to increased employment, investments and wages which achieving substantial, yet smaller than anticipated, reductions in energy use, for instance. Or, as Shrubsole *et al*. (2014) warn, a negative set of implications that extend as far as increases in feelings of social isolation if windows are too airtight and noise cannot infiltrate.

Eighth, *the spatial dynamics, and potential spatial and temporal asymmetries of policy interventions need to be acknowledged, and where possible, avoided*. This has implications across the three dimensions of innovations discussed in this book; emergence, diffusion and impact. In the emergence and diffusion phases, for instance, innovations are likely to have spatially and temporally distinct characteristics, playing out at national through to local scales. As an illustration, Chapter 5, Hopkins and Schwanen point to the replication of existing dynamics in the trialling of new technologies. Given the likely costs of emerging innovations (e.g. AVs, EVs) diffusion in wealthy suburbs – and policies to accelerate diffusion – are likely to have unequal impacts. Guiding visions and expectations are also often a-spatial; they homogenise places and overlook diversities in people, infrastructures, cultures, etc., with implications for public and political acceptance. Likewise, impacts may benefit some places, while disadvantaging others, at least in the short-medium term.

Ninth, and finally, *policies should aim to address all three facets of transitions*: The emergence of radical alternatives, the diffusion of new sociotechnical

Table 14.1 Six sociotechnical research debates and areas for future study

Theme	Research debate
Emergence	• The contribution of outsiders and incumbents to emergence • The role of visions and expectations in emergence • Geographies of innovation emergence and impacts for social equity
Diffusion	• Political will and contextual pressures for deliberately accelerating diffusions • Policy mixes for accelerated diffusion
Impact	• Rebound effects of low-energy innovations • Frameworks for evaluating broader impacts

Source: the authors, with reference to Geels *et al.*, Chapter 2.

systems and their ultimate impact on energy demand. Thus, it is critical that going forward, sustained and long-term policy to support sustainable innovation accounts for the various steps along the innovation pathway, to ensure not just successful emergence into the marketplace, but also that the innovation(s) is/are as impactful as possible in order to meet the challenge of energy demand reduction.

Conclusion: the future of energy demand research and policy

If there is one lesson we hope the readers of this book take away, it is the value of a sociotechnical approach in understanding transitions in energy demand. Looking beyond this, we hope that readers engage with each of the chapters that are relevant to their own practice and implement the recommendations given within. We return too, to the new areas of research reflected upon in Chapter 2 (and summarised and in one case, further developed in Table 14.1), which not only have the potential to build on each other to achieve radical, systematic change, but place this volume at the forefront of a new research agenda into the future of energy demand research and policy.

What is more, we reiterate the magnitude of this challenge and state once more, that reductions in energy demand will not be accomplished by following any one magical formula. This book's focus on complexity and context-dependence should be sufficient evidence of that. We have provided general principles, heuristics, cautionary tales and ideas, but transitions on-the-ground will always depend on the ingenuity, imagination and dogged effort of those who work on making them happen. To all engaged in meeting this challenge, we wish them good luck.

Note

1 The distribution of energy-efficient boilers, as one example.

References

Clarke, J.A., Johnstone, C.M., Kelly, N.J., Strachan, P.A. and Tuohy, P. (2008) The role of the built environment energy efficiency in a sustainable UK energy economy. *Energy Policy* 36(12): 4605–4609.

Climate Action Europe (2018) Off target: Ranking of EU Countries' Ambition and Progress in Fighting Climate Change. CAE, *Brussels, Belgium*. Available at: www.caneurope.org/publications/reports-and-briefings/1621-off-target-ranking-of-eu-countries-ambition-and-progress-in-fighting-climate-change.

Committee on Climate Change (2014) Meeting Carbon Budgets – 2014. Progress Report to Parliament. Committee on Climate Change, HM Government, London, UK.

Committee on Climate Change (2015) The Fifth Carbon Budget – The Next Step towards a Low-carbon Economy – November 2015. Committee on Climate Change, HM Government, London, UK. Available at: www.theccc.org.uk/wp-content/uploads/2015/11/Committee-on-Climate-Change-Fifth-Carbon-Budget-Report.pdf.

Committee on Climate Change (2016) Meeting Carbon Budgets – 2016 Progress Report to Parliament. Committee on Climate Change, HM Government, London, UK. Available at: www.theccc.org.uk/publication/meeting-carbon-budgets-2016-progress-report-to-parliament.

Committee on Climate Change (2018) Reduce UK Emissions: 2018 Progress Report to Parliament. Committee on Climate Change, London, UK. Available at: www.theccc.org.uk/publication/reducing-uk-emissions-2018-progress-report-to-parliament/.

Cooper, S.J.G. and Hammond, G.P. (2018) 'Decarbonising' UK industry: Towards a cleaner economy. Proceedings of the Institution of Civil Engineers. *Energy*: 1–11.

Dahlin, K.B. and Behrens, D.M. (2005) When is an invention really radical? Defining and measuring technological radicalness. *Research Policy* 34(5): 717–737.

Geels, F.W. (2004) From sectoral systems of innovation to socio-technical systems: Insights about dynamics and change from sociology and institutional theory. *Research Policy* 33(6–7): 897–920.

Geels, F.W. (2005) The dynamics of transitions in socio-technical systems: A multi-level analysis of the transition pathway from horse-drawn carriages to automobiles (1860–1930). *Technology Analysis and Strategic Management* 17(4): 445–476.

Geels, F.W., Schwanen, T., Sorrell, S., Jenkins, K. and Sovacool, B.K. (2018) Reducing energy demand through low carbon innovation: A sociotechnical transitions perspective and thirteen research debates. *Energy Research and Social Science* 40: 23–35.

Guertler, P. and Rosenow, J. (2016) Buildings and the 5th Carbon Budget. ACE, London, UK. Available at: www.ukace.org/wp-content/uploads/2016/09/ACE-RAP-report-2016-10-Buildings-and-the-5th-Carbon-Budget.pdf.

Hardt, L., Owen, A., Brockway, P., Heun, M.K., Barrett, J., Taylor, P.G. and Foxon, T.J. (2018) Untangling the drivers of energy reduction in the UK productive sectors: Efficiency or offshoring? *Applied Energy* 223(1): 124–133.

Hopkins, D. and Schwanen, T. (2018a) Automated mobility transitions: governing processes in the UK. *Sustainability* 10(956): 1–19. DOI: 10.3390/su10040956.

Hopkins, D. and Schwanen, T. (2018b) Governing the Race to Automation. In: Marsden, G. and Reardon, L. (Eds) *Governance of Smart Mobility*. Emerald, Bingley, UK.

Hughes, T.P. (1987) The Evolution of Large Technological Systems. In: Bijker, W.E., Hughes, T.P. and Pinch, T. (Eds) *The Social Construction of Technological Systems: New Directions in the Sociology and History of Technology*. The MIT Press, Cambridge, MA, 51–82.

Mallaburn, P.S. and Eyre, N. (2014) Lessons from energy efficiency policy and programmes in the UK from 1973 to 2013. *Energy Efficiency* 7(1): 23–41.

Mokyr, J. (2010) Chapter 2 – The Contribution of Economic History to the Study of Innovation and Technical Change: 1750–1914. *Handbook of the Economics of Innovation 1*, Elsevier, Oxford, UK, 11–50.

Rogelj, J., den Elzen, M., Hohne, N., Fransen, T., Fekete, H., Winkler, H., Schaeffer, R., Sha, F., Riahi, K. and Meinshausen, M. (2016) Paris Agreement climate proposals need a boost to keep warming well below 2 °C. *Nature* 534: 631–639.

Rosenow, J. and Cowart, R. (2017) Efficiency First: Reinventing the UK's Energy System Growing the Low-Carbon Economy, Increasing Energy Security, and Ending Fuel Poverty, The Regulatory Assistance Project. Available at: www.raponline.org/knowledge-center/efficiency-first-reinventing-uks-energy-system.

Rosenow, J., Guertler, P., Sorrell, S. and Eyre, N. (2018) The remaining potential for energy savings in UK households. *Energy Policy* 121: 542–552.

Shrubsole, C., Macmillan, A., Davies, M. and May, N. (2014) 100 unintended consequences of policies to improve energy efficiency of the UK housing stock. *Indoor and Built Environment* 23(3): 340–352.

Unruh, G.C. (2000) Understanding carbon lock-in. *Energy Policy* 28(12): 817–830.

Index

Page numbers in **bold** denote tables, those in *italics* denote figures.

acceleration of transitions 24–25, **29**, 177, 222–224
advocacy coalition framework (ACF) 225, **226–227**
agricultural modernisation, UK *178*, 179, 181–183, 188, 189, 190, 191
automobility 53, 56–58, 60–62, 64, 65, 67

behavioural economics 15, 16
BEIS *see* Department of Business, Energy and Industrial Strategy (BEIS), UK
Bonfield Review, UK 124, 208
Brexit 26, 78, 270
brokering and partnership management 248, **251**
Building Regulations, UK 112
business models, for residential retrofit *116*, 117–118, *117*

carbon dioxide emissions 1, 15, 53, 110, 235
Carbon Emissions Reduction Target (CERT), UK 112, 199, 200
carbon pricing 4–5, 15–16
Carbon Saving Multiplier 157–158, 166–168, **167**, 169–170
CCC *see* Committee on Climate Change (CCC), UK
Centre on Alternative Technology, UK 122
Centre on Innovation and Energy Demand (CIED) 8, 121, 215, 217, 228
CERT *see* Carbon Emissions Reduction Target (CERT), UK
CGE models *see* Computable General Equilibrium (CGE) models

children, fuel poverty and 40, 41
China, exergy efficiency 143, **143**, *143*
CIED *see* Centre on Innovation and Energy Demand (CIED)
Citizens Advice, UK 105
civic laboratories 76
Clean Growth Strategy, UK 112, 124, 196, 197, 203, 210
Climate Change Levy, UK 205
coal, Dutch transition to from *178*, 179, 183–185, 188, 189, 190, 191
co-construction 23–24
cognitive uncertainties 20
commercial buildings 197, 205, 208
Committee on Climate Change (CCC), UK 7, 63, 110, 196, 203, 264
Computable General Equilibrium (CGE) models 149, 158–159, 162–165, 170
condensing boiler regulations 25–26, 112, 196, 207–208
connected and autonomous vehicle (CAV) experimentation 72–73, 78–91; anticipated benefits of CAVs 73; benefits of experiments 79, 90–91; emergence of CAVs 78–79; framings and logics 81–82; Greenwich projects 81, 82, 83–89, **85**, 90; Oxford projects 80–83, **84**, 86, 87, 88, 89, 90; places and spaces 88–89; policy recommendations 91; practices 82–86, **84–85**; subjects 86–88; UK national policy 79–80
constant elasticity of substitution (CES) production functions 147, 148
consumption-based emissions 7
conversion efficiencies 3
cost optimal formulae 202–203
cost share theorem 136–137, 147, 148

CRC Energy Efficiency Scheme, UK 197, 205, 208
CSM *see* Carbon Saving Multiplier
cultural domination practices 40

DECC *see* Department of Energy and Climate Change (DECC), UK
Denmark, transition to district heating 178, 179, 185–187, 188, 189, 190, 191
Department of Business, Energy and Industrial Strategy (BEIS), UK 95, 96, 103, 106, 203, 205–206
Department of Energy and Climate Change (DECC), UK 96, 104, 105, 206
'Dieselgate' (Volkswagen emissions scandal) 79
diffusion of low-energy innovations 16, *19*, 22–25, **29**, **271**; *see also* energy efficiency policies, UK; policy mixes; political acceleration of sociotechnical transitions
discourse coalitions framework 225, **226–227**
distributional justice 37, 38–40, 100–102
district heating, Danish transition to 178, 179, 185–187, 188, 189, 190, 191
domestic retrofit *see* retrofitting residential buildings
DRIVEN Consortium project, Oxford, UK 82–83, **84**, 87

'Each Home Counts' report, UK 124, 208
ECO *see* Energy Company Obligation (ECO), UK
Eco Open Houses events, Brighton, UK 121, 122
ecological economics 134, 136–137
economic growth, energy consumption and 3, 4, 133–134, 135–138; *see also* exergy economics
economy-wide impacts of household energy efficiency improvements 156–170; policy recommendations 168–170; rebound effects 156, 157, 158, **160**, 161, **163**, 165–166, **167**; saving multiplier as alternative to rebound indicator 157–158, 165–168, **167**; Scottish case study 162–165, **163**; UK-wide case study 159–162, **160**
efficient markets hypothesis 198–199
electric vehicles and visions of future personal transport 17, 53–68; automobility 53, 56–58, 60–62, 64, 65, 67; central vision 60–62, *61*, 64, 65, 67, 68; factors affecting trajectories 62–63; policy recommendations 68; potential transitions to EVs 57–58; public acceptance issues 62–63; specific trajectories of interest 63–64; technical issues 62; theoretical background 54–56; transition pathways 55, 64–65; vision documents **59**, 60; visions as institutional work 56, 66–67
electricity: renewable energies 1–2; *see also* Smart Meter Implementation Programme (SMIP), UK
embedding of innovations 23–24
emergence of low-energy innovations 16, 19–22, *19*, **29**, **271**
endogenous momentum of innovations 23
Energiesprong retrofit initiative, Netherlands 117–118, *117*, 122, 126
Energy Cafés 43, 44
Energy Company Obligation (ECO), UK 44, 110, 112, 116, 118, 124, 200
energy consumption: economic growth and 3, 4, 133–134, 135–138; household 111–112, 196; net-zero 117; *see also* exergy economics
Energy Demand Research Project (ERDF), UK 98
energy efficiency policies, UK 7, 111–112, 195–211; cost optimal formulae use 202–203; counter-efficiencies of managing market efficiencies 203–205; efficient markets hypothesis and 198–199; energy supplier obligation (ESO) policies 44, 110, 112, 118, 124, 196–197, 199–201; EU Energy Performance of Buildings Directive and 202–203; Green Deal policy 7, 110, 112, 116, 119, 201–202; policy mixes 220, 223–224; policy recommendations 207–211; problem of institutionalising 205–207; reductions in public investment 196–197, 200–201; Scotland 209–210; sociological perspective 197–199; Warm Front scheme 44, 112, 118–119, 197; zero carbon homes policy 197, 203
energy justice 27, 34–44; distributional justice 37, 38–40, 100–102, 44–45; justice as recognition 37, 40–42, 104–105; policy recommendations 44–45; procedural justice 37, 41, 42–43, 44, 102–104; smart meter rollout 100–106; tenet frameworks 37–38
energy justice metrics 37

Index

Energy Performance Certificate (EPC) ratings, UK 39, 111, 112, 124, 210
energy poverty: distributional justice *see* fuel poverty
energy pricing and subsidies 44–45, 195
Energy Saving Multiplier 157–158, 166–168, **167**, 169–170
energy supplier obligation (ESO) policies 44, 110, 112, 118, 124, 196–197, 199–201
Energy Systems Catapult 209
energy–economy models 149, *150*
EPC *see* Energy Performance Certificate (EPC) ratings, UK
equity *see* energy justice
ESO *see* energy supplier obligation (ESO) policies
EU Energy Efficiency Directive 219–220
EU Energy Performance of Buildings Directive 202–203
EU Energy Poverty Observatory (EPOV) 39
EU Structural and Investment Funds 80
European Union: policy mixes 217, 219–220; smart meters 95, 106–107; UK's exit from 78
evaluation frameworks 27–28, **29**; *see also* energy justice; exergy economics
EVs *see* electric vehicles and visions of future personal transport
exergy economics 27–28, 133–151; exergy accounting 141–142, **141**; exergy efficiency 142–143, **143**, *143*, *144*; exergy intensity of national economies 144–147, *145*, *146*; policy recommendations 150–151; rebound effects 148; role of energy in economy 133–134, 135–138; useful exergy and economic production 139–140, *140*; useful exergy as production function 147–148; useful exergy concept 134, 138–139; useful exergy in energy–economy models 149, *150*
expectations and visions 20–21, 22, **29**, 55–56, 266; *see also* electric vehicles and visions of future personal transport

finance and investment-related uncertainties 20
financing mechanisms: need for 247–248, **251**; residential retrofit 114, 118–120, **123**, 125
Finland, policy mixes 220, 223–224, 230
fiscal incentives, for residential retrofit 120, 125

fuel poverty: distributional justice 38–40; education 41; justice as recognition 40–42; procedural justice 41, 42–43, 44; residential retrofit 118–119, 125
Fuel Poverty Strategy 2015; UK 43

GATEway Project, Greenwich, UK 83, **85**, 88
German Advisory Council on Global Change 24
Germany, KfW programme 119, 120, 125
global primary energy intensity 2, 133
government policies, UK: autonomous vehicles 79–80; policy mixes 220, 223–224; *see also* energy efficiency policies, UK
grassroots innovation approach 21
Green Building Council, UK 125
Green Building Partnership, UK 121
Green Deal Pioneering Places programme, UK 121
Green Deal policy, UK 7, 110, 112, 116, 119, 201–202
greenhouse gas emissions 1, 7, 15, 53, 110, 235
Greenwich automated vehicle projects, UK 81, 82, 83–89, **85**, 90

health: fuel poverty and 40–41; residential retrofit and 110
heat pumps 6
Home Energy Services Gateway 209
household energy consumption 111–112, 196
household energy efficiency *see* economy-wide impacts of household energy efficiency improvements; energy efficiency policies, UK; retrofitting residential buildings
hype-cycles 22

IHDs *see* in-home display units (IHDs)
impact of low-energy innovations 16, *19*, 25–28, **29**, **271**; *see also* energy justice; exergy economics; rebound effects of energy-saving innovations
incentives: fiscal incentives for residential retrofit 120, 125; split 114, 118–120
incumbent actors: power and resistance 267; relative role of 21, **29**
Industrial Strategy 53, 73, 80
in-home display units (IHDs) 96, 101, 102, 103, 104–105, 106
Innovate UK 120–121

Index 277

innovation funders 120–121
Institute of Directors, UK 99
institutional dynamics 56
institutionalisation of energy efficiency policy 205–207
International Energy Agency (IEA) 1, 7, 58, 157, 160, 216, 223, 235
International Renewable Energy Agency (IREA) 235
investment-related uncertainties 20

Jevons' paradox 156
justice *see* energy justice
justice as recognition 37, 40–42, 104–105

KfW programme, Germany 119, 120, 125

laboratories 73, 79, 86–87; civic 76; grassroots 76; strategic 74–76; urban 74–78, 90, 91n1
learning processes 21
Least Cost Planning 199
liberalisation of energy markets, UK 198, 199, 206
Local Enterprise Partnerships (LEPs), UK 80
local innovation strategies, UK 80
lock-in effects 17–18
Low Carbon Transition Plan 2009; UK 35
low-energy innovations 6–7, 15, 17; diffusion of 16, *19*, 22–25, **29**, **271**; emergence of 16, 19–22, *19*, **29**, **271**; impact of 16, *19*, 25–28, **29**, **271**

macroeconomic benefits of energy efficiency *see* economy-wide impacts of household energy efficiency improvements
MAcroeconomic Resource COnsumption (MARCO–UK) model 149, *150*
Managed Energy Services Agreement (MESA) 117–118, *117*
marginal abatement cost curve (MACC) concept 206–207
market efficiencies 198–199; counter-efficiencies of managing 203–205
MEES *see* minimum energy efficiency standards (MEES), UK
MESA *see* Managed Energy Services Agreement (MESA)
minimum energy efficiency standards (MEES), UK 112, 124
moral licensing 27
MOVE_UK project, Greenwich, UK 83, **85**

Multi-Level Perspective (MLP) 17–18, *19*, 56, 236
multiple benefits approach 157, 160
multiple streams approach 225, **226–227**
mutual learning 248, **251**

National Audit Office (NAO), UK 101, 200–201, 209
National Energy Action (NEA) 41
National Farmers' Union (NFU), UK 182, 183
National Health Service (NHS), UK 41, 110
National Infrastructure Commission (NIC), UK 209
natural gas, Dutch transition to *178*, 179, 183–185, 188, 189, 190, 191
neoclassical economics 15–16
Netherlands, transition from coal to natural gas *178*, 179, 183–185, 188, 189, 190, 191
net-zero energy consumption 117
new entrants, relative role of 21, **29**
niche innovations 18, *19*
non-recognition patterns 40, 41
normative approaches *see* energy justice

Office of Gas and Electricity Markets (Ofgem), UK 95
offshoring/outsourcing 7, 8, 133, 142
oil, Danish transition from *178*, 179, 185–187, 188, 189, 190, 191
outsiders, relative role of 21, **29**
OWL meter 101
Oxbotica 80, 81, **84**, 86, 87, 89
Oxford automated vehicle projects, UK 80–83, **84**, 86, 87, 88, 89, 90

PACE *see* property assessed clean energy finance (PACE), USA
Paris Agreement 1, 235, 261
Passive House Platform, Belgium 121
passive systems 3
peak car 57, 61
personal mobility *see* electric vehicles and visions of future personal transport
policy feedback theory 225, **226–227**, 228
policy mixes 24–25, **29**, 215–230, 267; acceleration of transitions by 24–25, 222–224; comprehensiveness 220–222; consistency and coherence 218–220, **219**; for creative destruction 218, 222–224; development processes 218–220, **219**; policy process theories

278 Index

policy mixes *continued*
    and 224–228, **226–227**; policy recommendations 229–230; *see also* energy efficiency policies, UK
policy process theories 224–228, **226–227**
policy recommendations 44–45, 68, 91, 122–126, **123**, 150–151, 168–170, 192, 207–211, 229–230, 254, 268–271
political acceleration of sociotechnical transitions 24–25, **29**, 177–192
political will 24, **29**; *see also* energy efficiency policies, UK; political acceleration of sociotechnical transitions
Portugal, exergy efficiency 142
poverty *see* fuel poverty
precautionary principle 203
privatisation of energy system, UK 198, 199, 206
procedural justice 37, 42–43; fuel poverty 41, 42–43, 44; smart meter rollout 102–104
production functions 136–137, 140, 147–148, 149
property assessed clean energy finance (PACE), USA 119–120
Public Accounts Committee, UK 101
public buildings 197, 208
punctuated equilibrium theory 225, **226–227**

rail transport, British transition from 178, *178*, 179–181, 188, 189, 190, 191
real-world experimentation *see* urban experimentation
rebound effects of energy-saving innovations 26–27, **29**, 156, 157, 158, **160**, 161, **163**, 165–166, **167**
reducing energy demand 2–5
RE:FIT programme, London, UK 126
regulations *see* standards and regulations
Remote Applications in Challenging Environments (RACE) facility, UK 86, 87
renewable energies 1–2
retrofit mortgages 119–120
retrofitting residential buildings 110–126; anticipated benefits 110–111, 113–114; business models 114–118, *116*, *117*; capital costs and split incentives 114, 118–120; financing 114, 118–120, **123**, 125; fiscal incentives 120, 125; fuel poverty and 118–119, 125;

intermediaries 120–122, **123**, 125–126; key challenges 113–115, *115*; new institutions **123**, 125–126; policy recommendations 122–126, **123**, 207–211; standards and regulations **123**, 124–125, 207–209; *see also* energy efficiency policies, UK
road transport: automobility 53, 56–58, 60–62, 64, 65, 67; British transition to 178, *178*, 179–181, 188, 189, 190, 191; peak car 57, 61; *see also* connected and autonomous vehicle (CAV) experimentation; electric vehicles and visions of future personal transport

Scottish Energy Strategy 210
second-order learning 74, 90, 91
Smart Energy GB 98, 103–104, 105, 106
Smart Meter Central Delivery Body (SMCDB), UK 98
Smart Meter Implementation Programme (SMIP), UK 94–107; anticipated benefits 96, **97–98**, **99**, 101–102, 208–209; costs 96, 101, 102; customer resistance 99, 102–104; distributional justice 100–102; energy justice 100–106; financial benefits 96, **99**; in-home display units (IHDs) 96, 101, 102, 103, 104–105, 106; justice as recognition 104–105; procedural justice 102–104; rollout issues 100–101, 209; rollout programme 98–99; security and privacy concerns 101, 103; smart meter technologies 95–96, *95*
Smart Mobility Living Lab project, Greenwich, UK 83, **85**, 88, 89
Smart Oxford Lab project, Oxford, UK 82–83, **84**
SMIP *see* Smart Meter Implementation Programme (SMIP), UK
SO *see* supplier obligation (SO) policies
social housing providers 121
social justice *see* energy justice
social network building 21
social psychology 15, 16
social uncertainties 20
societal embedding of innovations 23–24
sociotechnical landscape 18, *19*, 24, 53, 55, 57, 64–65, 67–68, 188
sociotechnical niche 18, *19*, 20–21, 23, 54–58, 65, 66, 67, 73–74, 79, 90, 91, 236–238, **237**, 263; *see also* political acceleration of sociotechnical transitions; systematic review of Strategic Niche Management research

Index

sociotechnical regime 17–18, *19*, 21, 24, 54–58, 63, 64–68, 118, 222–224, 236, **237**, 249, 252, 254; regime actors 54, 55–56, 57–58, 60, 64–68, 74, **237**; *see also* political acceleration of sociotechnical transitions
sociotechnical transitions approach 5–7, 16, 17–19, *19*, 262–264, **271**; diffusion of low-energy innovations 16, *19*, 22–25, **29**, **271**; emergence of low-energy innovations 16, 19–22, *19*, **29**, **271**; impact of low-energy innovations 16, *19*, 25–28, **29**, **271**
split incentives 114, 118–120
standards and regulations: condensing boilers 25–26, 112, 196, 207–208; minimum energy efficiency standards (MEES) 112, 124; residential retrofit **123**, 124–125, 207–209
Starship Technologies 83–86
stereotyping 40
strategic laboratories 76
Strategic Niche Management (SNM) approach 20, 21, 73–74, 236–238, **237**; *see also* systematic review of Strategic Niche Management research
supplier obligation (SO) policies 44, 110, 112, 118, 124, 196–197, 199–201
systematic review of Strategic Niche Management research 235–254; article analytical strategies **239**, 241, 249–250; article attitudes towards pace of transitions **239**, 241, 248–249; article author demographics **239**, 240, 242, *243*, 253, 254; article case studies **239**, 240, 245–246, 254; article methods **239**, 240, 244–245, *244*, 254; article policy recommendations **239**, 241–242, 250–253, **251**; article topics **239**, 240–241, 246–248, *247*, 253; content analysis methodology 238–242, **239**; key findings 253–254; recommendations for future research and policy 254; summary of Strategic Niche Management 236–238, **237**

techno-economic uncertainties 20
thermodynamics *see* exergy economics
total factor productivity (TFP) 135–136
transition theory 54–55
transmission losses 3
transport 17; automobility 53, 56–58, 60–62, 64, 65, 67; hype-cycles 22; peak car 57, 61; policy mixes 25; rebound effects 26–27; *see also* connected and autonomous vehicle (CAV) experimentation; electric vehicles and visions of future personal transport
Transport Research Laboratory (TRL) 83, **85**, 86, 88

United Kingdom 264–266; exergy efficiency 142, **143**, *143*, 149, *150*; household energy consumption 111–112, 196; policy mixes 220, 223–224; privatisation of energy system 198, 199, 206; transition from rail to road transport 178, *178*, 179–181, 188, 189, 190, 191; transition to modern agriculture *178*, 179, 181–183, 188, 189, 190, 191; *see also* connected and autonomous vehicle (CAV) experimentation; economy-wide impacts of household energy efficiency improvements; electric vehicles and visions of future personal transport; energy efficiency policies, UK; fuel poverty; retrofitting residential buildings; Smart Meter Implementation Programme (SMIP), UK
urban experimentation 72–91; benefits 79, 90–91; connected and autonomous vehicles (CAVs) 72–73, 78–91, **84–85**; framework for analysis 74–78, *75*; framings and logics 74–76, *75*, 81–82; functions 73–74; places and spaces 77–78, 88–89; policy recommendations 91; practices 76–77, 82–86, **84–85**; strategic, civic, and grassroots 76; strategic, civic and grassroots 86–87; subjects 77, 86–88; trial, enclave, demonstration and platform 77–78, 90
urban/rural divide 105

'valley of death' in innovation financing 20
vehicle automation 8, 72–91; *see also* connected and autonomous vehicle (CAV) experimentation
visions and expectations 20–21, 22, **29**, 55–56, 266; *see also* electric vehicles and visions of future personal transport

Warm Front scheme, UK 44, 112, 118–119, 197

zero carbon homes policy, UK 197, 203